P9-CKL-618

ON SOCIAL STRUCTURE AND SCIENCE

THE HERITAGE OF SOCIOLOGY

A Series Edited by Donald N. Levine

Morris Janowitz, *Founding Editor*

ROBERT K. MERTON

ON SOCIAL STRUCTURE AND SCIENCE

Edited and with an introduction by

PIOTR SZTOMPKA

THE UNIVERSITY OF CHICAGO PRESS

Chicago and London

the Jagiellonian
ert K. Merton: An
86).

The University of Chicago Press, Chicago 60637
The University of Chicago Press, Ltd., London

Printed in the United States of America

05 04 03 02 01 00 99 98 97 96 5 4 3 2 1

ISBN (cloth): 0-226-52070-6
ISBN (paper): 0-226-52071-4

Library of Congress Cataloging-in-Publication Data
Merton, Robert King, 1910–
On social structure and science / Robert K. Merton ; edited and with an introduction by Piotr Sztompka.
p. cm. — (Heritage of sociology)
Includes bibliographical references and index.
ISBN 0-226-52070-6. — ISBN 0-226-52071-4 (pbk.)
1. Sociology—Philosophy. 2. Social structure. 3. Science—Social aspects. I. Sztompka, Piotr. II. Title. III. Series.
HM24.M4719 1996
301′.01—dc20 96-928
CIP

∞The paper used in this publication meets the minimum requirements of the American National Standard for Information Sciences—Permanence of Paper for Printed Library Materials, ANSI Z39.48-1984.

Contents

Paradoxes of Social Process

III SCIENCE AS A SOCIAL STRUCTURE

The Sociology of Knowledge

The Social Institution of Science

CODA

Acknowledgments

I wish to acknowledge the substantial assistance provided by the series editor, Donald Levine, in the preparation of this volume. A reader for the Press, Charles Crothers, also offered a number of helpful suggestions regarding the selections and the content of the introduction. Special thanks are due to Robert Merton for graciously examining and authorizing a number of stylistic changes in the selections reprinted here.

Piotr Sztompka
Krakow, 1995

Introduction

Rarely has the heritage of sociology been so deeply savored and dramatically enriched as in the life and work of Robert K. Merton. Nourished by a bountiful range of classical authors and guided by a succession of remarkable mentors, Merton opened up fruitful areas of inquiry along lines that he and generations of others would pursue for decades. The self-fulfilling prophecy, focussed interview (whence 'focus groups') opportunity structure, middle-range theory, manifest and latent functions, role-sets and status-sets, social dysfunctions, locals and cosmopolitans, scientific paradigms, the Matthew effect, accumulation of advantage and disadvantage, self-exemplification of sociological ideas, strategic research site, reconceptualization, and the serendipity pattern in research are some of his ingenious generative concepts. Some of these concepts in turn have proven so useful that, once adopted in sociology, they have then entered the common parlance in ways that obscure their origins, a process Merton describes colorfully as "obliteration by incorporation."

A celebrated colleague, teacher, correspondent, author, lecturer, editor, and New Yorker, Merton gracefully expanded a knowing constituency for sociology and genially reshaped its mission—in ways that remain compelling. Merton's forte has been to establish connections. As sociologists again separate into theorists divorced from empirical research and researchers detached from theory, his work affirms the possibility and the promise of their fruitful interdependence. At a time when academic competition often devolves into a search for novel fads and narcissistic forensics, he upholds a search for continuities as the hallmark of scholarly enterprise. At a time of increased academic nationalism, he exemplifies a commitment to international communication and mutual respect. At a time of widening gaps between science and humanism, Merton exemplifies what Robert Redfield once described as the destiny of the social scientist, to exist midway between natural science and the humanities; indeed, the year that Merton became the first sociologist invited to the White House to receive the National Medal of Science, he also became the first sociologist invited by a consortium of humanists, the American Council of Learned Societies, to present the Haskins Lecture. At a time of surging subjectivisms in the intellectual world, Merton reminds us of the communal standards essential to a life of learning.

Published Work

Merton's bibliography includes a dozen books, another dozen edited or coedited volumes, some two hundred articles, and a hundred or so book reviews. True books, in the sense so dear to humanists and so alien to natural scientists, only appear early on: the revised doctoral dissertation, *Science, Technology and Society in Seventeenth Century England* (1938a) and *Mass Persuasion* (1946). The heart of his output takes the form of paradigmatic essays, long articles, introductions, reviews, commentaries—circumscribed pieces that sometimes become so long as to turn imperceptibly into a book, like the "Shandean Postscript" of 290 pages which became his favorite *On the Shoulders of Giants* (1965, 1985, 1993), or the "Episodic Memoir" of 150 pages, which traces the emergence and development of the sociology of science (1979), or the monograph of 160 pages introducing the "sociological theory of reference groups" embedded in the first memorable collection of his paradigmatic essays, *Social Theory and Social Structure* in three major editions—1949, 1957, and 1968. Many of his other papers are gathered together in later collections: *The Sociology of Science* (1973), *Sociological Ambivalence* (1976), and *Social Research and the Practicing Professions* (1982).

These writings exhibit a wide range of interests: from drug addicts to professionals, from anomie to social time, from friendship formation to role-conflicts, from functional analysis to scientific ethos, from medical education to multiple discoveries, from bureaucratic structure to the origins of medieval aphorisms. Merton seems to pick up topics wherever he finds them, then pursues them methodically, meticulously, and in depth, sometimes for many years.

One may distinguish five phases in Merton's productive career (Crothers 1987, 34–40; Clark 1990, 15). In the 1930s, as a graduate student at Harvard University, he prepared a doctoral dissertation on the effects of Protestant Puritanism and of economic and military events on the development of science and technology in seventeenth-century England that was published in 1938 (see chap. 18 in this volume). His interest in European sociology was expressed in two review articles, "Recent French Sociology" (1934a) and "Durkheim's Division of Labor in Society" (1934b). Continuing at Harvard as tutor and instructor after receiving his doctorate in 1936, he produced three paradigmatic essays: "The Unanticipated Consequences of Purposive Social Action" (1936a; chap. 15 in this volume), "Social Time" (with Pitirim Sorokin, 1937), and "Social Structure and Anomie" (1938b; chap. 12 in this volume)—a paper that became the most-quoted paper in the literature on deviance for the period 1955 to 1970 (Cole 1975).

After a brief stint at Tulane University Merton moved to Columbia University where he spent the rest of his career. In the 1940s he conducted a number of empirical projects with Columbia's Bureau of Applied Social Research, including a study of the first radio propaganda marathon, a war-bond drive, which he analyzed in *Mass Persuasion* (1946a). Another of his contributions (with Alice Kitt Rossi in 1950) was the reinterpretation of selected findings in war-time studies conducted by Samuel Stouffer and his team on the American soldier, which resulted in the first sociological analysis of "reference groups." At the same time, Merton worked on several methodological and theoretical topics. In 1948 his paradigmatic article on the self-fulfilling prophecy appeared (chap. 16 in this volume). In 1949 he brought together the results of several investigations in *Social Theory and Social Structure* (2d, rev. ed., 1957; further rev. in 1968), a volume which set forth the foundations of his frameworks for functional and structural analysis and established his reputation as a preeminent sociological theorist.

In the 1950s Merton's research was still linked to the Bureau of Applied Social Research, while his lectures and seminars became arenas of theoretical work through the medium of what he called "oral publication" (Merton 1980a). Collaboration with Paul Lazarsfeld resulted in an important article, "Friendship as a Social Process: A Substantive and Methodological Analysis" (1954). Empirical studies of medical education led to *The Student-Physician* (1957a). *The Focused Interview* (with M. Fiske and P. Kendall), which would later be recognized as the origin of "focus group research," appeared in 1956. Two theoretical papers joined the list that became canonized: "The Role-Set: Problems in Sociological Theory" (1957b; chap. 10 in this volume) and "Social Conformity, Deviation and Opportunity-Structures" (1959). In the latter, Merton returned after twenty years to the theory of anomie, which he continued to develop in "Anomie, Anomia and Social Interaction: Contexts of Deviant Behavior" in 1964, and again in 1994 in "Opportunity Structure" (excerpted in part as chap. 13 in this volume).

In the late 1950s and 1960s Merton returned to his "first love" (Lazarsfeld 1975, 43), the sociology of science, a field which he did so much to conceive, guide, and support. The first sign of this thematic shift came in 1957 with his widely hailed presidential address to the American Sociological Society, "Priorities in Scientific Discovery: A

Chapter in the Sociology of Science" (chap. 22 in this volume). This was followed by a series of paradigmatic essays on diverse problems in the sociology of science: "The Matthew Effect in Science: the Reward and Communication Systems of Science" (1968b; the concept was developed further in a 1988 paper, reproduced here in part as chap. 24), "Patterns of Evaluation in Science: Institutionalization, Structure, and Function of the Referee System" (with Harriet Zuckerman, 1971), "Insiders and Outsiders: a Chapter in the Sociology of Knowledge" (1972; chap. 19 in this volume), and others.

In 1965 Merton published *On the Shoulders of Giants* (2d ed., 1985; 3d, "post-Italianate" ed., 1993). This "prodigal brainchild," as he called it, offered a sociology of science essay in the guise of a historical search for the origins of the medieval metaphor commonly ascribed to Newton: "If I have seen further, it is by standing on the shoulders of giants." He collected his basic papers in his favorite subdiscipline in *The Sociology of Science: Theoretical and Empirical Investigations* (1973) and a few years later recorded the history of that field in *The Sociology of Science: An Episodic Memoir* (1979). In this period he also produced crucial new statements in general theory: the analysis of "sociological ambivalence" (with Elinor Barber, 1963), and an essay on structural analysis (1975; chaps. 11 and 9 in this volume).

In the 1980s and 1990s Merton continued to work in a variety of areas. He devoted himself to rewriting and editing earlier unpublished manuscripts and preparing further volumes of papers for print which included *Social Research and Practicing Professions* (1982). He also composed new articles of considerable importance: "Client Ambivalences in Professional Relationships" (with Vanessa Merton and Elinor Barber, 1983b), "Socially Expected Durations" (1984a; chap. 14 in this volume), and "The Fallacy of the Latest Word: The Case of Pietism and Science" (1984b). At the same time, he began to work with a new genre: reminiscences about his teachers (Talcott Parsons, George Sarton, Pitirim A. Sorokin, and those teachers-at-a-distance, Émile Durkheim, W. I. Thomas and Florian Znaniecki), collaborators (chiefly Paul Lazarsfeld), and students, including Peter Blau, James S. Coleman, Lewis A. Coser, Rose Laub Coser, Franco Ferrarotti, Louis Schneider, Alvin W. Gouldner, and Seymour Martin Lipset. In 1991, he collaborated with David L. Sills to compile and edit *Social Science Quotations: Who Said What, When, and Where* which appeared first as Volume 19 of the *International Encyclopedia of the Social Sciences* and then as a free-standing volume. In 1994, at the request of the largely humanistic American Council of Learned Societies, he turned his focus on himself to produce "A Life of Learning" (which appears as a coda to this volume).

Formative Influences

Merton long maintained that "the writings of classical authors in every field of learning can be read with profit time and again, additional ideas and intimations coming freshly into view with each rereading" (1965, 45). This is particularly so in sociology, which has been less successful than the natural sciences in "retrieving relevant cumulative knowledge of the past" (1968a, 35). On the other hand, he was fond of quoting A. N. Whitehead's adage about the dangers of refusing to venture into new directions: "A science which hesitates to forget its founders is lost" (1968a, 13).

These tenets generated a policy toward the classic writings that seems to imply three directives. First, approach the classics selectively, with an eye to sifting the core of their ideas from their inevitable marginal contributions, blind alleys, or outright mistakes. Second, enter into critical dialogue with the masters, interpreting their ideas in the light of new perspectives and later observations. Third, enrich, supplant, or reject past ideas to the extent that they appear incomplete, deficient, or obsolete: "The founding fathers are honored, not by zealous repetition . . . but by extensions, modifications and, often enough, by rejection of some of their ideas and findings" (1968a, 587).

This is the policy Merton adhered to as, time and again, he drew inspiration from a broad range of European and American classic authors. In the forefront stands his indebtedness to Émile Durkheim, an indebtedness that began during his undergraduate years at Temple College when he was inducted into the mysteries of sociology. There is a striking similarity in the dominant orientation of their sociologies: their common attempt to have sociology develop into a reasonably rigorous, "hard" science of a specifically social subject-matter, with explanations couched in terms of social processes. Merton's sociology also continues Durkheim's methodological approach: structural and functional analyses. Finally, there are remarkable substantive continuities: from Durkheim's study of suicide to Merton's analyses of anomie, from Durkheim's gloss on crime as functional for mechanical solidarity to Merton's view of crime as an "innovative" route to upward mobility, and from Durkheim's sociological interpretations of religion to Merton's sociological analysis of science. These parallels in theoretical outlook were recognized in the orange-colored paper band encircling the

French translation of *Social Theory and Social Structure* in 1965 which read: "*un Durkheim américain.*"

While Durkheim may have charted the course for Merton's intellectual trajectory, other classic authors influenced him in particular ways at certain times and places. Karl Marx was one of these. In contrast to the many sociologists who set Durkheim and Karl Marx in systematic opposition, Merton set them side by side when he first presented his paradigm for functional analysis in 1949. Merton arguably drew from Marx in placing such emphasis on social structural factors in the explanation of differential opportunities and in his attention to the contradictions, conflicts, and circularity of social processes. The Marxian idea of the existential determination of knowledge turned into the germinal idea of the social-structural influences on science: "few, if any, would contest [Marx's] basic influence upon widespread ways of construing the interaction between society and scientific thought" (Merton 1979, 8).

Max Weber influenced Merton's work chiefly at two junctures. These were the doctoral dissertation on the Puritan ethic and the origins of modern science, and the later analyses of bureaucratic structure. Otherwise, Merton engaged with Weber's problematics less than might be supposed, perhaps because Merton's structuralist approach implies the relative neglect of the theory of action, human motivation, or forms of rationality, and because his preference for the study of interactional and group phenomena entails the neglect of civilizational patterns and macropolitical structures. Georg Simmel came to influence Merton mainly in the 1950s, when he focused attention on Simmel in a two-year seminar and used Simmel to inspire a number of propositions on the structural properties of groups, and directed a number of graduate students to exploit Simmel more fully in their own work.

A particularly interesting influence derived from Bronislaw Malinowski. Introducing the Polish edition of *Social Theory and Social Structure* (1982b), Merton emphasizes the crucial role of Malinowski's functional anthropology in stimulating his own ideas on functional analysis. To some extent it was a negative influence, for it was by rejecting certain of Malinowski's tenets that Merton reached his own version of dynamic functional analysis, one that embraces the elements of conflict and change. As Merton observes: "It was not necessary to agree with Malinowski in order to learn from him" (Merton 1982b, 8).

Among American masters, Merton drew particular inspiration from W. I. Thomas and Florian Znaniecki. Their classic, *The Polish Peasant in Europe and America* (1918–21), influenced his image of social structure and his predilection for "human documents" in shaping his own uses of "content analysis." Merton described Thomas as the dean of American sociologists and celebrated Thomas's formulation about the definition of the situation by naming it "the Thomas theorem" (see chap. 16 in this volume). He also describes himself as "a self-selected student-at-a-distance of Znaniecki" (Merton 1983a, 124), whose *Social Role of the Man of Knowledge* (1940) resonated with some of his work in what was emerging as the sociology of science.

Of prime importance for his work on science was George Sarton, at the time of Merton's graduate studies "the acknowledged world dean among historians of science" (Merton 1985, 477). Beyond serving as an exemplar of historical scholarship regarding the development of science and the functioning of scientific communities, Sarton aided his protégé by providing an environment supportive of his doctoral work, offering *Isis,* the principal international journal of the history of science, for some of Merton's early publications, and publishing the revision of Merton's dissertation in Sarton's monograph series, *Osiris* (1938a).

The two most influential teachers under whom, and later with whom, Merton learned and worked were Pitirim Sorokin and Talcott Parsons. These were not entirely direct and wholly positive influences. Merton was evidently not an easy pupil. While admiring his teachers, he did not hesitate to criticize them and to build his own intellectual system partly in opposition to theirs.

Speaking for his entire generation of Harvard sociologists Merton observed, "Talcott was both cause and occasion for our taking sociological theory seriously" (Merton 1980b, 70). His influence on steering Merton's interest toward theoretical considerations was certainly immense. Even so, at what was to be for ever after a memorable meeting of the American Sociological Society in 1948, Parsons's abstract manner of theorizing was a target of Merton's critique, leading Merton to counter with the notion of "middle-range theory" (see chap. 2 in this volume). Similarly, the static and ahistoric "structural functionalism" proposed by Parsons was a subject of Merton's strong critique contributing to the birth of his own dynamic "functional analysis." But their theoretical debate always stayed within the borders of exemplary civility. As Merton recollects: "I remember the grace with which, some thirty years ago, he responded in a forum of this same Association to my mild-mannered but determined criticism of certain aspects of his theoretical orientation" (Merton 1980b, 70; coda to this volume). Years later Parsons came to acknowledge Merton's "major contribution to the understanding and clarification of the theoretical methodology of what he, I think quite appropriately, called 'functional analysis'" (Parsons 1975, 67).

Sorokin encouraged the young Merton by asking him to collaborate on two chapters (on science and technology) of his treatise *Social and Cultural Dynamics* (1937) and on an article, "Social Time" (1937). Decades later, he would praise *On the Shoulders of Giants* (1965) as a masterpiece. Even so, in a typically Sorokinian fulmination, he described Merton's paradigm of functional analysis as "heuristically sterile, empirically useless, and a logically cumbersome table of contents" and his theory of reference groups as "a codification of trivialities dressed up as scientific generalizations" (Sorokin 1966, 451–2). Sorokin's ambivalence recurs in the personal inscription in one of his books: "To my darned ennemy [*sic*] and dearest friend—Robert—from Pitirim."

Finally, one must recognize Merton's decades-long "improbable collaboration" (Merton 1994, 15) with Paul Lazarsfeld. Theirs was an extraordinary colleagueship, a rare case of continuously productive complementarity between two strong, independent individuals with divergent backgrounds, thought styles, and scholarly goals. Conjoining fundamentally different styles of research and theorizing, they cotaught an enormously stimulating graduate seminar, engaged in numerous research projects, and produced several coauthored publications.

The Image of Sociology: Its Goals and Scope

From the mosaic of Merton's substantive contributions, produced at various levels of generality in various areas over various periods of his long career, there emerges a coherent system of ideas. Some of them refer to sociology: its goals, orientation, and methods. Some refer to the distinctive mission of theoretical work. Some of them refer to society: its constitution and transformations. And in the background, there hovers a complex image of science: as a cognitive enterprise, a social institution, and a community of scholars.

Two of Merton's formulations come closest to his definition of what sociology is all about. The calling of a sociologist is described as "lucidly presenting claims to logically interconnected and empirically confirmed propositions about the structure of society and its changes, the behavior of man within that structure and the consequences of that behavior" (Merton 1968a, 70). And the goal of the whole discipline is characterized as follows: "In the large, sociology is engaged in finding out how man's behavior and fate are affected, if not minutely governed, by his place within particular kinds, and changing kinds, of social structure and culture" (Merton 1976, 184). The focus on social structure appears from the beginning as the defining trait of Merton's sociology.

This becomes all the more evident when we see how Merton handles the problem of demarcation, especially how he draws the boundary between sociology and psychology. The difference has to do with what aspects of phenomena are taken for granted or, conversely, treated as problematic. For the psychologist, the social-structural aspect is given, as the context of individual behavior, whereas individual behavior is problematic. For the sociologist, the opposite holds: "the analytical focus is upon the structure of the environment, with the attributes and psychological processes of individuals being regarded as given and not within the theoretical competence of the sociologist to analyze" (1982a, 169–170).

Merton evidently perceives society as a reality *sui generis,* possessing specific properties and displaying specific regularities, emergent in relation to the properties and regularities of individual human beings. The social reality is superindividual (since it is basically novel in relation to individuals) but it is not supraindividual (since it is not independent of individuals). Social structure involves people and their actions, but it is something more than an aggregate of people and activities. It is an inter-individual reality, produced by varied relationships among pluralities of individuals. However, it is also a systemic reality, insofar as it results in various social wholes consisting of individuals bound together by networks of interrelations. Such an ontological conception, a sort of "structural realism," avoids the pitfall of reification that is so typical of metaphysical holism. It also avoids the atomization of social reality that is found in various forms of theoretical (or "methodological") individualism.

On Theoretical Work

Merton is widely (and correctly) perceived as a theorist. The term often connotes either one who weaves highly abstract general theory or engages mainly in the scholastic exegesis of classic texts. By precept and example, Merton is neither of these. Ever engaged with fresh inquiry on a number of substantive fronts, he has propounded and modeled a set of specific practices that together constitute a distinctly Mertonian program for theoretical work in sociology. Following his own style of exposition, we append here a paradigm of the functions of theoretical work that Merton has advanced in various contexts. Most of these functions are represented by selections included in Part One of this volume; most of the others are discussed in an influential pair of essays first published in the late 1940s, "The Bearing of Sociological Theory on Empirical Research," and "The Bearing of Empirical Research on Sociological Theory" (1968a). These functions include:

- Problem finding—defining scientific questions, finding rationales for them, and specifying what must be done to answer them
- Conceptual articulation and reconceptualization—advancing from an early, rudimentary, particularized, and largely unexplicated idea (proto-concept) to a genuine concept—an idea that has been defined, generalized, and explicated to the point where it can effectively guide inquiry into seemingly diverse phenomena (1984a, 287; see also "Robert K. Merton as Teacher" by James S. Coleman in Clark 1990, 29)
- Conceptual clarification—making explicit the character of data subsumed under a concept and suggesting observable indices for conceptualized phenomena that cannot be directly observed
- Construction of middle-range generalizations—formulating generalizations that deal with limited aspects of phenomena or that apply to limited ranges of phenomena
- Functional analysis—specifying the consequences, positive and negative, of given social phenomena for the various social structures in which they are implicated
- Structural analysis—specifying the antecedent structural conditions that give rise to social phenomena
- Construction of typologies—systematizing the types of behavioral patterns found among actors in various domains[1]
- Codification—ordering the available empirical generalizations in a given domain, showing connections among generalizations in apparently different spheres of behavior, and tracing continuities within research traditions
- Construction of paradigms (in a sense that antedates Kuhn's usage of the term)—systematizing the concepts and problems of a given domain of inquiry in compact form[2]

1. Although Merton himself refers to his classifications of forms of deviant behavior and of attitudes in race relations as paradigms, they are more appropriately considered as typologies (Crothers 1987, 60).

2. Merton's own paradigms include those for functional analysis, structural analysis, and the sociology of knowledge, reprinted here in chapters 6, 9, and 17.

- Formalization—deriving the implications of theoretical assumptions and postulates for other investigable properties of social phenomena (yet remaining aware of the danger of pursuing logical consistency to the point of sterile theorizing)
- Recasting theory—extending theoretical formulations in new directions in response to the appearance of unanticipated, anomalous, and strategic facts (serendipity) or the repeated observation of facts previously neglected
- Specification of ignorance—expressly recognizing what is not yet known but what needs to be known in order to advance the pursuit of knowledge
- Location in theoretical space—understanding the implications of theoretical pluralism for a given perspective or research program, including the fact that perspectival differences may entail complementary or unconnected as well as contradictory theories
- Productive return to classics—mining the classics for crisper formulations, authoritative support, and critical rejoinders regarding current formulations, and for models of intellectual excellence

By specifying these practices as significant for the growth of sociology as a scientific discipline as well as by exemplifying them throughout his career, Merton succeeded in lending new dignity to the role of theory in sociology at a time when so much of its advancement seemed to hinge on the creation of new observational and analytic techniques. This is not to say that these theoretic functions had necessarily to be carried out by specialists in theory; Merton applauded conversations between practitioners of theory and of research that are carried out within the same person as well as between separate persons. What Merton stresses is the blindness of empirical research without theory and the emptiness of theory without empirical content. Coming at a time when the options for sociology seemed to lie among abstract grand theory, atheoretic empiricism, or humanistic scholarship, he upheld the vision of sociology as a genuinely scientific discipline. His work continues to remind us that, without the special steps that only theoretical work can provide, sociology's status as a cooperative and cumulative enterprise is endangered.

Social Structure and Its Vicissitudes

The tasks set by Merton for theoretical work in sociology pointedly exclude construction of a general, all-embracing theory of social phenomena: the search for a total system of sociological theory, he writes, "has the same exhilarating challenge and the same small promise as those many all-encompassing philosophical systems which have fallen into deserved disuse" (1968a, 45; see below, chap. 3). Yet, as Stinchcombe was perhaps the first to suggest, Merton's strictures against general theory have been misleading. They obscure the fact that his own writings appear to contain a coherent, plausible, and fruitful general theory of social structure. Despite Merton's occasional semantic inconsistencies (Levine 1985, chap. 4) and changing theoretic formulations as well as varying emphases among those who attempt to articulate this theory (e.g., Stinchcombe 1975, Sztompka 1986, Crothers 1987), certain key features of a general Mertonian theory of social structure can be readily identified.

To Merton's mind, social structure must be distinguished analytically from cultural structure, and it must be represented as multidimensional, full of tensions, and in continuous flux. In company with Sorokin, Parsons, and others, Merton distinguishes cultural structure—"the organized set of normative values"—from social structure, defined as "that organized set of social relationships in which members of the society or group are variously implicated" (1968a, 216). Although Merton offers a number of insightful propositions about culture, it cannot be said that he worked out a distinctive or well-developed general theory of culture. By contrast, those who have sought to articulate his implicit theory of social structure judge it to be sophisticated and original.

The notion of persons as structurally located, anchored in networks of social relationships, is at the core of Merton's general theory. The building block of social structure is the social status, defined as an identifiable "position in a social system occupied by designated individuals" (1968a, 368), e.g., a doctor, student, Muslim, Hispanic-American, mother. All persons occupy a complex of statuses—a "status-set"—and typically move through a succession of statuses in their lifetime—a "status-sequence." Social statuses are organized along two dimensions. First, a social status includes a normative aspect—the patterned set of expectations regarding the socially defined, appropriate behavior for an incumbent of that status. These expectations are diverse, for they are tendered by those regularly linked with the status as members of its attendant "role-set"—one of Merton's most fruitful theoretical constructions (see below, chap. 10). Second, each social status has an opportunity aspect that includes both the set of life-chances (Weber's term), options, resources, and facilities differentially accessible to the incumbent and the correlative opportunity structure which distributes and redistributes the probabilities of such differential access (chap. 13 in this volume). Together, normative structures and opportunity structures exert constraining and facilitating influences on the behaviors, beliefs, attitudes, and motivations of persons located in them. And the behaviors they induce have consequences that often reshape social institutions.

In addition to theorizing about relationships among statuses, Merton analyzes the properties of social groups built out of particular types of statuses. He examines the general characteristics of membership groups and reference groups, and goes on to codify twenty-six properties that groups are considered to possess (Merton 1968a, chaps. 10–11; Crothers 1987, 96–101). His image of social structure is rendered yet more complex by other considerations. Although he sees social relationships as patterned, regular, and repetitive, he does not assume their patterns to be coherent and harmonious. For one thing, there may be much dissensus regarding the normative expectations in a given situation. More important, the norms frequently entail contradictory expectations—a phenomenon Merton describes as "sociological ambivalence" (chap. 11 in this volume). Rose Laub Coser has gone so far as to assert that "Merton has stood Durkheim on his head; rather than having the individual confronted with ready-made social norms that are external, coming down in toto, so to speak, for Merton individuals have to find their own orientations among multiple, incompatible, and contradictory norms" (1975, 239). What is more, important aspects of social structure remain hidden to the actors themselves, a fact Merton represents by distinguishing between latent levels and manifest levels of social structure, just as early on he distinguished between manifest and latent functions (1976, 126; chap. 9 in this volume). In sum: "Even the seemingly simple social structure is extremely complex" (1968, 424).

Merton's image of social structure must also be described as dynamic. It attends routinely to forces that produce changes of two sorts: changes *within* social structures and changes *of* social structures. The first type of change consists of the ongoing adaptive processes that reproduce specified states of a social structure, or at least keep them within the limits that give that structure its identity. They follow a logic of negative feedback, correcting deviations and reducing the amount of conflict faced by occupants of a given status.

The second of these types of change, transformations of social structure, comes about through amplification rather than counteraction—through positive rather than negative feedback. Merton connects this mode of change with two general processes, the accumulation of dysfunctions and the accumulation of innovations. Accumulation of dysfunctions appears in many forms. Certain elements may be dysfunctional for the entire system, such as an unrestrained pattern of egoistic hedonism. The larger the number of such dysfunctional elements, and the more dysfunctional each of them, the more likely is the system to break down. Another case occurs when some elements are functional for the entire system, but carry certain dysfunctional side-effects: the competitive success orientation may benefit the economy but have dysfunctional effects for family structures. In this case, structural transformation comes when the dysfunctional side effects reach a certain threshold and outweigh the functional outcomes. A third pattern occurs when certain arrangements, such as progressive taxation, apartheid, social security, or affirmative action, are functional for certain groups or strata and dysfunctional for others. As the groups or strata affected adversely attain sufficient power, they tend to introduce changes in those structures.

Structural change also follows from an accumulation of innovations. Merton's chief illustration of this pattern appears when he analyzes the crescive change of normative structures, a process that begins with the private deviation from norms by those who still regard those norms as legitimate. When incidents of such deviation become widespread, public awareness of the deviation is awakened, and "successful rogues . . . become prototypes for others in their environment who, initially less vulnerable and less alienated, now no longer keep to the rules they once regarded as legitimate" (1964, 235). Widespread incentives to evade coupled with the widely shared belief that "everybody does it" along with the tendency to imitate successful evaders leads to the patterned evasion of norms in the face of their continuing claim to legitimacy. Tax evasions, cheating on exams, avoidance of customs duties and currency controls, adultery, and petty theft in business firms provide familiar examples.

The most transformative phase arrives when "a mounting frequency of deviant but 'successful' behavior tends to lessen and, as an extreme potentiality, to eliminate the legitimacy of the institutional norms for others in the system" (1968a, 234). This has occurred, for example, in the increasing obsolescence and finally transformation of laws governing divorce. More recently, it has appeared in the norms regarding homosexuality: private evasions of prevailing heterosexual norms were followed by public defiance—"coming out of the closet"—that finally gave way to a refiguring of public norms, to the extent that growing numbers of communities have come to affirm homosexual unions, according them positive rewards and social benefits, most recently, by American courts allowing unmarried homosexual and heterosexual couples to adopt children. Thus, institutionalized and other patterned evasions provide a distinct mechanism of social change. Such an amplifying, cumulative dynamic is also at work when predicted states that are not so at the moment become so as more and more people come to believe in them, as when runs on banks result in the very bank failures that depositors fear—the self-fulfilling prophecy (chap. 16 in this volume).

These considerations must prod careful readers to question certain notions about Merton's work that sprang up from the reaction against "functionalism" that swept sociology in the 1960s and 1970s. By pinning that label upon him and charging him with its associated sins, critics could assault his work for ignoring social conflict and evincing a static bias. Yet Merton had maintained all along that he was a proponent of functional analysis, not of "functionalism," and his functional and structural analyses consistently and systematically attend to conflictual situations and dynamic processes. As Lewis Coser has stated, "Merton's analytic strategies are characterized, *inter alia,* by his close attention to contradictions and conflicts within global structures, to ambivalence in the motivation of actors, and to ambiguities in the perceptual fields to which they orient themselves" (1975, 98). It is time for readers to engage afresh with the actual statements in which this illuminating approach to social structure and its vicissitudes is embodied.

Science as a Social Structure

While Merton decisively enriched both the vocation of sociological theory and contemporary discourse about social structure, he has been credited for establishing virtually single-handedly the sociology of science as a viable subdiscipline, an achievement marked by his election as founding president of the Society for Social Studies of Science in 1975. Merton's interest in the societal dimension of intellectual activity was lifelong. It started with his graduate work on the history of science for Sorokin and his dissertation on the emergence of modern science in seventeenth-century England, the latter stimulating long and continuing debates that have been recently summarized in a volume edited by I. B. Cohen (1990). This work was followed by his studies of the classical sociology of knowledge as represented by Marx, Scheler, and

Mannheim, studies that eventuated in a stock-taking paradigm for this field (chap. 17 in this volume).

At a certain juncture, Merton made a dramatic departure from the philosophical concerns of the sociology of knowledge and the largely descriptive concerns of the history of science to launch a systematic research program involving the empirical study of science as a social institution. The program rests on the assumption that even if the content of science is not much affected by the social organization of scientists, their work goes on within groups whose organizational characteristics demonstrably affect the production, selection, and distribution of scientific knowledge. This departure was heralded by a number of early pieces, including his 1942 piece on the ethos of science (chap. 20 in this volume), and his 1952 observations on the neglect of the sociology of science (in a foreword to Bernard Barber's *Science and the Social Order* [1973, chap. 10]) and, with Barber, a stock-taking bibliography of the field. The new program became fully manifest with his 1957 presidential address to the American Sociological Society, "Priorities in Scientific Discovery" (see chap. 22, this volume). In that pivotal presentation, Merton demonstrated the cognitive payoffs of regarding science as an institution pervaded by distinctive rules (the ethos of science) and as a community, like any other kind of human community, that brings persons together through systems of roles, interactions, and relationships. Work within the scientific community subjects one to conditions that generate disputes over priority. These include a normative emphasis on the value of creating new knowledge, a reward system that prods scientists to seek recognition of their work from peers, partly through citations (chap. 22 in this volume), and the realization that discoveries are often made by several scientists at much the same time (chap. 23 in this volume). Muting the intensity of the disputes over priority and other forms of social conflict is the norm in the ethos of science calling for the public expression of modesty and humility in the face of a vast shared ignorance.

In area after area, Merton proceeded to explore new components and dimensions of the scientific community. He extended his analyses of the evaluation and reward system of science in several directions, looking at the dynamics of the referee system that is so typical of modern science, and entailing a system of institutional vigilance, or "organized skepticism," which submits scientific productions to the scrutiny of peers. He examines the complex working of the "Matthew effect," according to which already eminent scientists gain disproportionate peer recognition and acclaim in cases of collaboration or independent multiple discoveries (chap. 24 in this volume). He investigates the stratification system of science, the hierarchical differentiation of scholars that results from these processes.

The communication system of science is also affected by the Matthew effect, which accounts for the special visibility enjoyed by works published by recognized scholars. Yet another pattern appears in the process of "obliteration by incorporation," in which long and widespread use of an idea leads scientists to forget both the original sources and its literal formulation. The communication system of science receives further attention in analyses of the networks of acquaintanceships, friendships, loyalties, and solidarities that make up an informal influence system among scientists. In this work Merton gives new prominence to the seventeenth-century notion of "invisible colleges" as redefined for modern usage by Derek de Solla Price, and retrieves the notion of "thought collective" from the long-forgotten work of Ludwik Fleck.

In work carried out with Harriet Zuckerman in the early 1970s, Merton turned to the phenomena of age, aging, and age structure in the scientific community (1973, chap. 22). This work linked the earlier conceptualizations of reward system, communication system, and stratification in science with age-related variations, investigating such questions as the roles that scientists occupy at different points in the life cycle, and how age differences affect patterns of collaboration, authority, visibility, and openness to new ideas. Renewed attention to Merton's early work promises to reopen concerns about the effects of social and economic factors on the selection of scientific problems, as the seventeenth-century needs for navigational devices boosted an interest in astronomy, and on the conditions in the societal environment that support or negate the enterprise of science (chap. 21 in this volume).

Science forms the arena where all of Merton's sociological interests and perspectives come together—his image of the discipline, his penchant for functional and structural analysis, and his middle-range theories of role-sets and status-sets, reference groups, sociological ambivalence, and the like. At the same time, concepts and hypotheses initially devised in his sociology of science have proven useful in other areas, exemplifying precisely the sort of disciplinary advance through the codification process that he advocated long before. A prime instance of this is the Matthew effect, which became generalized into the concept of the accumulation of advantage and disadvantage. The thesis that those who achieve an initial advantage in scientific competition secure more advantages than their accomplishments may strictly warrant has spread into other sociological areas, such as research on stratification and minorities, and even into adjoining disciplines, notably political science and economics. With Mertonian irony, one might even say that the fate of the Matthew-effect thesis is self-exemplifying.

The Mertonian Legacy

Perhaps no less consequential for the heritage of sociology than the body of Merton's published work has been his contribution to enhancing the enterprise of sociological scholarship. He made this contribution in many ways. As a preeminent figure at Columbia University during its heyday, he played a signal role, along with Paul Lazarsfeld, in powering what has been described amusingly as the "Columbia Sociology Machine" (Clark 1995). As an indefatigable editor, he compiled volumes, evaluated manuscripts, and decisively upgraded innumerable publications by colleagues and students (Caplovitz 1977). As a corespondent of unlimited enthusiasm, he disseminated invaluable criticism, stimulation, and support to hundreds of colleagues all over the world. As a public figure in the media, he was celebrated and consulted for statements about the state of sociology as a discipline. As an academic statesman, he tried—if not always successfully—to avoid internecine combat, out of concern about the distorting effects of public polemics among scientists. Indeed, he often served to conciliate and otherwise bring together disparate parties of the American and international sociological community. Above all, we have noted, he did much to save sociology from devolving into a pattern of atheoretic empiricism, fashioning with Lazarsfeld an amalgam of theory and empirical work that transformed sociology around the globe.

Although Merton's momentous contributions of this invisible sort reverberate in the quality of work of the legions who came under this influence, the visible portion of his legacy remains the body of published work to which this volume may serve as a new introduction. As perhaps the first sociological classic author to meditate in extended form on the issues involved in the utilization of classics, his life and work offer us guidelines for how to approach the part of the heritage of sociology that he has created. They tell us to approach that work in a spirit of appreciative openness and critical discernment. This means searching for lines of continuity between his interests and present concerns. It means working to clarify notions that he may have presented in confused or ambiguous form. It means integrating his work through processes of substantive codification and theoretical formalization. In so doing, we open the road to interesting discoveries.

As this essay and other commentators have argued, Merton's strictures against general theory notwithstanding, he promulgated an implicit system of general theory that arguably accounts for much of his appeal to sociologists with such diverse interests. In spite of the often dispersed, piecemeal, and fragmentary nature of Merton's contributions, they add up to a coherent system of thought. To be sure, the system is far from complete, and there are many lacunae. Yet its islands of enlightenment fit nicely into an overall topography whose shadowy areas offer invitations for further inquiry. To accept such invitations is to embrace the kind of life of learning that Robert K. Merton exemplifies.

References

Robert K. Merton's works appear in a separate bibliography at the end of this book.

Caplovitz, David. 1977. "Robert K. Merton as Editor: Review Essay." *Contemporary Sociology* 6:142–150.

Clark, Jon. 1990. "Robert Merton as Sociologist." In *Robert K. Merton: Consensus and Controversy,* ed. Jon Clark et al. 3–23. London: Falmer Press.

Clark, Terry. 1994. "Paul Lazarsfeld and the Columbia Sociology Machine." In *Proceedings of the Paul Lazarsfeld Colloquium,* ed. Bernard-Pierre Lecuyer. Forthcoming.

Cohen, I. Bernard, ed. 1990. *Puritanism and the Rise of Modern Science: The Merton Thesis.* New Brunswick: Rutgers University Press.

Cole, Stephen. 1975. "The Growth of Scientific Knowledge: Theories of Deviance as a Case Study." In *The Idea of Social Structure: Papers in Honor of Robert K. Merton,* ed. Lewis A. Coser, 175–220. New York: Harcourt Brace Jovanovich.

Coser, Lewis A. 1975. "Merton's Use of European Sociological Tradition." In *The Idea of Social Structure: Papers in Honor of Robert K. Merton,* ed. L. Coser, 85–100. New York: Harcourt Brace Jovanovich.

Coser, Rose Laub. 1975. "The Complexity of Roles as a Seedbed of Individual Autonomy." In *The Idea of Social Structure: Papers in Honor of Robert K. Merton,* ed. L. Coser, 237–63. New York: Harcourt Brace Jovanovich.

Crothers, Charles. 1987. *Robert K. Merton.* London: Tavistock.

Lazarsfeld, Paul. 1975. "Working with Merton." In *The Idea of Social Structure: Papers in Honor of Robert K. Merton,* ed. L. Coser, 35–66. New York: Harcourt Brace Jovanovich.

Levine, Donald N. 1985. *The Flight from Ambiguity.* Chicago: University of Chicago Press.

Parsons, Talcott. 1975. "The Present Status of 'Structural-Functional' Theory in Sociology." In *The Idea of Social Structure: Papers in Honor of Robert K. Merton,* ed. L. Coser, 67–83. New York: Harcourt Brace Jovanovich.

Sorokin, Pitirim A. 1966. *Sociological Theories of Today.* New York: Harper & Row.

Stinchcombe, Arthur L. 1975. "Merton's Theory of Social Structure." In *The Idea of Social Structure: Papers in Honor of Robert K. Merton,* ed. L. Coser, 11–53. New York: Harcourt Brace Jovanovich.

Sztompka, Piotr. 1986. *Robert K. Merton: An Intellectual Profile.* New York: St. Martin's Press.

Znaniecki, Florian. 1940. *The Social Role of the Man of Knowledge.* New York: Columbia University Press.

I

THEORETICAL WORK IN SOCIOLOGY

Varieties of Theoretical Work

1

The Uses and Abuses of Classical Theory (1967)

Humanistic and Scientific Aspects of Sociology

The contrast between the orientation of the sciences toward the great classical works and that of the humanities has often been noticed. It stems from profound differences in the kind of selective accumulation that takes place in civilization (which includes science and technology) and in culture (which includes the arts and value-configurations).[1] In the more exact sciences, the selective accumulation of knowledge means that classical contributions made by scientists of genius or great talent in the past are largely developed in later work, often by those of distinctly smaller talent.

The severest test of truly cumulative knowledge is that run-of-the-mill minds can solve problems today that great minds could not begin to solve earlier. An undergraduate student of mathematics knows how to identify and solve problems that defied the best powers of a Leibniz, Newton, or Cauchy.[2]

Because the theory and findings of the fairly remote past are largely incorporated into cumulative present knowledge in the more exact sciences, commemoration of the great contributors of the past is substantially reserved to the history of the discipline; scientists at their workbenches and in their papers make use primarily of the more recent contributions, which have developed these earlier discoveries. The re-

From "On the History and Systematics of Sociological Theory," in *On Theoretical Sociology* (New York: Free Press, 1967), 26–37. Reprinted by permission of The Free Press, an imprint of Simon and Schuster.

1. The distinction among processes of society, culture, and civilization was emphasized by Alfred Weber in "Prinzipielles zur Kultursoziologie: Gesellschaftsprozess, Zivilisationsprozess und Kulturbewegung," *Archiv für Sozialwissenschaft und Sozialpolitik* 47 (1920): 1–49.

2. Charles C. Gillispie, *The Edge of Objectivity: An Essay in the History of Scientific Ideas* (Princeton University Press, 1960), 8. "Every college freshman knows more physics than Galileo knew, whose claim is higher than any other's to the honor of having founded modern science, and more too than Newton did, whose mind was the most powerful ever to have addressed itself to nature."

sult of this practice is that earlier and often much weightier scientific contributions tend to be obliterated (though not without occasional and sometimes significant exceptions) by incorporation into later work.

In the humanities, by direct contrast, each classical work—each poem, drama, novel, essay, or historical work—tends to remain a part of the direct experience of succeeding generations of humanists. As Derek Price has put it in instructive imagery: "the cumulating structure of science has a texture full of short-range connexions like knitting, whereas the texture of a humanistic field of scholarship is much more of a random network with any point being just as likely to be connected with any other."[3] In short, firsthand acquaintance with classics plays a small role in the work of the physical and life scientists and a very large one in the work of humanistic scholars.

M. M. Kessler, another student of information systems in science, has put the point in deliberately provocative if not exasperating language:

> Even the masterpieces of scientific literature will in time become worthless except for historical reasons. This is a basic difference between the scientific and belletristic literature. It is inconceivable for a serious student of English literature, for example, not to have read Shakespeare, Milton and Scott. A serious student of physics, on the other hand, can safely ignore the original writings of Newton, Faraday and Maxwell.[4]

Kessler's language is designed to raise the hackles of the reader. And indeed, from the standpoint of humanism and the history of science, this statement appears to be an expression of latter-day barbarism. It is hard for many of us to distinguish our historical and commemorative interest in the pathbreaking works of science from our interest in advancing a contemporary science that requires little direct acquaintance with Newton's *Principia* or Lavoisier's *Traité*. Yet the same observation as Kessler's was eloquently advanced by one of the founding fathers of modern sociology. In language that personalizes the fateful process of incorporation and extension in science, Max Weber observes that

> in science, each of us knows that what he has accomplished will be antiquated in ten, twenty, fifty years. That is the fate to which science is subject; it is the very *meaning* of scientific work, to which it is devoted in a quite specific sense, as compared with other spheres of culture for which in general the same holds. Every scientific 'fulfillment' raises new 'questions'; it

3. Derek J. De Solla Price, "The Scientific Foundations of Science Policy," *Nature* 206 (17 April 1965): 233–38.

4. M. M. Kessler, "Technical Information Flow Patterns," *Proceedings,* Western Joint Computer Conference, 9 May 1961, 247–57.

> *asks* to be 'surpassed' and outdated. Whoever wishes to serve science has to resign himself to this fact. Scientific works certainly can last as 'gratifications' because of their artistic quality, or they may remain important as a means of training. Yet they will be surpassed scientifically—let that be repeated—for it is our common fate and, more, our common goal. We cannot work without hoping that others will advance further than we have. In principle, this progress goes on ad infinitum.[5]

Sociologists, poised between the physical and life scientists and the humanists, are subject to cross-pressures in their orientation toward the classic contributions and do not take easily to the commitment described by Weber. Only a few sociologists adapt to these pressures by acting wholly the scientific role suggested by Weber or the humanistic one. Perhaps the majority oscillate between the two, and a few try to consolidate them. These efforts to straddle scientific and humanistic orientations typically lead to merging the systematics of sociological theory with its history.

That the social sciences stand between the physical sciences and the humanities in their cumulation of knowledge is dramatically confirmed by so-called citation studies which compare the distributions of dates of publications *cited* in the several fields. The findings are notably consistent. In the physical sciences—represented by such journals as *The Physical Review* and the *Astrophysical Journal* some 60 to 70 percent of the citations refer to publications appearing within the preceding five years. In the humanities—represented by such journals as the *American Historical Review,* the *Art Bulletin,* and the *Journal of Aesthetics and Art Criticism*—the corresponding figures range from 10 to 20 percent. In between are the social sciences—represented by such journals as the *American Sociological Review,* the *American Journal of Sociology,* and the *British Journal of Psychology*—where from 30 to 50 percent of the citations refer to publications of the preceding five years. Other students of citation patterns testify that these findings are typical in their main outlines.

In one way, sociology adopts the orientation and practice of the physical sciences. Research moves from the frontiers advanced by the cumulative work of past generations: sociology is, in this precise sense, historically short-sighted, provincial, and effective. But in another way, sociology retains its kinship with the humanities. It is reluctant to abandon a firsthand acquaintance with the classical works of sociology and pre-sociology as an integral part of the experience of the sociologist

5. Max Weber, *From Max Weber: Essays in Sociology,* trans. and ed. H. H. Gerth and C. Wright Mills (New York: Oxford University Press, 1946), 138; the extract is, of course, from his enduring eloquent affirmation of "science as a vocation."

qua sociologist. Every contemporary sociologist with a claim to sociological literacy has had direct and repeated encounters with the works of the founding fathers: Comte, Marx, and Spencer, Durkheim, Weber, Simmel, and Pareto, Sumner, Cooley, and Veblen, and the rest of the short list of great talents who have left their indelible stamp on sociology today. Since I have long shared the reluctance to lose touch with the classics, even before finding a rationale for it, and since to a degree I continue to share it, this may be reason enough for speculating about its character and sources.

Erudition Versus Originality

No great mystery shrouds the affinity of sociologists for the works of their predecessors. There is a degree of immediacy about much of the sociological theory generated by the more recent members of this distinguished lineage, and current theory has a degree of resonance to many of the still unsolved problems identified by the forerunners.

However, interest in classical writings of the past has also given rise to intellectually degenerative tendencies in the history of thought. The first is an uncritical reverence toward almost any statement made by an illustrious ancestor. This has often been expressed in the dedicated but, for science, largely sterile exegesis of the commentator. It is to this practice that Whitehead refers in the epigraph to this chapter: "A science which hesitates to forget its founders is lost." The second degenerative form is banalization. For one way a truth can become a worn and increasingly dubious commonplace is simply by being frequently expressed, preferably in unconscious caricature, by those who do not understand it. (An example is the frequent assertion that Durkheim assigned a great place to coercion in social life by developing his conception of 'constraint' as one attribute of social facts.) Banalization is an excellent device for drying up a truth by sponging upon it.

In short, the study of classical writings can be either deplorably useless or wonderfully useful. It all depends on the form that study takes. For a vast difference separates the anemic practices of mere commentary or banalization from the active practice of following up and developing the theoretical leads of significant predecessors. It is this difference that underlies scientists' ambivalence toward extensive reading in past writings.

This ambivalence of scientists has historical and psychological roots. From the beginning of modern science, it was argued that scientists should know the work of their predecessors in order to build on what had gone before and to give credit where credit was due. Even the most vocal prophet of anti-scholasticism, Francis Bacon, took this for granted: "When a man addresses himself to discover something, he first seeks out and sees before him all that has been said about it by others; then he begins to meditate for himself. . . ."[6] This practice has since been institutionalized in the format of scientific papers which calls for a summary of the theory and investigations that bear upon the problems in hand. The rationale for this is as clear as it is familiar: ignorance of past work often condemns the scientist to discovering for himself what is already known. As Sorokin has put the case for our own field,

> Not knowing that a certain theory has been developed long ago, or that a certain problem has been carefully studied by many predecessors, a sociologist may easily devote his time and energy to the discovery of a new sociological America after it was discovered long ago. Instead of a comfortable crossing of the scientific Atlantic in the short period of time necessary for the study of what has been done before, such a sociologist has to undergo all the hardships of Columbus to find, only after his time and energy are wasted, that his discovery has been made long ago, and that his hardships have been useless. Such a finding is a tragedy for a scholar, and a waste of valuable ability for society and sociology.[7] [. . .]

Since the policy and in part the practice of searching the antecedent literature have been long institutionalized in science, they require no further documentation. But the counter-emphasis—little institutionalized yet often put into practice—requires extensive documentation if we are to understand the ambivalence of scientists toward erudition.

Through at least the last four centuries, eminent men of science[8] have warned of the alleged dangers of erudition. The historical roots of this attitude are embedded in the revolt against the scholasticism of the commentator and exegetist. Thus, Galileo gives his clarion call:

6. Francis Bacon, *Novum Organum* (London: George Routledge and Sons, n.d.), 1: aphorism LXXXII, 105.

7. Pitirim A. Sorokin, *Contemporary Sociological Theories* (New York: Harper, 1928), xviii–xix.

8. On the term "man of science," established since the sixteenth century, and its slow displacement by the nineteenth-century coinage, "scientist," see Robert K. Merton, "De-gendering 'Man of Science': The Genesis and Epicene Character of the Word *Scientist*," in *Sociological Visions,* ed. Kai Erikson (New Haven: Yale University Press, 1996). For Merton's earlier sociological analyses of this semantic shift, see "Le molteplici origini e il carattere epiceno del termine inglese *Scientist,*" in *Scientia: L'immagine e il mondo* (Milan: Scientia, 1989), 279–93 and "Ueber die vielfältigen Wurzeln und den geschlechtlosen Charakter des englischen Wortes 'Scientist'," in *Generationsdynamik und Innovation in der Groundlagenforschung,* ed. Peter Hans Hofshneider and Karl Ulrich Mayer (Munich: Max-Planck-Gesellschaft, 1990), 259–94.—*Ed.*

> . . . a man will never become a philosopher by worrying forever about the writings of other men, without ever raising his own eyes to nature's works in the attempt to recognize there the truths already known and to investigate some of the infinite number that remain to be discovered. This, I say, will never make a man a philosopher, but only a student of other philosophers and an expert in their works.[9] [. . .]

In due course, the choice between scholarship and original scientific work was converted by some into an ambivalence toward erudition. [. . .] For example, a Claude Bernard assumes that a man of science must know the work of his predecessors. But, he goes on to say, the reading of even such "useful scientific literature . . . must not be carried too far, lest it dry up the mind and stifle invention and scientific originality. What use can we find in exhuming worm-eaten theories or observations made without proper means of investigation?" In a word, "misconceived erudition has been, and still is, one of the greatest obstacles to the advancement of experimental science."[10] [. . .]

Other have dealt with their ambivalence by largely abandoning the effort to become versed in the antecedent literature in order to get on with their own work. The social sciences have their own complement of such adaptations. Long ago, Vico was ready to quote with pleasure Hobbes' observation that if he had read as much as other men he would have known as little.[11] Herbert Spencer—of whom it can be said that never before had anyone written so much with so little knowledge of what others before him had written on the same wide range of subjects—elevated both his hostility toward authority and his illness (he was dizzied by reading) into a philosophy of investigation that gave little room to acquaintance with predecessors.[12] And Freud, repeatedly and quite self-consciously, maintained the policy of working up his clinical data and theory without recourse to antecedent work. As he put it on one occasion, "I am really very ignorant about my predecessors. If we ever meet up above they will certainly treat me ill as a plagiarist. But it is such a pleasure to investigate the thing itself instead of reading the literature about it." And again: "In later years I have denied myself the very great pleasure of reading the works of Nietzsche from a

9. *Le Opere di Galileo Galilei,* Edizione Nazione, vol. 3, bk. 1 (Florence: G. Barbèra, 1892), 395.

10. Claude Bernard, *An Introduction to the Study of Experimental Medicine* (1865; reprint, New York: Henry Schuman, 1949), 145, 141.

11. *The Autobiography of Giambattista Vico,* trans. Max Harold Fisch and Thomas Goddard Bergin (Ithaca, New York: Great Seal Books, 1963).

12. *Autobiography of Herbert Spencer* (New York: D. Appleton and Co., 1904).

deliberate resolve not to be hampered in working out the impressions received in psychoanalysis by any sort of expectation derived from without. I have to be prepared, therefore—and am so, gladly—to forgo all claim to priority in the many instances in which laborious psychoanalytic investigation can merely confirm the truths which this philosopher recognized intuitively."[13]

It was a founding father of sociology who managed to carry this sort of adaptation to the tension between erudition and originality to its inept extreme. During the dozen years he devoted to writing the *Course of Positive Philosophy,* Comte followed the "principle of cerebral hygiene"—he washed his mind clean of everything but his own ideas by the simple tactic of not reading anything even remotely germane to his subject. As he proudly put it in a letter to A. B. Johnson, "For my part, I read nothing except the great poets ancient and modern. The cerebral hygiene is exceedingly salutary to me, particularly in order to maintain the originality of my peculiar meditations."[14] Thus we find Comte making the ultimate—and, at this extreme, absurd—distinction between the history and the systematics of sociology; as historian of science, he tried to reconstruct the development of science through a relatively extensive reading of the classics, while as originator of the positivist *system* of sociological theory, he devoutly ignored immediately antecedent ideas—not least, those of his onetime master, Saint-Simon—in order to achieve a Pickwickian kind of originality.

As we have seen, the historically recurring tension between erudition and originality is a problem yet to be solved. Since the seventeenth century, scientists have warned that erudition often encourages mere scholastic commentary on earlier writings instead of new empirical investigation and that a deep involvement with earlier ideas hobbles originality by producing inflexible sets of mind. But despite these dangers, great scientists have been able to combine erudition and original inquiry for the advancement of science either by reading only the immediately prior research devoted to their problem, which presumably incorporates the relevant cumulative knowledge of the past, or by exploring more remote sources only after their inquiry has been brought to a head. However, an extreme effort to emancipate oneself from an-

13. The first observation comes from Freud's letter to Pfister, 12 July 1909; the second from his "History of the Psychoanalytic Movement," *Collected Papers,* vol. 1 (London: Hogarth Press, 1949), 297.

14. The letter was addressed to Alexander Bryan Johnson and is printed in the new edition of his remarkable *Treatise on Language,* ed. David Rynin (Berkeley: University of California Press, 1959), 5–6.

tecedent ideas—as made by Comte—can deteriorate into the conscientious neglect of all the pertinent theory of the past and an artificial distinction between the history and systematics of theory.

The Functions of Classical Theory

Not even a founding father should be allowed to caricature the fundamental difference we have been investigating between authentic history and the systematics of sociological theory. For the distinction we have been emphasizing resembles Comte's little or not at all. A genuine *history* of sociological theory must extend beyond a chronologically ordered set of critical synopses of doctrine; it must deal with the interplay between theory and such matters as the social origins and statuses of its exponents, the changing social organization of sociology, the changes that diffusion brings to ideas, and their relations to the environing social and cultural structure. We want now to sketch out some distinctive functions for systematic theory of a thorough grounding in the classical formulations of sociological theory.

The condition of the physical and life sciences remains considerably different from that of the social sciences and of sociology in particular. Though the physicist *qua* physicist has no need to steep himself in Newton's *Principia* or the biologist *qua* biologist to read and reread Darwin's *Origin of Species,* the sociologist *qua* sociologist rather than as historian of sociology, has ample reason to study the works of a Weber, Durkheim, and Simmel and, for that matter, to turn back on occasion to the works of a Hobbes, Rousseau, Condorcet, or Saint-Simon.

The reason for this difference has been examined here in detail. The record shows that the physical and life sciences have generally been more successful than the social sciences in retrieving relevant cumulative knowledge of the past and incorporating it in subsequent formulations. This process of obliteration by incorporation[15] is still rare in sociology. As a result, previously unretrieved information is still there

15. Merton's conception of obliteration by incorporation (Merton, *On the Shoulders of Giants* [Chicago: University of Chicago Press, 1993], 312–13) identifies a distinct and theoretically important phenomenon in the history of thought: the forgetting of the sources or original formulations of ideas, methods, concepts, or findings by their incorporation in canonical knowledge, or even in the vernacular, so that only a few persons remain aware of their parentage. This process involves a paradox: on the one hand, it signifies acceptance of a scholar's contribution; on the other, it results in the loss of recognition for having made that contribution. As noted in the introduction, several of Merton's own contributions have become obliterated by incorporation.—*Ed.*

to be usefully employed as new points of departure. The present uses of past theory in sociology are still more complex as evidenced by the range of functions served by citations of classical theory.

One type of citation involves neither mere commentary on the classics nor the use of authority to establish credentials for current ideas. Instead this form of citation represents moments of affinity between our own ideas and those of our predecessors. More than one sociologist has had the self-deflating experience of finding that their independent discovery is an unwitting *rediscovery,* and, moreover, that the language of the classical prediscovery, long lost to view, is so crisp, so eloquent, or so implicative as to make their own version only second-best. In the ambivalent state of misery over having been preempted and joy at the beauty of the earlier formulation, the sociologist cites the classical idea.

Differing only by a nuance are citations to classical writings that come about when readers, stocked with their own ideas, find in the earlier book precisely what they already had in mind. The idea, still hidden from other readers, is noted precisely because it is congenial to the reader who has developed it independently. It is often assumed that to cite an earlier source *necessarily* means that the idea or finding in that citation first came to mind upon the reading of it. Yet the evidence often indicates that the earlier passage is noted only because it agrees with what the reader has already developed on his own. What we find here is that unlikely sounding event: a dialogue between the dead and the living. These do not differ much from dialogues between contemporary scientists in which each is delighted at the discovery that the other agrees with what was until then an idea held in solitude and perhaps even regarded as suspect. Ideas take on new validity when they are independently expressed by another, either in print or in conversation. The only advantage of coming upon it in print is that one knows there has been no inadvertent contagion between the book or article and one's own prior formulation of the same idea.

Sociologists conduct "dialogues" with classical formulations in still another way. A contemporary sociologist often comes upon a discussion in the classics questioning an idea that he was ready to affirm as sound. Reflections that ensue are sobering. The later theorist, forced to consider that he just might be mistaken, reexamines the idea in question, and if it is found in fact to be defective, that theorist reformulates it in a version that profits from the unrecorded dialogue.

A fourth function of the classics is that of providing a model for intellectual work. Exposure to such penetrating sociological minds as those of Durkheim and Weber helps us form standards of taste and judgment in identifying a *good* sociological problem—one that has significant implications for theory—and to learn what constitutes an apt theoretical solution to the problem. The classics are what Salvemini liked to call *libri fecondatori*—books that sharpen the faculties of exacting readers who give them their undivided attention. It is this process, presumably, that led the great and youthful Norwegian mathematician Niels Abel to record in his notebook: "It appears to me that if one wants to make progress in mathematics, one should study the masters and not the pupils."[16]

Finally, a classical sociological book or paper worth reading at all is worth rereading periodically. For part of what is communicated by the printed page changes as the result of an interaction between the dead author and the live reader. Just as the *Song of Songs* is different when it is read at age 17 and at age 70, so Weber's *Wirtschaft und Gesellschaft* or Durkheim's *Suicide* or Simmel's *Soziologie* differ when they are read at various times. For, just as new knowledge has a *retroactive effect* in helping us to recognize anticipations and adumbrations in earlier work, so changes in current sociological knowledge, problems, and foci of attention enable us to find *new* ideas in a work we had read before. The new context of recent developments in our own intellectual life or in the discipline itself brings into prominence ideas or hints of ideas that escaped notice in an earlier reading. Of course, this process requires intensive reading of the classics—the kind of concentration evidenced by that truly dedicated scholar (described by Edmund Wilson) who, interrupted at his work by a knock on the door, opened it, strangled the stranger who stood there, and then returned to his work.

As an informal check on the potentially creative function of rereading the classics, we need only examine the marginalia and notes we have taken on a classical work which has been read and then reread years later. If the book has precisely the same things to say to us the second time, we are suffering from severe intellectual stagnation or the classical work has less intellectual depth than has been attributed to it, or both unhappy conditions obtain.

What is a familiar experience in the intellectual life of the individual sociologist can become prevalent for entire generations of sociologists. For as each new generation accumulates its own repertoire of knowledge and thus becomes sensitized to new theoretical problems, it comes to see much that is "new" in earlier works, however often these works have been previously examined. There is much to be said for the reread-

16. The extract from Abel's notebook is recorded in Oystein Ore, *Niels Henrik Abel: Mathematician Extraordinary* (Minneapolis: University of Minnesota Press, 1957), 138.

ing of older works—particularly in an imperfectly consolidated discipline such as sociology—providing that this study consists of something more than that thoughtless mimicry through which mediocrity expresses its tribute to greatness. Rereading an older work through new spectacles allows contemporary sociologists to find fresh perceptions that were blurred in the course of firsthand research and, as a result, to consolidate the old, half-formed insight with newly developing inquiry.

All apart from reading the masters for the purposes of writing a history of sociological theory, then, acquaintance and reacquaintance with the classics have a variety of functions. These range from the direct pleasure of coming upon an aesthetically pleasing and more cogent version of one's own ideas, through the satisfaction of independent confirmation of these ideas by a powerful mind and the educative function of developing high standards of taste for sociological work, to the interactive effect of developing new ideas by turning to older writings within the context of contemporary knowledge. Each function derives from the imperfect retrieval of past sociological theory that has not yet been fully absorbed in subsequent thought. For that reason, sociologists in our time must continue to behave unlike their contemporaries in the physical and life sciences and devote more of themselves to close familiarity with their not-so-distant classical predecessors. But if they are to be effective rather than merely pious, if they are to *use* earlier formulations of theory rather than simply commemorate them, they must distinguish between the scholastic practice of commentary and exegesis and the scientific practice of extending antecedent theory.[17] And most important, sociologists must distinguish between the distinctive tasks of developing the history of sociological theory and developing its current systematics.

17. This, clearly, is what Whitehead meant by the dictum I adopted as an epigraph to *Social Theory and Social Structure* back in 1949: "A science which hesitates to forget its founders is lost."

2

Theoretical Pluralism (1981)

In its most general aspect, the concept of theoretical pluralism is of long standing and has turned up more or less independently in various disciplines. One of its chief exponents, Karl Popper, observed that its importance in the growth of scientific knowledge had been emphasized at the close of the nineteenth century by the geologist T. C. Chamberlin under the heading of "the method of multiple working hypotheses."[1] We need reach back no further than the 1940s to observe a pluralistic orientation being developed in sociology, economics, biology, and, for that matter, in literary criticism as well. The case for a plurality of theories, paradigms, or conceptual schemes was being developed in opposition to the quest for a single, all-embracing theory on grounds that have since been rather fully elucidated by philosophers of science.

In sociology during the 1940s, this division of opinion derived largely from pro-and-con responses to Talcott Parsons's mode of theorizing. He was anticipating and advocating a theoretical monism in which the then current theories advanced within "the professional group . . . should converge in the development of a single conceptual structure."[2] This monistic orientation was countered in a rather mild-mannered polemic:

> . . . when [Mr Parsons] suggests that our chief task is to deal with "theory" rather than with "theories," I must take strong exception.
>
> The fact is that the term "sociological theory," just as would be the case with the terms "physical theory" or "medical theory," is often misleading.

From "Foreword: Remarks on Theoretical Pluralism," in *Continuities in Structural Inquiry,* ed. Peter M. Blau and Robert K. Merton (London: Sage Publications, 1981), i–viii. Reprinted by permission of Sage Publications Ltd.

1. Karl R. Popper, "Replies to My Critics," in *The Philosophy of Karl Popper,* ed. Paul A. Schilpp (La Salle, Ill.: Open Court, 1974), 1187, 80n.

2. Talcott Parsons, "The Position of Sociological Theory," *American Sociological Review* 13 (April 1948): 156–64.

It suggests a tighter integration of diverse working theories than ordinarily obtains in any of these disciplines. Let me try to make clear what is here implied. Of course, every discipline has a strain toward logical and empirical consistency. Of course, the temporary co-existence of logically incompatible theories sets up a tension, resolved only when one or another of the theories is abandoned or so revised as to eliminate the inconsistency. Of course, also, every discipline has basic concepts, postulates and theorems which are the common resources of all theorists, irrespective of the special range of problems with which they deal. . . .

Of course, distinct theories often involve partly overlapping concepts and postulates. But the significant fact is that the progress of these disciplines consists in working out a large number of theories specific to certain types of phenomena and in exploring their mutual relations, and not in centering attention on "theory" as such. . . .

Sociological theory must advance on these interconnected planes: through special theories adequate to limited ranges of social data and through the evolution of a conceptual scheme adequate to *consolidate* groups of special theories.

To concentrate solely on special theories is to run the risk of emerging with ad hoc speculations consistent with a limited range of observations and inconsistent among themselves.

To concentrate solely on the major conceptual scheme for deriving all sociological theory is to run the risk of producing twentieth-century equivalents of the large philosophical systems of the past, with all their suggestiveness, all their architectonic splendor and all their scientific sterility.[3]

At much the same time, the economist Wassily Leontief was independently complementing the advocacy of theoretical pluralism *within* a discipline by advocacy of *inter-discipline* pluralism in the interpretation of historical development. There he was, in the late 1940s, noting the distinctive cognitive contributions of diverse disciplinary perspectives which, even when incommensurable in their internal logics, could be complementary:

. . . we face the choice between obdurate insistence on some *monistic interpretation*—which means overtaxing the analytical resources of one, chosen discipline and neglecting the capacities of all the others—or *practical pluralism.*

The pluralistic character of any single explanation reveals itself not in simultaneous application of essentially disparate types of considerations but rather in the ready shift from one type of interpretation to another. The justification of such *methodological eclecticism* lies—and this is the principal point of the argument that follows—in the limited nature of any type of

3. Robert K. Merton, "The Position of Sociological Theory," *American Sociological Review* 13 (April 1948): 164–68.

> interpretation or causation (I use the two terms interchangeably). Neither the economic, nor the anthropological, nor, say, geographical argument can, in the present state of the development of the respective disciplines, lead to the statement of uniquely defined necessities. Considering any given sequence of events *alternatively in the light of each one of such different approaches,* one can at best assign it to as many different ranges of "possibilities." Although the internal logics of the respective disciplines are *incommensurable,* the various ranges of possibilities thus derived are *comparable,* since all of them are described in terms of alternative developments of the same particular processes.[4]

Leontief thus centers on the idea central to theoretical pluralism: that though theoretical orientations may differ in their cogency when directed toward the same problems, their very difference of perspective typically leads them to focus on different rather than the same problems. As a result, the theories are often complementary or unconnected rather than contradictory. Failure to recognise this possibility often results in the mock controversies that recurrently pepper the history of the sciences. That this is not at all peculiar to sociology, or specifically to alternative modes of structural analysis, appears clear from the observations of the biologist-polymath Paul A. Weiss on the plurality of concepts, theories and orientations in various disciplines:

> Science has a good record of success in resolving *tenacious sham controversies by proving opposing tenets to be not mutually exclusive, but rather validly coexisting alternatives.* Scientific history abounds with scientific verdicts in which, on unassailably "objective" evidence, cases of supposedly irreconcilable contradictoriness were adjudicated by showing the conclusions of both contenders to have been valid. The complementarity principle of Bohr, affirming the right of coexistence of both a corpuscular and a wave concept of light; the duplicity theory of von Kries, establishing that both of two theories concerning the function of the retinal elements in color vision, formerly thought to be in conflict, were correct; the perennial fight between the embryological credos of preformation versus epigenesis—whether the whole array of organs in an adult organism is preformed as such in the egg in miniature form or whether all development is *de novo* to both concepts; all these are classical illustrations. . . .[5]

Nor is the concept of pluralism as involving complementary rather than inevitably contradictory theories limited to the sciences. In the domain of literary criticism during the 1940s and afterward, the Chicago

4. Wassily Leontief, "Notes on the Pluralistic Interpretation of History and the Problem of Interdisciplinary Cooperation," *The Journal of Philosophy* 45 (4 November 1948): 617–24.

5. Paul A. Weiss, *The Science of Life* (Mt. Kisco: Future Publishing Co., 1973), xii.

Neo-Aristotelians also argued the case for pluralism. As Ronald S. Crane, one of the chief architects of that school, put it in his introduction to one of their composite manifestos-and-exemplars:

> These essays . . . are as much an essential part of the program of the Chicago group as their essays in "Aristotelian" poetics. And *the only critical philosophy that underlies them all* is contained in the very un-Aristotelian attitude toward criticism, including the criticism of Aristotle, *which they have called "pluralism."* The term may be unfortunate: what they meant it to convey was simply their conviction that *there are and have been many valid critical methods, each of which exhibits the literary object in a different light, and each of which has its characteristic powers and limitations.*[6]

And in his exemplifying "Outline of Poetic Theory," Elder Olson, another of the same critical persuasion, went on to note:

> If a *plurality of valid and true kinds of criticism* is possible, choice must still be exercised, for it is impossible to employ all methods simultaneously. . . . Choice is determined by *the questions one wishes to ask and the form of answer one requires* and by the relative adequacy of given systems.[7]

This small sampling of pluralist perspectives could of course be easily enlarged—most easily by turning to the prime, disputatious discussions of pluralism and monism which have lately gathered force in the philosophy of science. But in these remarks, it must be enough only to refer to a few works by some of the principals in that ongoing energetic debate: the practically ubiquitous Popper and Kuhn, along with Naess, Lakatos, and the extreme formulations of Feyerabend, with some notice of the useful overviews of this clash of philosophical opinion which have been provided by Radnitsky and Klima.[8]

6. R. S. Crane, ed., *Critics and Criticism* (1952; reprint, Chicago: University of Chicago Press, 1970).

7. Elder Olson, "An Outline of Poetic Theory," (1952) in Crane, *Critics and Criticism*, 546–66.

8. Popper, "Replies to My Critics," in *Philosophy of Karl Popper;* Thomas S. Kuhn, *The Structure of Scientific Revolutions* (Chicago: University of Chicago Press, 1962, 1970) and Kuhn, *The Essential Tension: Selected Studies in Scientific Tradition and Change* (Chicago: University of Chicago Press, 1977); Arne Naess, *The Pluralist and Possibilist Aspect of the Scientific Enterprise* (Oslo: Universitetsforlager, 1972); Imre Lakatos, *The Methodology of Scientific Research Programmes* (Cambridge: Cambridge University Press, 1978); Paul Feyerabend, *Against Method: Outline of an Anarchistic Theory of Knowledge* (London: NLB, 1975); Gerald Radnitzky, "Theorienpluralismus-Theorienmonismus," ed. Alwin Diemer, *Der Methoden- und Theorienpluralismus in den Wissenschaften* (Meisenheim: Hain, 1971); Rolf Klima, "Theorienpluralismus in der Soziologie," in ibid., 198–219.

As Merton explains elsewhere, "The term 'theoretical pluralism' is adopted here in the broad sense of a plurality of hypotheses, ideas, or, for that matter, theories and par-

The few formulations by workers in various disciplines quoted here are perhaps enough to suggest that however much the disciplines differ in other respects, they are alike in perforce taking a single, all-embracing theoretical orientation which unifies the postulates of the discipline as an ultimate ideal rather than as a description of the actual state of the field. It is being proposed here that the plurality of current theories, paradigms, and thought-styles is not a mere happenstance, simply incidental to the development of each field of inquiry. Rather, it appears to be integral to the socially patterned, cognitive processes operating in the disciplines. With the institutionalization of science, the behavior of scientists oriented toward norms of organized skepticism and mutual criticism works to bring about such theoretical pluralism. As particular theoretical orientations come to be at the focus of attention of a sufficient number of workers in the field to constitute a thought collective, interactively engaged in developing a particular thought-style,[9] they give rise to a variety of new key questions requiring investigation. As the theoretical orientation is increasingly put to use, further implications become identifiable. And, in anything but a paradoxical sense, newly acquired knowledge produces newly acquired ignorance. For the growth of knowledge and understanding within a field of inquiry brings with it the growth of *specifiable and specified ignorance:* a new awareness of what is not yet known or understood and a rationale for its being worth the knowing. To the extent that current theoretical frameworks prove unequal to the task of dealing with some of the newly emerging key questions, there develops a composite social-and-cognitive pressure within the discipline for developing new or revised frameworks. But typically, the new does not completely crowd out the old, as earlier theoretical perspectives remain capable of dealing with problems distinctive of them.

This sort of process seems to have been at work to produce the diversity of theoretical orientations [in current structural analysis]. Each account of particular modes of structural inquiry is in effect both a cognitive and social announcement. Cognitively, it announces the critical

---

adigms involved in the growth of a scientific discipline. The term is not being employed in the special sense most emphatically and extensively used by Feyerabend and Klima, which not only advocates the 'proliferation of hypotheses' but, as Naess, Lakatos, and many another point out, argues a kind of methodological dadaism." ("Structural Analysis in Sociology," in *Sociological Ambivalence* [New York: Free Press, 1976], 137, 72n.)—*Ed.*

9. Ludwik Fleck, *Genesis and Development of a Scientific Fact,* ed. T. J. Trenn and Robert K. Merton (1935; reprint, Chicago: University of Chicago Press, 1979).

rationale of a more or less coherent set of structural problems, concepts, analytical procedures, and findings. Socially, it announces the commitment of a group of scholars to the particular theoretical orientation and, with or without the author's intent, it attempts to make a case for others in the discipline to adopt that orientation in their own work.

This twin cognitive-and-social character of such expositions is once again taken as integral, not incidental, to the development of a field of inquiry. The public adoption of one rather than another theoretical orientation is at the least a tacit (and often, an explicit) claim that it is more powerful and consequential in solving the cognitive problems it is designed to solve and more fruitful in raising new significant problems than rival orientations. After all, in terms of a cognitive rationale, why else would a scholar adopt a structural orientation to begin with and, within that generic frame of structural thought, why else adopt a Lévi-Straussian, Marxist, Weberian, or other orientation? [. . .]

It is also not sociologically incidental that theoretical pluralism should make for controversy and cognitive conflict. Sociologists are scarcely alone in appearing to be forever engaged in hot dispute. Whether they are more so than the generality of other scholars and scientists is not clear and, to the best of my knowledge, comparative rates of cognitive conflict among disciplines have yet to be investigated systematically. At any rate, the tribe of sociologists seems no more controversial than the articulate, internecine tribe of today's philosophers of science, with its clans engaged in vigorously announcing their own claims to sound knowledge while cheerfully denouncing the claims of others. To the extent that members of a thought-collective claim exclusive access to sound knowledge about a given region of phenomena they deny that there is truth in the ideas being advanced by cognitively opposed thought-collectives. This is a condition admirably calculated to produce increasingly passionate cognitive conflict and ensuing cognitive segregation. The process is somewhat as follows: deepened cognitive conflict leads to progressive alienation in which members of each thought-collective develop selective perceptions of the theoretical orientation being advanced by the others. These perceptions harden into self-confirming stereotypes. This, in turn, leads to reciprocal (and often well-founded) claims to having been misunderstood and, accordingly, misrepresented. In due course, full cognitive segregation sets in, with members of rival thought-collectives no longer making an active effort to examine the work of cognitively opposed collectives. No doubt this is the sort of thing which led Alfred Marshall, in properly ironic mood, to propose the "general rule that in discussions on method and scope, a man is nearly sure to be right when affirming the usefulness of his own procedure, and wrong when denying that of others."[10]

10. Alfred Marshall, *Principles of Economics* (1890; reprint, New York: Macmillan, 1961), 1:771n.

3

On Sociological Theories of the Middle Range ([1949] 1968)

Like so many words that are bandied about, the word "theory" threatens to become meaningless. Because its referents are so diverse—including everything from minor working hypotheses, through comprehensive but vague and unordered speculations, to axiomatic systems of thought—use of the word often obscures rather than creates understanding.

Strictly speaking, the term *sociological theory* refers to logically interconnected sets of propositions from which empirical uniformities can be derived. Here, the focus is on what I have called *theories of the middle range:* theories that lie between the minor but necessary working hypotheses that evolve in abundance during day-to-day research and the all-inclusive systematic efforts to develop a unified theory that will explain all the observed uniformities of social behavior, social organization, and social change.

Middle-range theory is principally used in sociology to guide empirical inquiry. It is intermediate to general theories of social systems which are too remote from particular classes of social behavior, organization, and change to account for what is observed and to those detailed orderly descriptions of particulars that are not generalized at all. Middle-range theory involves abstractions, of course, but they are close enough to observed data to be incorporated in propositions that permit empirical testing. Middle-range theories deal with delimited aspects of social phenomena, as is indicated by their labels. One speaks of a theory of reference groups, of social mobility, or of role-conflict and of the formation of social norms just as one speaks of a theory of prices, a germ theory of disease, or a kinetic theory of gases.

The seminal ideas in such theories are characteristically simple: consider Gilbert on magnetism, Boyle on atmospheric pressure, or Darwin

From "On Sociological Theories of the Middle Range," in *Social Theory and Social Structure,* 3d ed., rev. and enl. (New York: Free Press, 1968), 39–42, 45–48, 50–53. Reprinted by permission of The Free Press, an imprint of Simon and Schuster.

on the formation of coral atolls. Gilbert begins with the relatively simple idea that the earth may be conceived as a magnet; Boyle, with the simple idea that the atmosphere may be conceived as a "sea of air"; Darwin, with the idea that one can conceive of the atolls as upward and outward growths of coral over islands that had long since subsided into the sea. Each of these theories provides an image that gives rise to inferences. To take but one case: if the atmosphere is thought of as a sea of air, then, as Pascal had inferred, there should be less air pressure on a mountain top than at its base. The initial idea thus suggests specific hypotheses which are tested by seeing whether the inferences from them are empirically confirmed. The idea itself is tested for its fruitfulness by noting the range of theoretical problems and hypotheses that allow one to identify new characteristics of atmospheric pressure.

In much the same fashion, the theory of reference groups and relative deprivation starts with the simple idea, initiated by James, Baldwin, and Mead and developed by Hyman and Stouffer, that people take the standards of significant others as a basis for self-appraisal and evaluation. Some of the inferences drawn from this idea are at odds with common-sense expectations based upon an unexamined set of "self-evident" assumptions. Common sense, for example, would suggest that the greater the actual loss experienced by a family in a mass disaster, the more acutely it will feel deprived. This belief is based on the unexamined assumption that the magnitude of objective loss is related linearly to the subjective appraisal of the loss and that this appraisal is confined to one's own experience. But the theory of relative deprivation leads to quite a different hypothesis—that self-appraisals depend upon people's comparisons of their own situation with that of other people perceived as being comparable to themselves. This theory therefore suggests that, under specifiable conditions, families suffering serious losses will feel *less* deprived than those suffering smaller losses if they are in situations leading them to compare themselves to people suffering even more severe losses. For example, it is people in the area of greatest impact of a disaster who, though substantially deprived themselves, are most apt to see others around them who are even more severely deprived. Empirical inquiry supports the theory of relative deprivation rather than the common-sense assumptions: "the feeling of being relatively *better off* than others *increases with objective loss* up to the category of highest loss" and only then declines. This pattern is reinforced by the tendency of public communications to focus on "the *most extreme sufferers* [which] tends to fix them as a reference group against which even other sufferers can compare themselves favorably." As the inquiry develops, it is found that these patterns of self-appraisal in turn affect the distribution of morale in the community of survivors and their motivation to help others.[1] Within a particular class of behavior, therefore, the theory of relative deprivation directs us to a set of hypotheses that can be empirically tested. The confirmed conclusion can then be put simply enough: when few are hurt to much the same extent, the pain and loss of each seems great; where many are hurt in greatly varying degree, even fairly large losses seem small as they are compared with far larger ones. The probability that comparisons will be made is affected by the differing visibility of losses of greater and less extent.

The specificity of this example should not obscure the more general character of middle-range theory. Obviously, behavior of people confronted with a mass disaster is only one of an indefinitely large array of particular situations to which the theory of reference groups can be instructively applied, just as is the case with the theory of change in social stratification, the theory of authority, the theory of institutional interdependence, or the theory of anomie. But it is equally clear that such middle-range theories have not been logically derived from a single all-embracing theory of social systems, though once developed they may be consistent with one. Furthermore, each theory is more than a mere empirical generalization—an isolated proposition summarizing observed uniformities of relationships between two or more variables. A theory comprises a set of assumptions from which empirical generalizations have themselves been derived.

Another case of middle-range theory in sociology may help us to identify its character and uses. The theory of role-sets[2] begins with an image of how social status is organized in the social structure. This image is as simple as Boyle's image of the atmosphere as a sea of air or Gilbert's image of the earth as a magnet. As with all middle-range theories, however, the proof is in the using, not in the immediate response to the originating ideas as obvious or odd, as derived from more general theory or conceived of to deal with a particular class of problems.

Despite the diverse meanings attached to the concept of "social status," one sociological tradition consistently uses it to refer to a position in a social system, with its distinctive array of designated rights and obligations. In this tradition, as exemplified by Ralph Linton, the related concept of "social role" refers to the behavior of status-occupants that is oriented toward the patterned expectations of others (who ac-

1. Allen Barton, *Social Organization Under Stress: A Sociological Review of Disaster Studies* (Washington: National Academy of Sciences–National Research Council, 1963).

2. See "The Role-Set," chap. 10 in this volume.—*Ed.*

cord the rights and exact the obligations). Linton, like others in this tradition, went on to state the long recognized and basic observations that each person in society inevitably occupies multiple statuses and that each of these statuses has its associated role.

It is at this point that the imagery of role-set theory departs from this long-established tradition. The difference is initially a small one—some might say so small as to be insignificant—but the shift in the angle of vision leads to successively more fundamental theoretical differences. Role-set theory begins with the concept that each social status involves not a single associated role, but an array of roles. This feature of social structure gives rise to the concept of role-set: that complement of social relationships in which persons are involved simply because they occupy a particular social status. Thus, a person in the status of medical student plays not only the role of student vis-à-vis the correlative status of his teachers, but also an array of other roles relating him diversely to others in the system: other students, physicians, nurses, social workers, medical technicians, and the like. Again, the status of school teacher has its distinctive role-set which relates the teacher not only to the correlative status, pupil, but also to colleagues, the school principal and superintendent, the Board of Education, professional associations and, in the United States, local patriotic organizations.

Notice that the role-set differs from what sociologists have long described as "multiple roles." The latter term has traditionally referred not to the complex of roles associated with a single social status but to the various social statuses (often, in different institutional spheres) in which people find themselves—for example, one person might have the diverse statuses of physician, husband, father, professor, church elder, Conservative Party member, and army captain. (This complement of distinct statuses of a person, each with its own role-set, is a status-set.)[3] [. . .]

Total Systems of Sociological Theory

The quest for theories of the middle range exacts a distinctly different commitment from the sociologist than does the quest for an all-embracing unified theory. The pages that follow assume that this search for a total system of sociological theory, in which observations about every aspect of social behavior, organization, and change promptly find

3. The analysis of "status-sets" is carried forward in Merton, *Social Theory and Social Structure,* 422–38.—*Ed.*

their preordained place, has the same exhilarating challenge and the same small promise as those many all-encompassing philosophical systems which have fallen into deserved disuse. The issue must be fairly joined. Some sociologists still write as though they expect, here and now, formulation of *the* general sociological theory broad enough to encompass the vast ranges of precisely observed details of social behavior, organization, and change, and fruitful enough to direct the attention of research workers to a flow of problems for empirical research. This I take to be a premature and apocalyptic belief. We are not ready. Not enough preparatory work has been done.

A historical sense of the changing intellectual contexts of sociology should be sufficiently humbling to liberate these optimists from this extravagant hope. For one thing, certain aspects of our historical past are still too much with us. We must remember that early sociology grew up in an intellectual atmosphere in which vastly comprehensive systems of philosophy were being introduced on all sides. Any philosopher of the eighteenth and early nineteenth centuries worth his salt had to develop his own philosophical system—of these, Kant, Fichte, Schelling, Hegel were only the best known. Each system was a personal bid for the definitive overview of the universe of matter, nature, and man.

These attempts of philosophers to create total systems became a model for the early sociologists, and so the nineteenth century was a century of sociological systems. Some of the founding fathers, like Comte and Spencer, were imbued with the *esprit de système,* which was expressed in their sociologies as in the rest of their wider-ranging philosophies. Others, such as Gumplowicz, Ward, and Giddings, later tried to provide intellectual legitimacy for this still "new science of a very ancient subject." This required that a general and definite framework of sociological thought be built at once rather than developing special theories designed to guide the investigation of specific sociological problems within an evolving and provisional framework.

Within this context, almost all the pioneers in sociology tried to fashion their own systems. The multiplicity of systems, each claiming to be the genuine sociology, led naturally enough to the formation of schools, each with its cluster of masters, disciples and epigoni. Sociology not only became differentiated from other disciplines, but it also became internally differentiated. This differentiation, however, was not in terms of specialization, as in the sciences, but rather, as in philosophy, in terms of total systems, typically held to be mutually exclusive and largely at odds. As Bertrand Russell noted about philosophy, this total sociology did not seize "the advantage, as compared with the [sociologies] of the system-builders, of being able to tackle its problems one at a time, instead of having to invent at one stroke a block theory of the whole [sociological] universe."[4]

Another route has also been followed by sociologists in their quest to establish the intellectual legitimacy of their discipline: they have taken as their prototype systems of scientific theory rather than systems of philosophy. This path too has sometimes led to the attempt to create total systems of sociology—a goal that is often based on one or more of three basic misconceptions about the sciences.

The first misconception assumes that systems of thought can be effectively developed before a great mass of basic observations has been accumulated. According to this view, Einstein might follow hard on the heels of Kepler, without the intervening centuries of investigation and systematic thought about the results of investigation that were needed to prepare the terrain. The systems of sociology that stem from this tacit assumption are much like those introduced by the system-makers in medicine over a span of 150 years: the systems of Stahl, Boissier de Sauvages, Broussais, John Brown, and Benjamin Rush. Until well into the nineteenth century, eminent personages in medicine thought it necessary to develop a theoretical system of disease long before the antecedent empirical inquiry had been adequately developed.[5] These garden paths have since been closed off in medicine but this sort of effort still turns up in sociology. It is this tendency that led the biochemist and avocational sociologist, L. J. Henderson, to observe the following:

> A difference between most system-building in the social sciences and systems of thought and classification in the natural sciences is to be seen in their evolution. In the natural sciences both theories and descriptive systems grow by adaptation to the increasing knowledge and experience of the scientists. *In the social sciences, systems often issue fully formed from the mind of one man.* Then they may be much discussed if they attract attention, but *progressive adaptive modification as a result of the concerted efforts of great numbers of men is rare.*[6]

The second misconception about the physical and biological sciences rests on a mistaken assumption of historical contemporaneity—*that all cultural products existing at the same moment of history have the same degree of maturity*. In fact, to perceive differences here would

4. Bertrand Russell, *A History of Western Philosophy* (New York: Simon and Schuster, 1945), 834.

5. Wilfred Trotter, *Collected Papers* (New York: Oxford University Press, 1941), 150.

6. Lawrence J. Henderson, *The Study of Man* (Philadelphia: University of Pennsylvania Press, 1941), 19–20 [italics added].

be to achieve a sense of proportion. The fact that the discipline of physics and the discipline of sociology are both identifiable in the mid-twentieth century does not mean that the achievements of the one should be the measure of the other. True, social scientists today live at a time when physics has achieved comparatively great scope and precision of theory and experiment, a great aggregate of tools of investigation, and an abundance of technological by-products. Looking about them, many sociologists take the achievements of physics as the standard for self-appraisal. They want to compare biceps with their bigger brothers. They, too, want to *count.* And when it becomes evident that they neither have the rugged physique nor pack the murderous wallop of their big brothers, some sociologists despair. They begin to ask: is a science of society really possible unless we institute a total system of sociology? But this perspective ignores the fact that between twentieth-century physics and twentieth-century sociology stand billions of hours of sustained, disciplined, and cumulative research. Perhaps sociology is not yet ready for its Einstein because it has not yet found its Kepler—to say nothing of its Newton, Laplace, Gibbs, Maxwell, or Planck.

Third, sociologists sometimes misread the actual state of theory in the physical sciences. This error is ironic, for physicists agree that they have not achieved an all-encompassing system of theory, and most see little prospect of it in the near future. What characterizes physics is an array of special theories of greater or less scope, coupled with the historically grounded hope that these will continue to be brought together into families of theory. As one observer puts it, "though most of us hope, it is true, for an all-embracive future theory which will unify the various postulates of physics, we do not wait for it before proceeding with the important business of science."[7] More recently, the theoretical physicist Richard Feynman reported without dismay that "today our theories of physics, the laws of physics, are a multitude of different parts and pieces that do not fit together very well."[8] But perhaps most telling is the observation by that most comprehensive of theoreticians who devoted the last years of his life to the unrelenting and unsuccessful search "for a unifying theoretical basis for all these single disciplines, consisting of a minimum of concepts and fundamental relationships, from which all the concepts and relationships of the single disciplines might be derived by logical process." Despite his own profound and lonely commitment to this quest, Einstein observed that

7. Henry Margenau, "The Basis of Theory in Physics," unpublished ms., 1949, 5–6.

8. Richard Feynman, *The Character of Physical Law* (London: Cox and Wyman Ltd., 1965), 30.

> The greater part of physical research is devoted to the development of the various branches in physics, in each of which the object is the theoretical understanding of more or less restricted fields of experience, and in each of which the laws and concepts remain as closely as possible related to experience.[9]

These observations might be pondered by those sociologists who expect a sound general system of sociological theory in our time—or soon after. If the science of physics, with its centuries of enlarged theoretical generalizations, has not managed to develop an all-encompassing theoretical system, then *a fortiori* the science of sociology, which has only begun to accumulate empirically grounded theoretical generalizations of modest scope, would seem well advised to moderate its aspirations for such a system. [. . .]

Total Systems of Theory and Theories of the Middle Range

From all this it would seem reasonable to suppose that sociology will advance insofar as its major (but not exclusive) concern is with developing theories of the middle range, and it will be retarded if its primary attention is focussed on developing total sociological systems. So it is that in his inaugural address at the London School of Economics, T. H. Marshall put in a plea for sociological "stepping-stones in the middle distance."[10] Our major task today is to develop special theories applicable to limited conceptual ranges—theories, for example, of deviant behavior, the unanticipated consequences of purposive action, social perception, reference groups, social control, the interdependence of social institutions—rather than to seek immediately the total conceptual structure that is adequate to derive these and other theories of the middle range.

Sociological theory, if it is to advance significantly, must proceed on these interconnected planes: (1) by developing special theories from which to derive hypotheses that can be empirically investigated and (2) by evolving, not suddenly revealing, a progressively more general conceptual scheme that is adequate to consolidate groups of special theories.

To concentrate entirely on special theories is to risk emerging with

9. Albert Einstein, "The Fundamentals of Theoretical Physics," in *Great Essays by Nobel Prize Winners*, ed. L. Hamalian and E. L. Volpe (New York: Noonday Press, 1960), 219–30 at 220.

10. The inaugural lecture was delivered February 21, 1946. It is printed in T. H. Marshall, *Sociology at the Crossroads* (London: Heinemann, 1963), 3–24.

specific hypotheses that account for limited aspects of social behavior, organization, and change but that remain mutually inconsistent.

To concentrate entirely on a master conceptual scheme for deriving all subsidiary theories is to risk producing twentieth century sociological equivalents of the large philosophical systems of the past, with all their varied suggestiveness, their architectonic splendor, and their scientific sterility. The sociological theorist who is *exclusively* committed to the exploration of a total system with its utmost abstractions runs the risk that, as with modern decor, the furniture of his mind will be bare and uncomfortable.

The road to effective general schemes in sociology will only become clogged if, as in the early days of sociology, each charismatic sociologist tries to develop a general system of theory. The persistence of this practice can only make for the balkanization of sociology, with each principality governed by its own theoretical system. Though this process has periodically marked the development of other sciences—conspicuously, chemistry, geology, and medicine—it need not be reproduced in sociology if we learn from the history of science. We sociologists can look instead toward progressively comprehensive theory which, instead of proceeding from the head of one sociologist, gradually consolidates theories of the middle range, so that these become special cases of more general formulations.

Developments in sociological theory suggest that emphasis on this orientation is needed. Note how few, how scattered, and how unimpressive are the specific sociological hypotheses *derived* from a master conceptual scheme. [. . .] A large part of what is now described as sociological theory consists of *general orientations toward data, suggesting types of variables which theories must somehow take into account, rather than clearly formulated, verifiable statements of relationships between specified variables.* We have many concepts but fewer confirmed theories; many points of view, but few theorems; many "approaches" but few arrivals. Perhaps some further changes in emphasis would be all to the good.

Consciously or unconsciously, people allocate their scant resources as much in the production of sociological theory as they do in the production of plumbing supplies, and their allocations reflect their underlying assumptions. Our discussion of middle-range theory in sociology is intended to make explicit a policy decision faced by all sociological theorists. Which shall have the greater share of our collective energies and resources: the search for confirmed theories of the middle range or the search for an all-inclusive conceptual scheme? I believe—and beliefs are of course notoriously subject to error—that theories of the middle range hold the largest promise, provided that the search for them is coupled with a pervasive concern with consolidating special theories into more general sets of concepts and mutually consistent propositions. Even so, we must adopt the provisional outlook of our big brothers and of Tennyson:

Our little systems have their day;
They have their day and cease to be.

[. . .]

4

Specified Ignorance (1987)

[. . .] It was Francis Bacon who made "the advancement of learning" a watchword in the culture of science emerging in the seventeenth century. From then till now, efforts to understand how science develops have largely centered on the modes of replacing ignorance with knowledge, with little attention paid to the formation of a useful kind of ignorance, as distinct from the manifestly dysfunctional kind. Karl Popper provides the distinctive contemporary exception that illuminates the rule, most powerfully in his analytical essay "On the Sources of Knowledge and of Ignorance."[1] The general inattention to the formation of useful ignorance has long obtained as well in the sociology of scientific knowledge [but now see Smithson[2] along with the early collateral paper on the functions of ignorance in social life by Moore and Tumin[3]].

These retrospective notes focus on the dynamic cognitive role played by the particular form of ignorance I describe as "specified ignorance": "the express recognition of what is not yet known but needs to be known in order to lay the foundation for still more knowledge."[4] "As the history of thought, both great and small, attests, *specified* ig-

From "Three Fragments from a Sociologist's Notebooks," *Annual Review of Sociology* 13 (1987): 6–10. Reprinted by permission of Annual Reviews, Inc.

1. Karl Popper, *Conjectures and Refutations: The Growth of Scientific Knowledge* (1960; reprint, London: Routledge and Kegan Paul, 1962).

2. M. Smithson, "Toward a Social Theory of Ignorance," *Journal of the Theory of Social Behavior* 15 (1985): 149–70.

3. Wilbert E. Moore and Melvin M. Tumin, "Some Social Functions of Ignorance," *American Sociological Review* 14 (1949): 787–95.

4. Robert K. Merton, "The Precarious Foundations of Detachment in Sociology" in *The Phenomenon of Sociology*, ed. E. A. Tiryakian (New York: Appleton-Century-Crofts, 1971), 191.

norance is often a first step toward supplanting that ignorance with knowledge."[5]

The concept of specified ignorance hints at various other kinds and shades of acknowledged ignorance in science. The familiar kind of a general, rote, and vague admission of ultimate ignorance serves little direct cognitive purpose though it may have symbolic significance in reminding us of our limitations. This kind, however, does not issue in definite questions requiring further inquiry. And vague questions evoke dusty answers. After all, in the domain of science, it takes no great courage or skill to acknowledge a general want of knowledge. It is not merely that Socrates set an ancient pattern of announcing one's ignorance. Beyond that, the values of modern science have long put a premium on the public admission of one's limitations or the expression of humility in the face of the vast unknown. Scientists of epic stature have variously insisted on how little they have come to know and to understand in the course of their lives. We remember Galileo teaching himself and his pupils to reiterate "I do not know." And then, inevitably, one recalls the "memorable sentiment" reportedly uttered by Newton "a short time before his death":

> I do not know what I may appear to the world, but to myself I seem to have been only like a boy playing on the seashore, and diverting myself in now and then finding a smoother pebble or prettier shell than ordinary, whilst the great ocean of truth lay all undiscovered before me.[6]

Or again, Laplace—the French Newton—is said to have put much the same sentiment in a typically Gallic epigram: "What we know is not much; what we do not know is immense."[7] What the mathematician Bell describes elsewhere as "a common and engaging trait of the truly eminent scientist [found] in his frequent confession of how little he knows"[8] can be identified sociologically as the living up to a normative expectation of ultimate humility in a community of sometimes egocentric scientists. It is not simply that a goodly number of scientists happen to express these self-belittling sentiments; they are applauded for doing so.

But of course these paradigmatic figures in science do not confine

5. Robert K. Merton, *Social Theory and Social Structure* (Glencoe, Ill.: Free Press, 1957), 417.

6. David Brewster, *Memoirs of the Life, Writings, and Discoveries of Sir Isaac Newton,* 2 vols. (Edinburgh: Thomas Constable, 1855), 407.

7. E. T. Bell, *Men of Mathematics* (New York: Simon and Schuster, 1937), 172.

8. E. T. Bell, "Mathematics and Speculation," *Scientific Monthly* 32 (1931): 193–209, at 204.

themselves to such generic confessions of ignorance as may reinforce the norm of a decent humility without directly shaping the growth of scientific knowledge. They repeatedly adopt the cognitively consequential practice of specifying this or that piece of ignorance derived from having acquired the added degree of knowledge that made it possible to identify definite portions of the still unknown. In workaday science, it is not enough to confess one's ignorance; the point is to specify it. That, of course, amounts to instituting, or finding, a new, worthy, and soluble scientific problem.

Thus, as I have had occasion to propose, the process of successive specification of our ignorance in light of newfound knowledge provides a recurrent sociocognitive pattern:

> As particular theoretical orientations come to be at the focus of a sufficient number of workers in the field to constitute a thought collective, interactively engaged in developing a distinctive thought style (L. Fleck, *Genesis and Development of a Scientific Fact,* [Chicago: University of Chicago Press, 1979]) they give rise to a variety of key questions requiring investigation. As the theoretical orientation is put to increasing use, further implications become identifiable. In anything but a paradoxical sense, newly acquired knowledge produces newly acquired ignorance. For the growth of knowledge and understanding within a field of inquiry brings with it the growth of *specifiable and specified ignorance:* a new awareness of what is not yet known or understood and a rationale for its being worth the knowing. To the extent that current theoretical frameworks prove unequal to the task of dealing with some of the newly emerging key questions, there develops a composite social-and-cognitive pressure within the discipline for new or revised frameworks. But typically, the new does not wholly crowd out the old, as [long as] earlier theoretical perspectives remain capable of dealing with problems distinctive of them.[9]

It requires a newly informed theoretical eye to detect long-obscured pockets of ignorance as a prelude to newly focussed inquiry. Each theoretical orientation or paradigm has its own problematics, its own sets of specified questions. As these questions about selected aspects of complex phenomena are provisionally answered, the new knowledge leads some scientists both within and without the given thought collective to become aware of other, newly identified aspects of the phenomena. There then develops a succession of specified ignorance.

As a case in point, consider the sociological theory of deviant behav-

9. Robert K. Merton, "Remarks on Theoretical Pluralism" in *Continuities in Structural Inquiry,* ed. Peter M. Blau and Robert K. Merton (London: Sage, 1981), v–vi. (See chap. 2 in this volume.—Ed.)

ior as it was developed in four thought collectives.[10] Initiated in the 1920s, E. H. Sutherland's theory of differential association centered on the problem of the *social transmission* of deviant behavior.[11] Its key question therefore inquired into the modes of socialization through which patterns of deviant behavior are learned from others. But as the brilliant philosopher of literature, Kenneth Burke, has reminded us, "A way of seeing is also a way of not seeing—a focus upon object A involves a neglect of object B."[12] In this case, Sutherland's focus on the acquisition of these deviant patterns left largely untouched specifiable ignorance about the ways in which the patterns emerge in the first place.

Upon identifying that pocket of theoretical neglect, Merton proposed the theory of anomie-and-opportunity structures,[13] that rates of various types of deviant behavior tend to be high among people so located in the social structure as to have little access to socially legitimate pathways for achieving culturally induced personal goals. The Sutherland and Merton theories were consolidated and extended by Cohen, who proposed that delinquency subcultures arise as adaptations to this disjunction between culturally induced goals and the legitimate opportunity structure, and by Cloward and Ohlin, who proposed that the social structure also provides differential access to *illegitimate* opportunities.[14] Since that composite of theories centered on socially structured *sources* of deviant behavior, it had next to nothing to say about how these patterns of misbehavior are transmitted or about how these initial departures from the social rules sometimes crystallize into deviant careers, yet another sphere of specifiable ignorance.

That part of the evolving problematics was taken up in labeling (or societal reaction) theory as initiated by Lemert and Becker and advanced by Erikson, Cicourel, and Kitsuse.[15] It centered on the processes

10. I draw upon the summary in Robert K. Merton, "The Sociology of Social Problems" in *Contemporary Social Problems,* ed. Robert K. Merton and R. A. Nisbet (New York: Harcourt Brace Jovanovich, 1976), 3–43.

11. Albert K. Cohen et al., eds., *The Sutherland Papers* (Bloomington: Indiana University Press, 1956).

12. Kenneth Burke, *Permanence and Change* (New York: New Republic, 1935), 70.

13. Robert K. Merton, "Social Structure and Anomie," *American Sociological Review* 3 (1938): 672–82. (See chap. 12 in this volume.—Ed.)

14. Cf. Robert K. Merton, "Opportunity Structure: The Emergence, Diffusion, and Differentiation of a Sociological Concept" in *The Legacy of Anomie Theory,* ed. Freda Adler and William S. Laufer (New Brunswick, New Jersey: Transaction Publishers, 1995), 3–78. (Chap. 13 in this volume.)—*Ed.*

15. Edwin M. Lemert, *Social Pathology* (New York: McGraw-Hill, 1951); Howard S. Becker, *Outsiders* (1963; reprint, New York: Free Press, 1973); Kai T. Erikson,

through which some people are assigned a social identity by being labeled as "delinquents," "criminals," "psychotics," and the like and how, by responding to such stigmatization, they enter upon careers as deviants. In Becker's words, "Treating a person as though he were generally rather than specifically deviant produces a self-fulfilling prophecy.[16] It sets in motion several mechanisms which conspire to shape the person in the image people have of him."[17] With this problem as its focus, labeling theory has little to say about the sources of primary deviance or the making of societal rules defining deviance. As Lemert specified this ignorance, "When attention is turned to the rise and fall of moral ideas and the transformation of definitions of deviance, labeling theory and ethnomethodology do little to enlighten the process."[18]

It is precisely this problem that the conflict theory of deviance took as central. Its main thrust, as variously set forth by Turk[19] and Quinney,[20] for example, holds that a more or less homogeneous power elite incorporates its interests in the making and imposing of legal rules. It thus addresses questions neglected by the earlier theories: How do legal rules get formulated, how does this process affect their substance, and how are they differentially administered?

The case of deviance theory indicates how a dimly felt sense of sociological ignorance was successively specified for one class of social phenomena. But it is not yet known whether scientific disciplines differ in the practice of specifying ignorance—in the extent to which their practitioners state what it is about an established phenomenon that is not yet known and *why it matters* for generic knowledge that it become known.[21] Such specified ignorance is at a far remove from the familiar

"Notes on the Sociology of Deviance" in *The Other Side: Perspectives on Deviance,* ed. Howard S. Becker (Glencoe, Ill.: Free Press, 1964), 9–21; Aaron V. Cicourel, *The Social Organization of Juvenile Justice* (New York: Wiley, 1968); John I. Kitsuse, "Societal Reaction to Deviant Behavior" in *The Other Side,* ed. Becker, 87–102.

16. See chap. 16 in this volume.—*Ed.*

17. Becker, ed., *Outsiders,* 34.

18. Edwin M. Lemert, "Beyond Mead: The Societal Reaction to Deviance" *Social Problems* 21:462.

19. Austin Turk, *Criminality and the Legal Order* (Chicago: Rand McNally, 1969).

20. Richard Quinney, *The Social Reality of Crime* (Boston: Little, Brown, 1970).

21. Mathematics, of course, has a long tradition of publishing fundamental problems (long ago, in the form of challenges). Upon reading this portion of the chapter, my colleagues, Joshua Lederberg and Eugene Garfield, informed me of their episodic interest in institutionalizing what amounts to the specification of ignorance. For one expression of that interest in print, see Garfield's "The Unanswered Questions of Science" (*Current Contents,* 5 June 1974:5–6). Lederberg has made me the beneficiary of his 1974 permuterm bibliography entitled "Unsolved Problems" in the various sciences and has referred me to a specimen volume entitled *100 Problems in Environmental Health: A Col-*

rote sentence which concludes not a few scientific papers to the effect that "more research is needed." Serendipity aside, questions not asked are questions seldom answered. The specification of ignorance amounts to problem-finding as a prelude to problem-solving.

It is being proposed that the socially defined role of the scientist calls for both the augmenting of knowledge and the specifying of ignorance. Just as yesterday's uncommon knowledge becomes today's common knowledge, so yesterday's unrecognized ignorance becomes today's specified ignorance.[22] As new contributions to knowledge bring about a new awareness of something else not yet known, the sum of manifest human ignorance increases along with the sum of manifest human knowledge.

---

lection of Promising Research Problems (Jack E. McKee et al. [Washington, D.C.: Jones Composition, 1961]). My attention was also redirected to that superb and lively anthology I had misplaced, *The Scientist Speculates: An Anthology of Partly Baked Ideas* (Irving J. Good [New York: Basic Books, 1962]), which is designed "to raise more questions than it answers." Of particular interest is the piece in the anthology happily entitled "Ignoratica" by one Félix Serratosa who ascribes the essential idea of a "science of unknowns" to the explosive imagination of that prolific and often paradoxical Florentine critic, novelist, poet, and journalist Giovanni Papini. However that may all be, it can be said in self-exemplifying style: That the specification of ignorance is indispensable to the advancement of knowledge, I do not doubt; whether disciplines do differ notably in the practice of such specification, I do not know. Since the phenomenon is not yet established, I do not undertake to explain such possible variation. But one can still speculate . . .

22. Merton, *Social Theory and Social Structure,* 417; Karl Popper, *Conjectures and Refutations: The Growth of Scientific Knowledge* (1960; reprint, London: Routledge and Kegan Paul, 1962); Piotr Sztompka, *Robert K. Merton: An Intellectual Profile* (New York: St. Martin's Press, 1986), 97–98.

5

Paradigms: The Codification of Sociological Theory (1949)

A major concern of this book is the codification of substantive theory and of procedures of qualitative analysis in sociology. As construed here, codification is the orderly and compact arrangement of fruitful procedures of inquiry and the substantive findings that result from this use. This process entails identification and organization of what has been implicit in work of the past rather than the invention of new strategies of research.

The following chapters, dealing with functional and structural analysis, set forth paradigms as a basis for codifying work in these fields.[1] I believe that such paradigms have great propaedeutic value. For one thing, they bring out into the open the array of assumptions, concepts, and basic propositions employed in a sociological analysis. They thus reduce the inadvertent tendency to hide the hard core of analysis behind a veil of random, though possibly illuminating, comments and thoughts. Despite the appearance of propositional inventories, sociology still has few formulas—that is, highly abbreviated symbolic expressions of relationships between sociological variables. Consequently, sociological interpretations tend to be discursive. The logic of procedure, the key concepts, and the relationships between them often

From "Paradigms: The Codification of Sociological Theory," in *Social Theory and Social Structure* (Glencoe, Ill.: Free Press, 1949), 12–16. Reprinted by permission of The Free Press, an imprint of Simon and Schuster. Also from "Appendix: A Note on the Use of Paradigms in Qualitative Analysis," in *Sociological Ambivalence and Other Essays* (New York: Free Press, 1976), 213–16. Reprinted by permission of The Free Press, an imprint of Simon and Schuster.

1. I have set forth other paradigms: on deviant social behavior [see chap. 12 in this volume]; on the sociology of knowledge [see chap. 17 in this volume]; on racial intermarriage in "Intermarriage and the Social Structure," *Psychiatry* 4 (1941), 361–74; on racial prejudice and discrimination in "Discrimination and the American Creed" in *Discrimination and National Welfare*, ed. R. M. MacIver (New York: Harper and Brothers, 1948).

become lost in an avalanche of words. When this happens, critical readers must laboriously glean for themselves the implicit assumptions of the author. The paradigm reduces this tendency for the theorist to employ tacit concepts and assumptions.

Contributing to the tendency for sociological exposition to become lengthy rather than lucid is the tradition—inherited slightly from philosophy, substantially from history, and greatly from literature—of writing sociological accounts vividly and intensely to convey all the rich fullness of the human scene. The sociologist who does not disavow this handsome but alien heritage becomes intent on searching for the exceptional constellation of words that will best express the *particularity* of the sociological case in hand, rather than on seeking out the objective, generalizable concepts and relationships it exemplifies—the core of a science, as distinct from the arts. Too often, this misplaced use of genuine artistic skills is encouraged by the plaudits of a lay public, gratefully assuring the sociologist that he writes like a novelist and not like an overly domesticated and academically henpecked Ph.D. Not infrequently, the sociologist pays for this popular applause, for the closer one approaches eloquence, the farther one retreats from methodical sense. It must be acknowledged, however, as St. Augustine suggested in mild rebuttal long ago, that ". . . a thing is not necessarily true because badly uttered, nor false because spoken magnificently." [. . .]

Since sound sociological interpretation inevitably implies some theoretical paradigm, it seems the better part of wisdom to bring it out into the open. If true art consists in concealing all signs of art, true science consists in revealing its scaffolding as well as its finished structure.

Without pretending that this tells the whole story, I suggest that paradigms for qualitative analysis in sociology have at least five closely related functions.[2]

First, paradigms have a notational function. They provide a compact arrangement of the central concepts and their interrelations that are utilized for description and analysis. Setting out concepts in sufficiently small compass to allow their *simultaneous* inspection is an important aid in the self-correction of one's successive interpretations—a goal hard to achieve when the concepts are scattered throughout discursive exposition. (As the historian of mathematics Florian Cajori reminds us, this is one of the important functions of mathematical sym-

2. For a critical appraisal of this discussion, see Don Martindale, "Sociological Theory and the Ideal Type," in *Symposium on Sociological Theory,* ed. Llewellyn Gross (Evanston: Row, Peterson, 1959), 57–91, at 77–80.

bols: they provide for the simultaneous inspection of all terms entering into the analysis.)

Second, paradigms lessen the likelihood of inadvertently introducing hidden assumptions and concepts, for each new assumption and each new concept must be either logically derived from previous components of the paradigm or explicitly introduced into it. The paradigm thus provides a guide for avoiding ad hoc (i.e., logically irresponsible) hypotheses.

Third, paradigms advance the cumulation of theoretical interpretation. In effect, the paradigm is the foundation upon which the house of interpretations is built. If a new story cannot be built directly upon this foundation, then it must be treated as a new wing of the total structure, and the foundation of concepts and assumptions must be extended to support this wing. Moreover, each new story that *can* be built upon the original foundation strengthens our confidence in its substantial quality just as every new extension, precisely because it requires an additional foundation, leads us to suspect the soundness of the original substructure. A paradigm worthy of great confidence will in due course support an interpretative structure of skyscraper dimensions, with each successive story testifying to the well-laid quality of the original foundation, while a defective paradigm will support only a rambling one-story structure, in which each new set of uniformities requires a new foundation to be laid, since the original cannot bear the weight of additional stories.

Fourth, paradigms, by their very arrangement, suggest the systematic cross-tabulation of significant concepts and can thus sensitize the analyst to empirical and theoretical problems which might otherwise be overlooked.[3] Paradigms promote analysis rather than the description of concrete details. They direct our attention, for example, to the components of social behavior, to possible strains and tensions among these components, and thereby to sources of departures from the behavior which is normatively prescribed.

Fifth, paradigms make for the codification of qualitative analysis in a way that approximates the logical if not the empirical rigor of quantitative analysis. The procedures for computing statistical measures and their mathematical bases are codified as a matter of course; their assumptions and procedures are open to critical scrutiny by all. By con-

3. Although they express doubts about the uses of systematic theory, Joseph Bensman and Arthur Vidich have admirably exhibited this heuristic function of paradigms in their instructive paper, "Social Theory in Field Research," *American Journal of Sociology* 65 (1960): 577–84.

trast, the sociological analysis of qualitative data often resides in a private world of penetrating but unfathomable insights and ineffable understandings. Indeed, discursive expositions not based upon paradigms often include perceptive interpretations. As the cant phrase has it, they are rich in "illuminating insights." But it is not always clear just which operations on which analytic concepts were involved in these insights. In some quarters, even the suggestion that these intensely private experiences must be reshaped into publicly certifiable procedures if they are to be incorporated into the science of society is taken as a sign of blind impiety. Yet the concepts and research procedures of even the most perceptive of sociologists must be reproducible and the results of their insights testable by others. Science, and this includes sociological science, is public, not private. It is not that we ordinary sociologists wish to cut all talents to our own small stature; it is only that the contributions of the great and small alike must be codified if they are to advance the development of sociology.

All virtues can easily become vices merely by being carried to excess, and this applies to the sociological paradigm. It is a temptation to mental indolence. Equipped with a paradigm, sociologists may shut their eyes to strategic data not expressly called for by the paradigm. Thus it can be turned from a sociological field-glass into a sociological blinder. Misuse results from absolutizing the paradigm rather than using it as a tentative point of departure. But if they are recognized as provisional and changing, destined to be modified in the immediate future as they have been in the recent past, these paradigms are preferable to sets of tacit assumptions.

Paradigms in Qualitative Analysis

Upon proposing this conception of paradigms in sociology back in the 1940s, I discovered that it was regarded as an unusual, not to say bizarre, usage. One candid friend went so far as to inform me that the notion of a "paradigm" was appropriate only as an exemplar for declension or conjugation that exhibits all the inflectional forms of a class of words. In rebuking me, he of course managed to put aside Plato's idea of paradeigmata as well as centuries-long usage of the word in the extended sense of pattern or exemplar. Over the past quarter-century, the notion of paradigm in the indicated sense became thoroughly domesticated, not alone in sociology and psychology but in other social and behavioral disciplines as well.

With the appearance in 1962 of Thomas S. Kuhn's vastly consequential book, *The Structure of Scientific Revolutions*, the term "paradigm" has acquired a substantially different set of meanings and far wider usage. In a recent overview,[4] Raymond Firth instructively summarizes the differences and, to some degree, the conceptual relations between the two usages, in a passage which can be instructively quoted at length:

> Some might characterize the present situation in social anthropology as the paradigmatic phase. Paradigm has become the key word for a lot of interpretation. Paradigm, a word derived from classical sources, has been in use in English since at least the seventeenth century to mean a pattern to follow, an exemplar. It also has a hint of providing the basic components underlying any variation which a phenomenon might assume. In this sense the term was used by Robert Merton long ago when he was arguing for stricter methodology and greater awareness of the theoretical framework of sociological analysis. Merton used what he called the device of the analytical paradigm to present in a succinct way "codified materials" on concepts, procedures, and inferences over a range of problems from the requirements of functional analysis to social pressures leading to deviant behavior. Merton pointed out that any sound sociological analysis inevitably implies some theoretical paradigm, and he held that an explicit statement of such analytical model allows assumptions which would otherwise be hidden to be brought to the surface and laid open to scrutiny. He also argued that such analytical paradigms suggest systematic cross-tabulation of concepts, help to give more rigor in codifying qualitative data, and generally aid the symbolic expression of relationships between sociological variables. (The synoptic charts used by Malinowski for study of Trobriand agriculture or African culture change are examples of analogous paradigms primarily directed to research in the field.)
>
> Modern social anthropologists use the notion of paradigm rather differently, borrowing it not from Merton but from de Saussure and Thomas Kuhn. Moreover, one tendency is to attribute the paradigm not to the analytical observer but to the people whose behavior and ideas are being analyzed. In this setting a paradigm is a framework of ideas which the people have for envisaging and dealing with a specific set of circumstances and problems relating thereto—a kind of mental map of a sector of the natural and social world. [. . .]
>
> Paradigms tend to be concerned with type rather than with instance, with thought rather than with action. Paradigmatic structures are essentially structures of ideas rather than structures of social relationships. As such, they can be conceived as having creative power for the actor. For the

4. Raymond Firth, "An Appraisal of Modern Social Anthropology," *Annual Review of Anthropology* 4 (1975): 1–25, at 12, 15. For other recent discussions of related kind, see Robert W. Friedrichs, "Dialectical Sociology: An Exemplar for the 1970s," *Social Forces* 50 (1973): 447–55; Raymond Boudon, "Notes sur la notion de théorie dans les sciences sociales," *Archives Européennes de Sociologie* 11 (1970): 201–51; S. B. Barnes, "Paradigms—Scientific and Social," *Man* 4 (1969): 94–102.

analyst, they can appear to have a higher power of comparison and prediction through their more abstract, analogic quality. But sharp-edged instruments need more care in handling. Sometimes the notion of paradigm is used in almost a mechanistic sense: one is given the impression of people being fitted out with portfolios of paradigms which encode their circumstances and experiences—pulling out the appropriate blueprint for use as occasions present themselves. And the notion of collectivity inherent in most definitions of paradigm carries with it the well-known difficulties of abstraction—in the last resort, whose paradigm is it, the actor's or the analyst's? Long ago, Merton pointed to some of these dangers of possible abuse of the sociological paradigm. He called it roundly a temptation to mental indolence, to shutting one's eyes to strategic data not expressly called for in the paradigm, to using it as a sociological blinker rather than a sociological fieldglass. "Misuse results from absolutizing the paradigm rather than using it tentatively, as a point of departure." [. . .] So, granted the interpretative value as well as the intellectual excitement of analysis in terms of paradigm and allied concepts, there is also a good case for analysis in more direct behavioral and social terms. A focus on the world of ideas should complement, not replace, a focus on the world of action.[5]

5. In "Culture, Cultural System and Science" (in *Essays in Memory of Imre Lakatos* [Dordrecht: D. Reidel, 1976], 106), Yehuda Elkana summarizes the varied history of the much-abused concept of paradigm. "Wittgenstein used it in his Cambridge lectures. Merton wrote a "Paradigm for the Sociology of Knowledge" in 1945 (reprinted in *Robert K. Merton: The Sociology of Science,* ed. N. Storer [Chicago: University of Chicago Press, 1961], 7–40). Toulmin used it in his *Foresight and Understanding* (Bloomington: Indiana University Press, 1961) meaning "ideals of natural order" (p. 56). Thomas Kuhn made it the core concept of his *Structure of Scientific Revolutions* (Chicago: University of Chicago Press, 1962). Yet, only Merton's and Kuhn's usage became fundamental analytical tools for a whole discipline. Thus, there is a Mertonian paradigm and a Kuhnian paradigm." For more on the history of the usage of this concept, see S. Toulmin, *Human Understanding* (Oxford: Clarendon Press, 1972), 106–7.

Analytic Frameworks for Sociology

6

Paradigm for Functional Analysis in Sociology (1949)

Functional analysis is at once the most promising and possibly the least codified of contemporary orientations to problems of sociological interpretation. Having developed on many intellectual fronts at the same time, it has grown in shreds and patches rather than in depth. The accomplishments of functional analysis are sufficient to suggest that its large promise will progressively be fulfilled, just as its current deficiencies testify to the need for periodically overhauling the past, the better to build for the future. At the very least, occasional reassessments bring into open discussion many of the difficulties which otherwise remain tacit and unspoken.

Like all interpretative schemes, functional analysis depends upon a triple alliance between theory, method, and data. Of the three allies, method is by all odds the weakest. Many of the major practitioners of functional analysis have been devoted to theoretic formulations and to the clearing up of concepts; some have steeped themselves in data directly relevant to a functional frame of reference; but few have broken the prevailing silence regarding how one goes about the business of functional analysis. Yet the plenty and variety of functional analyses force the conclusion that *some* methods have been employed and awaken the hope that much may be learned from their inspection.

Although methods can be profitably examined without reference to theory or substantive data—methodology or the logic of procedure of course has precisely that as its assignment—empirically oriented disciplines are more fully served by inquiry into procedures if this takes due account of their theoretic problems and substantive findings. For the use of "method" involves not only logic but, unfortunately perhaps for those who must struggle with the difficulties of research, also the practical problems of aligning data with the requirements of theory. At

From "Manifest and Latent Functions," in *Social Theory and Social Structure,* 3d ed., rev. and enl. (New York: Free Press, 1968), 73–91, 104–8.

least that is our premise. Accordingly, we shall interweave our account with a systematic review of some of the chief conceptions of functional theory.

The Vocabularies of Functional Analysis

From its very beginnings, the functional approach in sociology has been caught up in terminological confusion. *Too often, a single term has been used to symbolize different concepts, just as the same concept has been symbolized by different terms.* Clarity of analysis and adequacy of communication are both victims of this frivolous use of words. At times, the analysis suffers from the unwitting shift in the conceptual content of a given term, and communication with others breaks down when the essentially same content is obscured by a battery of diverse terms. We have only to follow, for a short distance, the vagaries of the concept of "function" to discover how conceptual clarity is effectively marred and communication defeated by competing vocabularies of functional analysis.

Single Term, Diverse Concepts

The word "function" has been preempted by several disciplines and by popular speech with the not unexpected result that its connotation often becomes obscure in sociology proper. By confining ourselves to only five connotations commonly assigned to this one word, we neglect numerous others. There is, first, popular usage, according to which function refers to some public gathering or festive occasion, usually conducted with ceremonial overtones. It is in this connection, one must assume, that a newspaper headline asserts, "Mayor Tobin Not Backing Social Function," for the news account goes on to explain that "Mayor Tobin announced today that he is not interested in any social function, nor has he authorized anyone to sell tickets or sell advertising for any affair." Common as this usage is, it enters into the academic literature too seldom to contribute any great share to the prevailing chaos of terminology. Clearly, *this* connotation of the word is wholly alien to the functional analysis in sociology.

A second usage makes the term function virtually equivalent to the term occupation. Max Weber, for example, defines occupation as "the mode of specialization, specification, and combination of the functions of an individual so far as it constitutes for him the basis of a continual opportunity for income or for profit."[1] This is a frequent, indeed al-

1. Max Weber *Theory of Social and Economic Organization*, ed. Talcott Parsons (London: William Hodge and Co., 1947), 230.

most a typical, usage of the term by some economists who refer to the "functional analysis of a group" when they report the distribution of occupations in that group. Since this is the case, it may be expedient to follow the suggestion of Sargant Florence, that the more nearly descriptive phrase "occupational analysis" be adopted for such inquiries.[2]

A third usage, representing a special instance of the preceding one, is found both in popular speech and in political science. Function is often used to refer to the activities assigned to the incumbent of a social status, and more particularly, to the occupant of an office or political position. This gives rise to the term functionary, or official. Although function in this sense overlaps the broader meaning assigned the term in sociology and anthropology, it had best be excluded since it diverts attention from the fact that functions are performed not only by the occupants of designated positions, but by a wide range of standardized activities, social processes, culture patterns, and belief-systems found in a society.

Since it was first introduced by Leibniz, the word function has its most precise significance in mathematics, where it refers to a variable considered in relation to one or more other variables in terms of which it may be expressed or on the value of which its own value depends. This conception, in a more extended (and often more imprecise) sense, is expressed by such phrases as "functional interdependence" and "functional relations," so often adopted by social scientists. When Mannheim observes that "every social fact is a function of the time and place in which it occurs," or when a demographer states that "birth-rates are a function of economic status," they are manifestly making use of the mathematical connotation, though the first is not reported in the form of equations and the second is. The context generally makes it clear that the term function is being used in this mathematical sense, but social scientists not infrequently shuttle back and forth between this and another related, though distinct, connotation, which also involves the notion of "interdependence," "reciprocal relation," or "mutually dependent variations."

It is this fifth connotation which is central to functional analysis as this has been practiced in sociology and social anthropology. Stemming in part from the native mathematical sense of the term, this usage is more often explicitly adopted from the biological sciences, where the term function is understood to refer to the "vital or organic processes considered in the respects in which they contribute to the maintenance

2. P. Sargant Florence, *Statistical Method in Economics* (New York: Harcourt, Brace and Co., 1929), 357, 58n.

of the organism."[3] With modifications appropriate to the study of human society, this corresponds rather closely to the key concept of function as adopted by the anthropological functionalists, pure or tempered.[4]

A. R. Radcliffe-Brown is the most often explicit in tracing his working conception of social function to the analogical model found in the biological sciences. After the fashion of Durkheim, he asserts that "the function of a recurrent physiological process is thus a correspondence between it and the needs (i.e., the necessary conditions of existence) of the organism." And in the social sphere where individual human beings, "the essential units," are connected by networks of social relations into an integrated whole, "the function of any recurrent activity, such as the punishment of a crime, or a funeral ceremony, is the part it plays in the social life as a whole and therefore the contribution it makes to the maintenance of the structural continuity."[5]

Though B. Malinowski differs in several respects from the formulations of Radcliffe-Brown, he joins him in making the core of functional analysis the study of "the part which [social or cultural items] play in the society." "This type of theory," Malinowski explains in one of his early declarations of purpose, "aims at the explanation of anthropological facts at all levels of development by *their function, by the part which they play within the integral system of culture, by the manner in which they are related to each other within the system*. . . ."[6]

As we shall presently see in some detail, such recurrent phrases as "the part played in the social or cultural system" tend to blur the important distinction between the concept of function as "interdependence" and as "process." Nor need we pause here to observe that the

3. See, for example, Ludwig von Bertalanffy, *Modern Theories of Development* (New York: Oxford University Press, 1933), 9ff., 184ff.

4. Lowie makes a distinction between the "pure functionalism" of a Malinowski and the "tempered functionalism" of a Thurnwald. Sound as the distinction is, it will soon become apparent that it is not pertinent for our purposes. Robert H. Lowie, *The History of Ethnological Theory* (New York: Farrar and Rinehart, 1937).

5. A. R. Radcliffe-Brown, "On the Concept of Function in Social Science," *American Anthropologist* 37 (1935): 395–96. See also his later presidential address before the Royal Anthropological Institute where he states, "I would define the social function of a socially standardized mode of activity, or mode of thought, as its relation to the social structure to the existence and continuity of which it makes some contribution. Analogously, in a living organism, the physiological function of the beating of the heart, or the secretion of gastric juices, is its relation to the organic structure." "On Social Structure," *The Journal of the Royal Anthropological Institute of Great Britain and Ireland* 70, pt. 1 (1940): 9–10.

6. Bronislaw Malinowski, "Anthropology," *Encyclopaedia Britannica*, First Supplementary Volume (London and New York: 1926), 132–33 [italics added].

postulate which holds that every item of culture has *some* enduring relations with other items, that it has *some* distinctive place in the total culture scarcely equips the field-observer or the analyst with a specific guide to procedure. All this had better wait. At the moment, we need only recognize that more recent formulations have clarified and extended this concept of function through progressive specifications. Thus, Kluckhohn: "... a given bit of culture is "functional" insofar as it defines a mode of response which is adaptive from the standpoint of the society and adjustive from the standpoint of the individual."[7]

From these connotations of the term "function," and we have touched upon only a few drawn from a more varied array, it is plain that many concepts are caught up in the same word. This invites confusion. And when many different words are held to express the same concept, there develops confusion worse confounded.

Single Concept, Diverse Terms

The large assembly of terms used indifferently and almost synonymously with "function" presently includes use, utility, purpose, motive, intention, aim, consequences. Were these and similar terms put to use to refer to the same strictly defined concept, there would of course be little point in noticing their numerous variety. But the fact is that the undisciplined use of these terms, with their ostensibly similar conceptual reference, leads to successively greater departures from tight-knit and rigorous functional analysis. The connotations of each term which differ from rather than agree with the connotation that they have in common are made the (unwitting) basis for inferences that become increasingly dubious as they become progressively remote from the central concept of function. One or two illustrations will bear out the point that a shifting vocabulary makes for the multiplication of misunderstandings.

In the following passage drawn from one of the most sensible of treatises on the sociology of crime, one can detect the shifts in meaning of nominally synonymous terms and the questionable inferences which depend upon these shifts. (The key terms are italicized to help in picking one's way through the argument.)

> *Purpose* of Punishment. Attempts are being made to determine the *purpose or function* of punishment in different groups at different times. Many investigators have insisted that some one *motive* was the *motive* in punish-

7. Clyde Kluckhohn, *Navaho Witchcraft,* Papers of the Peabody Museum of American Archaeology and Ethnology, Harvard University, vol. 22, no. 2 (Cambridge: Peabody Museum, 1944), 47a.

ment. On the other hand, the *function* of punishment in restoring the solidarity of the group which has been weakened by the crime is emphasized. Thomas and Znaniecki have indicated that among the Polish peasants the punishment of crime is *designed primarily* to restore the situation which existed before the crime and renew the solidarity of the group, and that revenge is *a secondary consideration*. From this point of view punishment *is concerned primarily* with the group and only *secondarily* with the offender. On the other hand, expiation, deterrence, retribution, reformation, income for the state, and other things have been posited as *the function* of punishment. In the past as at present it is not clear that any one of these is *the motive;* punishments seem to grow from *many motives* and to perform *many functions*. This is true both of the individual victims of crimes and of the state. Certainly the laws of the present day are not consistent in *aims or motives;* probably the same condition existed in earlier societies.[8]

We should attend to the list of terms ostensibly referring to the same concept: purpose, function, motive, design, secondary consideration, primary concern, aim. Through inspection, it becomes clear that *these terms group into quite distinct conceptual frames of reference*. At times, some of these terms—motive, design, aim and purpose—clearly refer to the *explicit ends-in-view of the representatives of the state*. Others—motive, secondary consideration—refer to the *ends-in-view of the victim of the crime*. And both of these sets of terms are alike in referring to the *subjective anticipations of the results of punishment*. But the concept of function involves the standpoint of *the observer,* not necessarily that of the participant. Social function refers to *observable objective consequences,* and not to *subjective dispositions* (aims, motives, purposes). And the failure to distinguish between the objective sociological consequences and the subjective dispositions inevitably leads to confusion in functional analysis:

The extreme of unreality is attained in the discussion of the so-called "functions" of the family. The family, we hear, performs important *functions* in society; it provides for the perpetuation of the species and the training of the young; it performs economic and religious functions, and so on. Almost we are encouraged to believe that *people marry and have children because* they are eager to perform these needed societal functions. In fact, people marry *because* they are in love, or for other less romantic but no less personal reasons. The *function* of the family, *from the viewpoint of individuals,* is to satisfy their wishes. The *function* of the family or any other social institution is *merely what people use it for*. Social "*functions*" are mostly *rationalizations of established practices; we* act first, explain afterwards; *we*

8. Edwin H. Sutherland, *Principles of Criminology,* 3d ed. (Philadelphia: J. B. Lippincott, 1939), 349–50.

> act for *personal reasons,* and justify *our* behavior by social and ethical *principles.* Insofar as these *functions* of institutions have any real basis, it must be stated in terms of the social processes in which people engage *in the attempt* to satisfy their wishes. Functions arise from the interaction of concrete human beings and concrete *purposes.*[9]

This passage is an interesting medley of small islets of clarity in the midst of vast confusion. Whenever it mistakenly identifies (subjective) motives with (objective) functions, it abandons a lucid functional approach. For it need not be assumed, as we shall presently see, that the *motives* for entering into marriage ("love," "personal reasons") are identical with the *functions* served by families (socialization of the child). Again, it need not be assumed that the *reasons* advanced by people for their behavior ("*we* act for personal reasons") are one and the same as the observed consequences of these patterns of behavior. The subjective disposition may coincide with the objective consequence, but again, it may not. The two vary independently. When, however, it is said that people are motivated to engage in behavior which may give rise to (not necessarily intended) functions, there is offered escape from the troubled sea of confusion.

This brief review of competing terminologies and their unfortunate consequences may be something of a guide to later efforts at codification of the concepts of functional analysis. There will plainly be occasion to limit the use of the sociological concept of function, and there will be need to distinguish clearly between subjective categories of disposition and objective categories of observed consequences. Else the substance of the functional orientation may become lost in a cloud of hazy definitions.

Prevailing Postulates in Functional Analysis

Chiefly—but not solely—in anthropology, functional analysts have commonly adopted three interconnected postulates which, it will now be suggested, have proved to be debatable and unnecessary to the functional orientation.

Substantially, these postulates hold that first, standardized social activities or cultural items are functional for the *entire* social or cultural system; second, *all* such social and cultural items fulfill functions; and third, these items are consequently *indispensable.* Although these three articles of faith are ordinarily seen only in one another's company, they

9. Willard Waller, *The Family* (New York: Cordon Company, 1938), 26 [italics added].

had best be examined separately, since each gives rise to its own distinctive difficulties.

Postulate of the Functional Unity of Society

It is Radcliffe-Brown who characteristically puts this postulate in explicit terms:

> The function of a particular social usage is the contribution it makes to the *total social life* as the functioning of the *total social system*. Such a view implies that a social system (*the total social structure* of a society together with the totality of social usages, in which that structure appears and on which it depends for its continued existence) has a certain kind of unity, which we may speak of as a functional unity. We may define it as a condition in which all parts of the social system work together with a sufficient degree of harmony or internal consistency, i.e., without producing persistent conflicts which can neither be resolved nor regulated.[10]

It is important to note, however, that Radcliffe-Brown goes on to describe this notion of functional unity as a hypothesis which requires further test.

It would at first appear that Malinowski was questioning the empirical acceptability of this postulate when he notes that "the sociological school" (into which he thrusts Radcliffe-Brown) "exaggerated the social solidarity of primitive man" and "neglected the individual."[11] But it is soon apparent that Malinowski does not so much abandon this dubious assumption as he succeeds in adding another to it. He continues to speak of standardized practices and beliefs as functional "for culture as a whole," and goes on to assume that they are *also* functional for every member of the society. Thus, referring to primitive beliefs in the supernatural, he writes,

> Here the functional view is put to its acid test. . . . It is bound to show in what way belief and ritual work for social integration, technical and economic efficiency, for *culture as a whole*—indirectly *therefore* for the biological and mental welfare *of each individual member.*[12]

If the one unqualified assumption is questionable, this twin assumption is doubly so. Whether cultural items do uniformly fulfill functions for the society viewed as a system and for all members of the society is presumably an empirical question of fact rather than an axiom.

Kluckhohn evidently perceives the problem inasmuch as he extends

10. Radcliffe-Brown, "On the Concept of Function," 397 [italics added].

11. See Malinowski, "Anthropology," 132 and "The Group and the Individual in Functional Analysis," *American Journal of Sociology* 44 (1939): 938–64, at 939.

12. Malinowski, "Anthropology," 135 [italics added].

the alternatives to include the possibility that cultural forms "are adjustive or adaptive . . . for the members of the society *or* for the society considered as a perduring unit."[13] This is a necessary first step in allowing for variation in the *unit* which is subserved by the imputed function. Compelled by the force of empirical observation, we shall have occasion to widen the range of variation in this unit even further.

It seems reasonably clear that the notion of functional unity is *not* a postulate beyond the reach of empirical test; quite the contrary. The degree of integration is an empirical variable,[14] changing for the same society from time to time and differing among various societies. That all human societies must have *some* degree of integration is a matter of definition—and begs the question. But not all societies have that *high* degree of integration in which *every* culturally standardized activity or belief is functional for the society as a whole and uniformly functional for the people living in it. Radcliffe-Brown need in fact have looked no further than to his favored realm of analogy in order to suspect the adequacy of his assumption of functional unity. For we find significant variations in the degree of integration even among individual biological organisms, although the commonsense assumption would tell us that here, surely, all the parts of the organism work toward a "unified" end. Consider only this:

> One can readily see that there are *highly integrated organisms* under close control of the nervous system or of hormones, the loss of any major part of which will strongly affect the whole system, and frequently will cause death, but, on the other hand, there are the lower *organisms much more loosely correlated,* where the loss of even a major part of the body causes only temporary inconvenience pending the regeneration of replacement tissues. Many of these more loosely organized animals are *so poorly integrated that different parts may be in active opposition to each other.* Thus, when an ordinary starfish is placed on its back, part of the arms may attempt to turn the animal in one direction, while others work to turn it in the opposite way. . . . On account of its *loose integration,* the sea anemone may move off and leave a portion of its foot clinging tightly to a rock, so that the animal suffers serious rupture.[15]

If this is true of single organisms, it would seem *a fortiori* the case with complex social systems.

13. Kluckhohn, *Navaho Witchcraft,* 46b [italics added].

14. It is the merit of Sorokin's early review of theories of social integration that this important fact is recognized. Cf. P. A. Sorokin, "Forms and Problems of Culture-Integration," *Rural Sociology* 1 (1936): 121–41, 344–74.

15. G. H. Parker, *The Elementary Nervous System,* quoted by William C. Allee, *Animal Aggregation* (University of Chicago Press, 1931), 81–82.

One need not go far afield to show that the assumption of the complete functional unity of human society is repeatedly contrary to fact. Social usages or sentiments may be functional for some group and dysfunctional for others in the same society. Anthropologists often cite "increased solidarity of the community" and "increased family pride" as instances of functionally adaptive sentiments. Yet, as Bateson[16] among others has indicated, an increase of pride among individual families may often serve to disrupt the solidarity of a small local community. Not only is the postulate of functional unity often contrary to fact, but it has little heuristic value, since it diverts the analyst's attention from possible disparate consequences of a given social or cultural item (usage, belief, behavior pattern, institution) for diverse social groups and for the individual members of these groups.

If the body of observation and fact which negates the assumption of functional unity is as large and easily accessible as we have suggested, it is interesting to ask how it happens that Radcliffe-Brown and others who follow his lead have continued to abide by this assumption. A possible clue is provided by the fact that this conception, in its recent formulations, was developed by *social anthropologists,* that is, by scholars primarily concerned with the study of nonliterate societies. In view of what Radin has described as "the highly integrated nature of the majority of aboriginal civilizations," this assumption may be tolerably suitable for some, if not all, nonliterate societies. But one pays an excessive intellectual penalty for moving this possibly useful assumption from the realm of small nonliterate societies to the realm of large, complex, and highly differentiated literate societies. [. . .]

Unity of the total society cannot be usefully posited in advance of observation. It is a question of fact, and not a matter of opinion. The theoretic framework of functional analysis must expressly require that there be *specification* of the *units* for which a given social or cultural item is functional. It must expressly allow for a given item having diverse consequences, functional and dysfunctional, for individuals, for subgroups, and for the more inclusive social structure and culture.

Postulate of Universal Functionalism

Most succinctly, this postulate holds that all standardized social or cultural forms have positive functions. As with other aspects of the functional conception, Malinowski advances this in its most extreme form:

16. Gregory Bateson, *Naven* (Cambridge: Cambridge University Press, 1936), 31–32.

> The functional view of culture *insists* therefore upon the principle that in *every type of civilization, every custom, material object, idea, and belief fulfills some vital function* . . .[17]

Although, as we have seen, Kluckhohn allows for variation in the unit subserved by a cultural form, he joins with Malinowski in postulating functional value for all surviving forms of culture. ("My basic postulate . . . is that *no* culture forms survive unless they constitute responses which are adjustive or adaptive, in some sense . . ."[18]) This universal functionalism may or may not be a heuristic postulate; that remains to be seen. But one should be prepared to find that it too diverts critical attention from a range of nonfunctional consequences of existing cultural forms.

In fact, when Kluckhohn seeks to illustrate his point by ascribing "functions" to seemingly functionless items, he falls back upon a type of function which would be found, *by definition* rather than by inquiry, to be served by all persisting items of culture. Thus, he suggests that

> The at present mechanically useless buttons on the sleeve of a European man's suit subserve the "function" of preserving the familiar, of maintaining a tradition. People are, in general, more comfortable if they feel a continuity of behavior, if they feel themselves as following out the orthodox and socially approved forms of behavior.[19]

This would appear to represent the marginal case in which the imputation of function adds little or nothing to direct description of the culture pattern or behavior form. It may well be assumed that all *established* elements of culture (which are loosely describable as "tradition") have the minimum, though not exclusive, function of "preserving the familiar, of maintaining a tradition." This is equivalent to saying that the "function" of conformity to *any* established practice is to enable the conformist to avoid the sanctions otherwise incurred by deviating from the established practice. This is no doubt true but hardly illuminating. It serves, however, to remind us that we shall want to explore the *types of functions* which the sociologist imputes. At the moment, it suggests the provisional assumption that, although any item of culture or social structure *may* have functions, it is premature to hold unequivocally that every such item *must* be functional.

The postulate of universal functionalism is of course the historical product of the fierce, barren and protracted controversy over "survi-

17. Malinowski, "Anthropology," 132 [italics added].
18. Kluckhohn, *Navaho Witchcraft,* 46 [italics added].
19. Ibid., 47.

vals" which raged among the anthropologists during the early part of the century. The notion of a social survival, that is, in the words of W. H. R. Rivers, of "a custom . . . [which] cannot be explained by its present utility but only becomes intelligible through its past history,"[20] dates back at least to Thucydides. But when the evolutionary theories of culture became prominent, the concept of survival seemed all the more strategically important for reconstructing "stages of development" of cultures, particularly for nonliterate societies which possessed no written record. For the functionalists who wished to turn away from what they regarded as the usually fragmentary and often conjectural "history" of nonliterate societies, the attack on the notion of survival took on all the symbolism of an attack on the entire and intellectually repugnant system of evolutionary thought. In consequence, perhaps, they overreacted against this concept central to evolutionary theory and advanced an equally exaggerated "postulate" to the effect that "every custom [everywhere] . . . fulfills some vital function."

It would seem a pity to allow the polemics of the anthropological forefathers to create splendid exaggerations in the present. Once discovered, ticketed, and studied, social survivals cannot be exorcised by a postulate. And if no specimens of these survivals can be produced, then the quarrel dwindles of its own accord. It can be said, furthermore, that even when such survivals are identified in contemporary literate societies, they seem to add little to our understanding of human behavior or the dynamics of social change. Not requiring their dubious role as poor substitutes for recorded history, sociologists of literate societies may neglect survivals with no apparent loss. But they need not be driven, by an archaic and irrelevant controversy, to adopt the unqualified postulate that all culture items fulfill vital functions. For this, too, is a problem for investigation, not a conclusion in advance of investigation. Far more useful as a directive for research would seem the provisional assumption that persisting cultural forms have a *net balance of functional consequences* either for the society considered as a unit or for subgroups sufficiently powerful to retain these forms intact, by means of direct coercion or indirect persuasion. This formulation at once avoids the tendency of functional analysis to concentrate on positive functions and directs the attention of the research worker to other types of consequences as well.

20. W. H. R. Rivers, "Survival in Sociology," *The Sociological Review* 6 (1913): 293–305. See also Edward B. Tylor, *Primitive Culture* (New York: Harper, 1874), esp. 1, 70–159; and for a more recent review of the matter, Lowie, *The History of Ethnological Theory,* 44 ff., 81 f. For a sensible and restrained account of the problem, see Émile Durkheim, *Rules of Sociological Method,* chap. 5, esp. at 91.

Postulate of Indispensability

The last of this trio of postulates common among functional social scientists is, in some respects, the most ambiguous. The ambiguity becomes evident in the aforementioned manifesto by Malinowski to the effect that

> in every type of civilization, every custom, material object, idea, and belief fulfills some *vital* function, has some task to accomplish, represents an *indispensable part* within a working whole.[21]

From this passage, it is not at all clear whether he asserts the indispensability of the *function,* or of the *item* (custom, object, idea, belief) fulfilling the function, or *both.*

This ambiguity is quite common in the literature. Thus, in their account of the role of religion, Kingsley Davis and Wilbert E. Moore seem at first to maintain that it is the *institution* which is indispensable: "The reason why religion is necessary . . ."; ". . . religion . . . plays a unique and indispensable part in society."[22] But it soon appears that it is not so much the institution of religion which is regarded as indispensable but rather the functions which religion is taken typically to perform. For Davis and Moore regard religion as indispensable only insofar as it functions to make the members of a society adopt "certain ultimate values and ends in common." These values and ends, it is said,

> must . . . appear to the members of the society to have some reality, and it is the role of religious belief and ritual to supply and reinforce this appearance of reality. Through ritual and belief the common ends and values are connected with an imaginary world symbolized by concrete sacred objects, which world in turn is related in a meaningful way to the facts and trials of the individual's life. Through the worship of the sacred objects and the beings they symbolize, and the acceptance of *supernatural prescriptions* that are at the same time codes of behavior, a powerful control over human conduct is exercised, guiding it along lines sustaining the institutional structure and conforming to the ultimate ends and values.[23]

The alleged indispensability of religion, then, is based on the assumption of fact that it is through "worship" and "supernatural pre-

21. Malinowski, "Anthropology," 132 [italics added].

22. Kingsley Davis and Wilbert E. Moore, "Some Principles of Stratification," *American Sociological Review* 10 (April 1945): 244, 246. See the more recent review of this matter by Davis in his introduction to William J. Goode, *Religion Among the Primitives* (Glencoe, Ill.: Free Press, 1951) and the instructive functional interpretations of religion in that volume.

23. Kingsley Davis and Wilbert E. Moore, "Principles of Stratification," 244–45 [italics added].

scriptions" *alone* that the necessary minimum of "control over human conduct" and "integration in terms of sentiments and beliefs" can be achieved.

In short, the postulate of indispensability as it is ordinarily stated contains two related, but distinguishable, assertions. First, it is assumed that there are certain *functions* which are indispensable in the sense that, unless they are performed, the society (or group or individual) will not persist. This, then, sets forth a concept of *functional prerequisites* or *preconditions functionally necessary* for a society, and we shall have occasion to examine this concept in some detail. Second, and this is quite another matter, it is assumed that *certain cultural or social forms* are indispensable for fulfilling each of these functions. This involves a concept of specialized and irreplaceable structures, and gives rise to all manner of theoretic difficulties. For not only can this be shown to be manifestly contrary to fact, but it entails several subsidiary assumptions that have plagued functional analysis from the very outset. It diverts attention from the fact that alternative social structures (and cultural forms) have served, under conditions to be examined, the functions necessary for the persistence of groups. Proceeding further, we must set forth a major theorem of functional analysis: *just as the same item may have multiple functions, so may the same function be diversely fulfilled by alternative items.* Functional needs are here taken to be permissive, rather than determinant, of specific social structures. Or, in other words, there is a range of variation in the structures that fulfill the function in question. (The limits upon this range of variation involve the concept of structural constraint, of which more presently.)

In contrast to this implied concept of indispensable cultural forms (institutions, standardized practices, belief-systems, etc.), there is, then, the concept of *functional alternatives,* or *functional equivalents,* or *functional substitutes.* This concept is widely recognized and used, but it should be noted that it cannot rest comfortably in the same theoretical system which entails the postulate of indispensability of particular cultural forms. Thus, after reviewing Malinowski's theory of "the functional necessity for such mechanisms as magic," Parsons is careful to make the following statement:

> . . . wherever such uncertain elements enter into the pursuit of emotionally important goals, if not magic, at least *functionally equivalent* phenomena could be expected to appear.[24]

24. Talcott Parsons, *Essays in Sociological Theory, Pure and Applied* (Glencoe, Ill.: Free Press, 1949), 58.

This is a far cry from Malinowski's own insistence that

> Thus magic fulfills *an indispensable function* within culture. It satisfies a definite need *which cannot be satisfied by any other factors of primitive civilization.*[25]

This twin concept of the indispensable function and the irreplaceable belief-and-action pattern flatly excludes the concept of functional alternatives.

In point of fact, the concept of functional alternatives or equivalents has repeatedly emerged in every discipline which has adopted a functional framework of analysis. It is, for example, widely utilized in the psychological sciences, as a paper by Horace B. English admirably indicates.[26] And in neurology, Karl S. Lashley has pointed out, on the basis of experimental and clinical evidence, the inadequacy of the "assumption that individual neurons are specialized for particular functions," maintaining instead that a particular function may be fulfilled by a range of alternative structures.[27]

Sociology and social anthropology have all the more occasion for avoiding the postulate of indispensability of given structures, and for systematically operating with the concept of functional alternatives and functional substitutes. For just as laymen have long erred in assuming that the "strange" customs and beliefs of other societies were "mere superstitions," so functional social scientists run the risk of erring in the other extreme, first, by being quick to find functional or adaptive value in these practices and beliefs, and second, by failing to see which alternative modes of action are ruled out by cleaving to these ostensibly functional practices. Thus, there is not seldom a readiness among some functionalists to conclude that magic or certain religious rites and beliefs are functional, because of their effect upon the state of mind or self-confidence of the believer. Yet it may well be in some instances that these magical practices obscure and take the place of accessible secular and more adaptive practices. As F. L. Wells has observed,

> To nail a horseshoe over the door in a smallpox epidemic may bolster the morale of the household but it will not keep out the smallpox; such beliefs and practices will not stand the secular tests to which they are sus-

25. Malinowski, "Anthropology," 136 [italics added].

26. Horace B. English, "Symbolic versus Functional Equivalents in the Neuroses of Deprivation," *Journal of Abnormal and Social Psychology* 32 (1937): 392–94

27. Karl S. Lashley, "Basic Neural Mechanisms in Behavior," *Psychological Review* 37 (1930): 1–24.

ceptible, and the sense of security they give is preserved only while the real tests are evaded.[28]

Those functionalists who are constrained by their theory to attend to the effects of such symbolic practices *only* upon the individual's state of mind and who therefore conclude that the magical practice is functional, neglect the fact that these very practices may on occasion take the place of more effective alternatives. And those theorists who refer to the indispensability of standardized practices or prevailing institutions because of their observed function in reinforcing common sentiments must look first to functional substitutes before arriving at a conclusion, more often premature than confirmed.

Upon review of this trinity of functional postulates, several basic considerations emerge which must be caught up in our effort to codify this mode of analysis. In scrutinizing, first, *the postulate of functional unity,* we found that one cannot simply assume full integration of all societies, but that this is an empirical question in which we should be prepared to find a range of degrees of integration. And in examining the special case of functional interpretations of religion, we were alerted to the possibility that, though human nature may be of a piece, it does not follow that the structure of nonliterate societies is uniformly like that of highly differentiated, "literate" societies. A difference in degree between the two—say, the existence of several disparate religions in the one and not in the other—may make hazardous the passage between them. From critical scrutiny of this postulate, it developed that a theory of functional analysis must call for *specification* of the social units subserved by given social functions, and that items of culture must be recognized to have multiple consequences, some of them functional and others, perhaps, dysfunctional.

Review of the second *postulate of universal functionalism,* which holds that all persisting forms of culture are inevitably functional, resulted in other considerations that must be met by a codified approach to functional interpretation. It appeared not only that we must be prepared to find dysfunctional as well as functional consequences of these forms but that theorists will ultimately be confronted with the difficult problem of developing an organon for assessing the net balance of consequences if their research is to have bearing on social technology. Clearly, expert advice based only on the appraisal of a limited, and perhaps arbitrarily selected, range of consequences to be expected as a re-

28. F. L. Wells, "Social Maladjustments: Adaptive Regression," in *Handbook of Social Psychology,* ed. Carl A. Murchison (N.p.: Clark University Press, 1935), 880.

sult of contemplated action, will be subject to frequent error and will be properly judged as having small merit.

The postulate of indispensability, we found, entails two distinct propositions: the one alleging the indispensability of certain functions, and this gives rise to the concept of *functional necessity* or *functional prerequisites;* the other alleging the indispensability of existing social institutions, culture forms, or the like, and this, when suitably questioned, gives rise to the concept of *functional alternatives, equivalents, or substitutes.*

Moreover, the currency of these three postulates, singly and in concert, is the source of the common charge that functional analysis inevitably involves certain ideological commitments. [. . .]

A Paradigm for Functional Analysis in Sociology

As an initial and admittedly tentative step in the direction of codifying functional analysis in sociology, we set forth a paradigm of the concepts and problems central to this approach. It will soon become evident that the chief components of this paradigm have progressively emerged in the foregoing pages as we have critically examined the vocabularies, postulates, concepts, and ideological imputations now current in the field. The paradigm brings these together in compact form, thus permitting simultaneous inspection of the major requirements of functional analysis and serving as an aid to self-correction of provisional interpretations, a result difficult to achieve when concepts are scattered and hidden in page after page of discursive exposition. The paradigm presents the hard core of concept, procedure, and inference in functional analysis.

Above all, it should be noted that the paradigm does not represent a set of categories introduced *de novo,* but rather a *codification* of those concepts and problems which have been forced upon our attention by critical scrutiny of current research and theory in functional analysis. (Reference to the preceding sections of this chapter will show that the groundwork has been prepared for all the categories embodied in the paradigm.)

1. *The item(s) to which functions are imputed.*

The entire range of sociological data can be, and much of it has been, subjected to functional analysis. The basic requirement is that the object of analysis represent a *standardized* (i.e., patterned and repetitive) item, such as social roles, institutional patterns, social processes, cultural patterns, culturally patterned emotions, social norms, group organization, social structure, mechanisms of social control, etc.

BASIC QUERY: What must enter into the protocol of observation of the given item if it is to be amenable to systematic functional analysis?

2. *Concepts of subjective dispositions (motives, purposes)*

At some point, functional analysis invariably assumes or explicitly operates with some conception of the motivation of individuals involved in a social system. As the foregoing discussion has shown, these concepts of subjective disposition are often and erroneously merged with the related, but different, concepts of the objective consequences of attitude, belief, and behavior.

BASIC QUERY: In which types of analysis is it sufficient to take observed motivations as *data,* as given, and in which are they properly considered as *problematical,* as derivable from other data?

3. *Concepts of objective consequences (functions, dysfunctions)*

We have observed two prevailing types of confusion enveloping the several current conceptions of "function":

(1) The tendency to confine sociological observations to the *positive* contributions of a sociological item to the social or cultural system in which it is implicated; and

(2) The tendency to confuse the subjective category of *motive* with the objective category of *function.*

Appropriate conceptual distinctions are required to eliminate these confusions.

The first problem calls for a concept of *multiple consequences* and *a net balance of an aggregate of consequences.*

Functions are those observed consequences that make for the adaptation or adjustment of a given system; and *dysfunctions,* those observed consequences that lessen the adaptation or adjustment of the system. There is also the empirical possibility of *nonfunctional* consequences, which are simply irrelevant to the system under consideration.

In any given instance, an item may have both functional and dysfunctional consequences, giving rise to the difficult and important problem of evolving canons for assessing the net balance of the aggregate of consequences. (This of course, is most important in the use of functional analysis for guiding the formation and enactment of policy.)

The second problem (arising from the easy confusion of motives and functions) requires us to introduce a conceptual distinction between the cases in which the subjective aim-in-view coincides with the objective consequence, and the cases in which they diverge.

Manifest functions are those objective consequences contributing to the adjustment or adaptation of the system which are intended and recognized by participants in the system;

Latent functions, correlatively, being those which are neither intended nor recognized.[29]

BASIC QUERY: What are the effects of the transformation of a previously latent function into a manifest function (involving the problem of the role of knowledge in human behavior and the problems of "manipulation" of human behavior)?

4. *Concepts of the unit subserved by the function*

We have observed the difficulties entailed in *confining* analysis to functions fulfilled for "the society" since items may be functional for some individuals and subgroups and dysfunctional for others. It is necessary, therefore, to consider a *range* of units for which the item has designated consequences: individuals in diverse statuses, subgroups, the larger social system and culture systems. (Terminologically, this implies the concepts of psychological function, group function, societal function, cultural function, etc.)

5. *Concepts of functional requirements (needs, prerequisites)*

Embedded in every functional analysis is some conception, tacit or expressed, of the functional requirements of the system under observation. As noted elsewhere,[30] this remains one of the cloudiest and empirically most debatable concepts in functional theory. As utilized by sociologists, the concept of functional requirement tends to be tautological or *ex post;* it tends to be confined to the conditions of "survival" of a given system; it tends, as in the work of Malinowski, to include biological as well as social "needs."

This involves the difficult problem of establishing *types* of functional requirements (universal versus specific); procedures for validating the assumption of these requirements; etc.

BASIC QUERY: What is required to establish the validity of such a

29. The relations between the "unanticipated consequences" of action and "latent functions" can be clearly defined, since they are implicit in the foregoing section of the paradigm. The unintended consequences of action are of three types:

(1) those which are functional for a designated system, and these comprise the latent functions;

(2) those which are dysfunctional for a designated system, and these comprise the latent dysfunctions; and

(3) those which are irrelevant to the system which they affect neither functionally nor dysfunctionally, i.e., the pragmatically unimportant class of nonfunctional consequences.

For a preliminary statement, see Robert K. Merton, "The Unanticipated Consequences of Purposive Social Action," *American Sociological Review* 1 (1936): 894–904; for a tabulation of these types of consequences see Goode, *Religion Among the Primitives,* 32–33.

30. Robert K. Merton, "Discussion of Parsons' 'Position of Sociological Theory,'" *American Sociological Review* 13 (1949): 164–68.

variable as "functional requirement" in situations where rigorous experimentation is impracticable?

6. *Concepts of the mechanisms through which functions are fulfilled*

Functional analysis in sociology, as in other disciplines like physiology and psychology, calls for a "concrete and detailed" account of the mechanisms that operate to perform a designated function. This refers, not to psychological, but to social mechanisms (e.g., role-segmentation, insulation of institutional demands, hierarchic ordering of values, social division of labor, ritual and ceremonial enactments, etc.).

BASIC QUERY: What is the presently available inventory of social mechanisms corresponding, say, to the large inventory of psychological mechanisms? What methodological problems are entailed in discerning the operation of these social mechanisms?

7. *Concepts of functional alternatives (functional equivalents or substitutes)*

As we have seen, once we abandon the gratuitous assumption of the functional indispensability of particular social structures, we immediately require some concept of functional alternatives, equivalents, or substitutes. This focuses attention on the *range of possible variation* in the items which can, in the case under examination, subserve a functional requirement. It unfreezes the identity of the existent and the inevitable.

BASIC QUERY: Since scientific proof of the equivalence of an alleged functional alternative ideally requires rigorous experimentation, and since this is not often practicable in large-scale sociological situations, which practicable procedures of inquiry most nearly approximate the logic of experiment?

8. *Concepts of structural context (or structural constraint)*

The range of variation in the items that *can* fulfill designated functions in a social structure is not unlimited (and this has been repeatedly noted in our foregoing discussion). The interdependence of the elements of a social structure limits the effective possibilities of change or functional alternatives. The concept of structural constraint corresponds, in the area of social structure, to Goldenweiser's "principle of limited possibilities" in a broader sphere. Failure to recognize the relevance of interdependence and attendant structural restraints leads to utopian thought in which it is tacitly assumed that certain elements of a social system can be eliminated without affecting the rest of that system. This consideration is recognized by both Marxist social scientists (e.g., Karl Marx) and by non-Marxists (e.g., Malinowski).

BASIC QUERY: How narrowly does a given structural context limit the range of variation in the items which can effectively satisfy functional requirements? Do we find, under conditions yet to be determined, an area of indifference, in which any one of a wide range of alternatives may fulfill the function?

9. *Concepts of dynamics and change*

We have noted that functional analysts *tend* to focus on the statics of social structure and to neglect the study of structural change.

This emphasis upon statics is not, however, *inherent* in the theory of functional analysis. It is, rather, an adventitious emphasis stemming from the concern of early anthropological functionalists to counteract earlier tendencies to write conjectural histories of nonliterate societies. This practice, useful at the time it was first introduced into anthropology, has disadvantageously persisted in the work of some functional sociologists.

The concept of dysfunction, which implies the concept of strain, stress, and tension on the structural level, provides an analytical approach to the study of dynamics and change. How are observed dysfunctions contained within a particular structure, so that they do not produce instability? Does the accumulation of stresses and strains produce pressure for change in such directions as are likely to lead to their reduction?

Basic Query: Does the prevailing concern among functional analysts with the concept of *social equilibrium* divert attention from the phenomena of *social disequilibrium?* Which available procedures permit the sociologist most adequately to gauge the accumulation of stresses and strains in a social system? To what extent does knowledge of the structural context permit the sociologist to anticipate the most probable directions of social change?

10. *Problems of validation of functional analysis*

Throughout the paradigm, attention has been called repeatedly to the *specific* points at which assumptions, imputations, and observations must be validated.[31] This requires, above all, a rigorous statement

31. By this point, it is evident that we are considering functional analysis as a method for the interpretation of sociological data. This is not to gainsay the important role of the functional orientation in sensitizing sociologists to the collection of types of data which might otherwise be neglected. It is perhaps unnecessary to reiterate the axiom that one's concepts do determine the inclusion or exclusions of data, that, despite the etymology of the term, data are not "given" but are "contrived" with the inevitable help of concepts. In the process of evolving a functional interpretation, the sociological analyst invariably finds it necessary to obtain data other than those initially contemplated. Interpretation and the collection of data are thus inextricably bound up in the array of concepts and propositions relating these concepts.

of the sociological procedures of analysis that most nearly approximate the *logic* of experimentation. It requires a systematic review of the possibilities and limitations of *comparative* (cross-cultural and cross-group) *analysis.*

BASIC QUERY: To what extent is functional analysis limited by the difficulty of locating adequate *samples of social systems* which can be subjected to comparative (quasi-experimental) study? [32]

11. *Problems of the ideological implications of functional analysis*

It has been emphasized that functional analysis has no intrinsic commitment to an ideological position. This does not gainsay the fact that *particular* functional analyses and *particular* hypotheses advanced by functionalists may have an identifiable ideological role. This, then, becomes a specific problem for the sociology of knowledge: to what extent does the social position of the functional sociologist (e.g., vis-à-vis a particular "client" who has authorized a given research) evoke one rather than another formulation of a problem, affect his assumptions and concepts, and limit the range of inferences drawn from the data?

BASIC QUERY: How does one detect the ideological tinge of a functional analysis and to what degree does a particular ideology stem from the basic assumptions adopted by the sociologist? Is the incidence of these assumptions related to the status and research role of the sociologist?

32. George P. Murdock's *Social Structure* (New York: Macmillan, 1949) is enough to show that procedures such as those involved in the cross-cultural survey hold large promise for dealing with certain methodological problems of functional analysis. See also the procedures of functional analysis in George C. Homans and David M. Schneider, *Marriage, Authority, and Final Causes* (New York: Free Press, 1955).

7

Manifest and Latent Functions (1949)

As stated in the previous chapter, the distinction between manifest and latent functions was devised to preclude the inadvertent confusion, often found in the sociological literature, between motivations for social behavior and its objective consequences. Our scrutiny of current vocabularies of functional analysis has shown how easily, and how unfortunately, the sociologist may identify motives with functions. It was further indicated that motive and function vary independently and that the failure to register this fact in an established terminology has contributed to the unwitting tendency among sociologists to confuse the subjective categories of motivation with the objective categories of function. This, then, is the central purpose of our succumbing to the not always commendable practice of introducing new terms into the rapidly growing technical vocabulary of sociology, a practice regarded by many laymen as an affront to their intelligence and an offense against common intelligibility.

As will be readily recognized, I have adapted the terms "manifest" and "latent" from their use in another context by Freud (although Francis Bacon had long ago spoken of "latent process" and "latent configuration" in connection with processes that are below the threshold of superficial observation).

The distinction itself has been repeatedly drawn by observers of human behavior at irregular intervals over a span of centuries. Indeed, it would be disconcerting to find that a distinction which we have come to regard as central to functional analysis has not been made by any of that numerous company who have in effect adopted a functional orientation. We need mention only a few of those who have, in recent decades, found it necessary to distinguish in their specific interpretations of behavior between the end-in-view and the functional consequences of action:

From "Manifest and Latent Functions," in *Social Theory and Social Structure,* 3d ed., rev. and enl. (New York: Free Press, 1968), 114–26. Reprinted by permission of The Free Press, an imprint of Simon and Schuster.

George H. Mead: "... that attitude of hostility toward the law-breaker has the unique advantage [read: latent function] of uniting all members of the community in the emotional solidarity of aggression. While the most admirable of humanitarian efforts are sure to run counter to the individual interests of very many in the community, or fail to touch the interest and imagination of the multitude and to leave the community divided or indifferent, the cry of thief or murderer is attuned to profound complexes, lying below the surface of competing individual efforts, and citizens who have [been] separated by divergent interests stand together against the common enemy."[1]

Émile Durkheim's[2] similar analysis of the social functions of punishment is also focused on its latent functions (unifying consequences for the community) rather than confined to manifest functions (deterrent consequences for the criminal).

W. G. Sumner: "... from the first acts by which men try to satisfy needs, each act stands by itself, and looks no further than the immediate satisfaction. From recurrent needs arise habits for the individual and customs for the group, but these results are consequences which were never conscious and never foreseen or intended. They are not noticed until they have long existed, and it is still longer before they are appreciated." Although this fails to locate the latent functions of standardized social actions for a designated social structure, it plainly makes the basic distinction between ends-in-view and objective consequences.[3]

R. M. MacIver: In addition to the direct effects of institutions, "there are further effects by way of control which lie outside the direct purposes of men ... this type of reactive form of control ... may, though unintended, be of profound service to society."[4]

W. I. Thomas and F. Znaniecki: "Although all the new [Polish peasant cooperative] institutions are thus formed with the definite purpose of satisfying certain specific needs, their social function is by no means limited to

1. George H. Mead, "The Psychology of Punitive Justice," *American Journal of Sociology* 23 (1918): 577–602, esp. 591.

2. Durkheim adopted a functional orientation throughout his work, and he often operates, albeit without explicit notice, with concepts equivalent to that of latent function in his researches. The reference in the text at this point is to his "Deux lois de l'évolution pénale," *L'année sociologique* 4 (1899–1900): 55–95, as well as to his *Division of Labor in Society* (1893; reprint, Glencoe, Ill.: Free Press, 1947).

3. This one of his many such observations is of course from W. G. Sumner's *Folkways* (Boston: Ginn and Co., 1906), 3. His collaborator, Albert G. Keller, retained the distinction in his own writings; see, for example, his *Social Evolution* (New York: Macmillan, 1927), at 93–95.

4. This is advisedly drawn from one of MacIver's earlier works, *Community* (London: Macmillan, 1915). The distinction takes on greater importance in his later writings, becoming a major element in his *Social Causation* (Boston: Ginn and Co., 1942), esp. at 314–21, and informs the greater part of his *More Perfect Union,* (New York: Macmillan, 1948).

their explicit and conscious purpose . . . every one of these institutions—commune or agricultural circle, loan and savings bank, or theater—is not merely a mechanism for the management of certain values but also an association of people, each member of which is supposed to participate in the common activities as a living, concrete individual. Whatever is the predominant, official common interest upon which the institution is founded, the association as a concrete group of human personalities unofficially involves many other interests; the social contacts between its members are not limited to their common pursuit, though the latter, of course, constitutes both the main reason for which the association is formed and the most permanent bond which holds it together. Owing to this combination of an abstract political, economic, or rather rational mechanism for the satisfaction of specific needs with the concrete unity of a social group, the new institution is also the best intermediary link between the peasant primary-group and the secondary national system."[5]

These and numerous other sociological observers have, then, from time to time distinguished between categories of subjective disposition ("needs, interests, purposes") and categories of generally unrecognized but objective functional consequences ("unique advantages," "never conscious" consequences, "unintended . . . service to society," "function not limited to conscious and explicit purpose").

Since the occasion for making the distinction arises with great frequency, and since the purpose of a conceptual scheme is to direct observations toward salient elements of a situation and to reduce inadvertent oversight of these elements, it would seem justifiable to designate this distinction by an appropriate set of terms. That is the rationale for the distinction between manifest functions and latent functions; the first referring to those objective consequences for a specified unit (person, subgroup, social or cultural system) which contribute to its adjustment or adaptation and were so intended; the second referring to unintended and unrecognized consequences of the same order.

There are indications that the christening of this distinction may serve a heuristic purpose by becoming incorporated into an explicit conceptual apparatus, thus aiding both systematic observation and later analysis. In recent years, for example, the distinction between manifest and latent functions has been utilized in analyses of racial inter-

5. The single excerpt quoted in the text is one of scores that have led to *The Polish Peasant in Europe and America* (2 vols. [1918–20; reprint, New York: Knopf, 1927]) being deservedly described as a "sociological classic." See pages 1426–27 and 1523ff. As noted later in this chapter, the many others like it in point of richness of content, were forgotten or never noticed by those industrial sociologists who recently came to develop the notion of "formal organization" in industry.

marriage,[6] social stratification,[7] Veblen's sociological theories,[8] prevailing American orientations toward Russia,[9] propaganda as a means of social control,[10] Malinowski's anthropological theory,[11] Navajo witchcraft,[12] problems in the sociology of knowledge,[13] fashion,[14] the dynamics of personality,[15] national security measures,[16] the internal social dynamics of bureaucracy,[17] and a great variety of other sociological problems.

The very diversity of this subject matter suggests that the theoretic distinction between manifest and latent functions is not bound up with a limited and particular range of human behavior. But there still remains the large task of ferreting out the specific uses to which this conceptual distinction can be put, and it is to this large task that we devote the remaining pages of this chapter.

6. Robert K. Merton, "Intermarriage and the Social Structure," *Psychiatry* 4 (1941): 361–74.

7. Kingsley Davis, "A Conceptual Analysis of Stratification," *American Sociological Review* 7 (1942): 309–21.

8. A. K. Davis, "Thorstein Veblen's Social Theory" (Ph.D. diss., Harvard University, 1941); A. K. Davis, "Veblen on the Decline of the Protestant Ethic," *Social Forces* 22 (1944): 282–86; and Louis Schneider, *The Freudian Psychology and Veblen's Social Theory* (New York: King's Crown Press, 1948), esp. chap. 2.

9. A. K. Davis, "Some Sources of American Hostility to Russia," *American Journal of Sociology* 53 (1947): 174–83.

10. Talcott Parsons, "Propaganda and Social Control," *Essays in Sociological Theory* (Glencoe, Ill.: Free Press, 1949).

11. Clyde Kluckhohn, "Bronislaw Malinowski, 1884–1942," *Journal of American Folklore* 56 (1943): 208–19.

12. Clyde Kluckhohn, *Navaho Witchcraft,* Papers of the Peabody Museum of American Archaeology and Ethnology, Harvard University, vol. 22, no. 2 (Cambridge: Peabody Museum, 1944), esp. at 46–47ff.

13. Robert K. Merton, chap. 17 in this volume.

14. Bernard Barber and L. S. Lobel, "'Fashion' in Women's Clothes and the American Social System," *Social Forces* 31 (1952): 124–31.

15. O. H. Mowrer and C. Kluckhohn, "Dynamic Theory of Personality," in *Personality and the Behavior Disorders,* ed. J. M. Hunt (New York: Ronald Press, 1944), 1:69–135, esp. at 72.

16. Marie Jahoda and S. W. Cook, "Security Measures and Freedom of Thought: An Exploratory Study of the Impact of Loyalty and Security Programs," *Yale Law Journal* 61 (1952): 296–333.

17. Philip Selznick, *TVA and the Grass Roots* (Berkeley: University of California Press, 1949); A. W. Gouldner, *Patterns of Industrial Bureaucracy* (Glencoe, Ill.: Free Press, 1954); P. M. Blau, *The Dynamics of Bureaucracy* (Chicago: University of Chicago Press, 1955); A. K. Davis, "Bureaucratic Patterns in Navy Officer Corps," *Social Forces* 27 (1948): 142–53.

Heuristic Purposes of the Distinction

Clarifies the analysis of seemingly irrational social patterns. In the first place, the distinction aids the sociological interpretation of many social practices which persist even though their manifest purpose is clearly not achieved. The time-worn procedure in such instances has been for diverse, particularly lay, observers to refer to these practices as "superstitions," "irrationalities," "mere inertia of tradition," etc. In other words, when group behavior does not—and, indeed, often cannot—attain its ostensible purpose there is an inclination to attribute its occurrence to lack of intelligence, sheer ignorance, survivals, or so-called inertia. Thus, the Hopi ceremonials designed to produce abundant rainfall may be labeled a superstitious practice of primitive folk and that is assumed to conclude the matter. It should be noted that this in no sense accounts for the group behavior. It is simply a case of name-calling; it substitutes the epithet "superstition" for an analysis of the actual role of this behavior in the life of the group. Given the concept of latent function, however, we are reminded that this behavior *may* perform a function for the group, although this function may be quite remote from the avowed purpose of the behavior.

The concept of latent function extends the observer's attention beyond the question of whether or not the behavior attains its avowed purpose. Temporarily ignoring these explicit purposes, it directs attention *toward* another range of consequences: those bearing, for example, upon the individual personalities of Hopi involved in the ceremony and upon the persistence and continuity of the larger group. Were one to consider only the problem of whether a manifest (purposed) function occurs, it becomes a problem, not for the sociologist, but for the meteorologist. And to be sure, our meteorologists agree that the rain ceremonial does not produce rain; but this is hardly to the point. It is merely to say that the ceremony does not have this technological use; that this purpose of the ceremony and its actual consequences do not coincide. But with the concept of latent function, we continue our inquiry, examining the consequences of the ceremony not for the rain gods or for meteorological phenomena, but for the groups which conduct the ceremony. And here it may be found, as many observers indicate, that the ceremonial does indeed have functions—but functions which are non-purposed or latent.

Ceremonials fulfill the latent function of reinforcing the group identity by providing a periodic occasion on which the scattered members of a group assemble to engage in a common activity. As Durkheim among others has long since indicated, such ceremonials are a means by which collective expression is afforded the sentiments which, in a further analysis, are found to be a basic source of group unity. Through the systematic application of the concept of latent function, therefore, apparently irrational behavior may at times be found to be positively functional for the group. Operating with the concept of latent function, we are not too quick to conclude that if an activity of a group does not achieve its nominal purpose, then its persistence can be described only as an instance of "inertia," "survival," or "manipulation by powerful subgroups in the society."

In point of fact, some conception like that of latent function has very often, almost invariably, been employed by social scientists observing *a standardized practice designed to achieve an objective which one knows from accredited physical science cannot be thus achieved.* This would plainly be the case, for example, with Pueblo rituals dealing with rain or fertility. *But with behavior which is not directed toward a clearly unattainable objective, sociological observers are less likely to examine the collateral or latent functions of the behavior.*

Directs attention to theoretically fruitful fields of inquiry. The distinction between manifest and latent functions serves further to direct the attention of sociologists to precisely those realms of behavior, attitude, and belief where they can most fruitfully apply their special skills. For what is their task if they confine themselves to the study of manifest functions? They are then concerned very largely with determining whether a practice instituted for a particular purpose does, in fact, achieve this purpose. They will then inquire, for example, whether a new system of wage payment achieves its avowed purpose of reducing labor turnover or of increasing output. They will ask whether a propaganda campaign has indeed gained its objective of increasing "willingness to fight" or "willingness to buy war bonds," or "tolerance toward other ethnic groups." Now, these are important, and complex, types of inquiry. But, so long as sociologists *confine* themselves to the study of manifest functions, their inquiry is set for them by practical men of affairs (whether a captain of industry, a trade union leader, or, conceivably, a Navaho chieftain, is for the moment immaterial) rather than by the theoretic problems which are at the core of the discipline. By dealing primarily with the realm of manifest functions, with the key problem of whether deliberately instituted practices or organizations succeed in achieving their objectives, sociologists become converted into industrious and skilled recorders of altogether familiar patterns of behavior. *The terms of appraisal are fixed and limited by the question, put to them by non-theoretic business interests,* e.g., has the new wage payment program achieved such-and-such purposes?

But armed with the concept of latent function, sociologists extend their inquiry in those very directions which promise most for the theoretic development of the discipline. They examine the familiar (or planned) social practice to ascertain the latent, and hence generally unrecognized, functions (as well, of course, as the manifest functions). They consider, for example, the consequences of the new wage plan for, say, the trade union in which the workers are organized or the consequences of a propaganda program, not only for increasing its avowed purpose of stirring up patriotic fervor, but also for making large numbers of people reluctant to speak their minds when they differ with official policies, etc. In short, it is suggested that the *distinctive* intellectual contributions of the sociologist are found primarily in the study of unintended consequences (among which are latent functions) of social practices, as well as in the study of anticipated consequences (among which are manifest functions).[18]

There is much evidence that it is precisely at the point where the research attention of sociologists has shifted from the plane of manifest to the plane of latent functions that they have made their distinctive and major contributions. [. . .]

The discovery of latent functions represents significant increments in sociological knowledge. There is another respect in which inquiry into latent functions represents a distinctive contribution of the social scientist. It is precisely the latent functions of a practice or belief which are *not* common knowledge, for these are unintended and generally unrecognized social and psychological consequences. As a result, findings concerning latent functions represent a greater increment in knowledge than findings concerning manifest functions. They represent, also, greater departures from "common-sense" knowledge about social life. Inasmuch as the latent functions depart, more or less, from the avowed manifest functions, the research which uncovers latent functions often produces "paradoxical" results. The seeming paradox arises from the sharp modification of a familiar popular preconception which regards a standardized practice or belief *only* in terms of its manifest functions by indicating some of its subsidiary or collateral latent functions. The introduction of the concept of latent function in

18. For an illustration of this general proposition, see Robert K. Merton, Marjorie Fiske, and Alberta Curtis, *Mass Persuasion* (New York: Harper, 1946), 185–89; Jahoda and Cook, "Security Measures."

social research leads to conclusions which show that "social life is not as simple as it first seems." For as long as people confine themselves to severely limited consequences (e.g., manifest consequences), it is comparatively simple for them to pass moral judgments upon the practice or belief in question. Moral evaluations, generally based on these manifest consequences, tend to be polarized, in terms of black or white. But the perception of further (latent) consequences often complicates the picture. Problems of moral evaluation (which are not our immediate concern) and problems of social policy (which are our concern[19]) both take on the additional complexities usually involved in responsible social decisions. [. . .]

The discovery of latent functions does not merely render conceptions of the functions served by certain social patterns more precise (as is the case also with studies of manifest functions), but introduces a *qualitatively different increment in the previous state of knowledge.*

Precludes the substitution of naive moral judgments for sociological analysis. Since moral evaluations in a society tend to be largely in terms of the manifest consequences of a practice or code, we should be prepared to find that analysis in terms of latent functions at times runs counter to prevailing moral evaluations. For it does not follow that the latent functions will operate in the same fashion as the manifest consequences which are ordinarily the basis of these judgments. Thus, in large sectors of the American population, the political machine or the "political racket" is judged to be unequivocally "bad" and "undesirable." The grounds for such moral judgment vary somewhat, but they consist substantially in pointing out that political machines violate moral codes: political patronage violates the code of selecting personnel on the basis of impersonal qualifications rather than on grounds of party loyalty or contributions to the party war chest; bossism violates the code that votes should be based on individual appraisal of the qualifications of candidates and of political issues, and not on abiding loyalty to a feudal leader; bribery and "honest graft" obviously offend the proprieties of property; "protection" for crime clearly violates the law and the mores; and so on.

In view of the manifold respects in which political machines, in varying degrees, run counter to the mores and at times to the law, it becomes pertinent to inquire how they manage to continue in operation. The

19. This is not to deny that social engineering has direct moral implications or that technique and morality are inescapably intertwined, but I do not intend to deal with this range of problems in the present chapter. For discussion of these problems see Merton, Fiske, and Curtis, *Mass Persuasion,* chap. 7.

familiar explanations for the continuance of the political machine are not here in point. To be sure, it may well be that if "respectable citizenry" would live up to their political obligations, if the electorate were to be alert and enlightened; if the number of elective officers were substantially reduced from the dozens, even hundreds, which the average voter is now expected to appraise in the course of town, county, state and national elections; if the electorate were activated by the "wealthy and educated classes without whose participation," as the not always democratically oriented Bryce put it, "the best-framed government must speedily degenerate";—if these and a plethora of similar changes in political structure were introduced, perhaps the "evils" of the political machine would indeed be exorcized.[20] But it should be noted that these changes are often not introduced, that political machines have had the phoenix-like quality of arising strong and unspoiled from their ashes, that, in short, this structure has exhibited a notable vitality in many areas of American political life.

Proceeding from the functional view, therefore, that we should *ordinarily* (not invariably) expect persistent social patterns and social structures to perform positive functions *which are at the time not adequately fulfilled by other existing patterns and structures,* the thought occurs that perhaps this publicly maligned organization is, under present conditions, satisfying basic latent functions.[21] [. . .]

20. These "explanations" are "causal" in design. They profess to indicate the social conditions under which political machines come into being. In so far as they are empirically confirmed, these explanations of course add to our knowledge concerning the problem: How is it that political machines operate in certain areas and not in others? How do they manage to continue? But these causal accounts are not sufficient. The functional consequences of the machine, as we shall see, go far toward supplementing the causal interpretation.

21. I trust it is superfluous to add that this hypothesis is not "in support of the political machine." The question whether the dysfunctions of the machine outweigh its functions and the question whether alternative structures are not available which may fulfill its functions without necessarily entailing its social dysfunctions still remain to be considered at an appropriate point. We are here concerned with documenting the statement that moral judgments based *entirely* on an appraisal of manifest functions of a social structure are "unrealistic" in the strict sense, i.e., they do not take into account other actual consequences of that structure, consequences which may provide basic social support for the structure. Be it noted that "social reforms" or "social engineering" that ignore latent functions do so on pain of suffering acute disappointments and boomerang effects.

The original text continues with an extended analysis of the functions of the political machine, work that spawned a distinct and still continuing tradition of research in political science.—*Ed.*

8

Social Dysfunctions (1976)

The study of social problems requires sociologists to consider the dysfunctions of patterns of behavior, belief, and organization rather than to focus primarily on their functions. A social dysfunction[1] is any process that undermines the stability or survival of a social system. The presence of this concept in sociology curbs any tendency toward adopting the doctrine that everything in society works for "harmony" and "the good."

We can briefly state the theoretical connection of social dysfunctions to social disorganization. *Social disorganization* refers to the whole composite of defects in the operation of a social system. A *social dysfunction* is a specific inadequacy of a particular part of the system for meeting a particular functional requirement. Social disorganization can thus be thought of as the resultant of various social dysfunctions.

Four general points will serve to keep our thinking straight about the concept of social dysfunction as a tool for analyzing social problems.

Specifying Dysfunctions

Social dysfunctions need to be specified. A full analysis of a social dysfunction provides a *designated* set of consequences of a *designated*

From "Social Disorganization and Deviant Behavior," in *Contemporary Social Problems, an Introduction to the Sociology of Deviant Behavior and Social Disorganization,* 4th ed., ed. Robert K. Merton and Robert A. Nisbet (New York: Harcourt Brace, 1976), 37–40. Reprinted by permission of Harcourt Brace.

1. It will be noticed that I have adopted the etymological barbarism, *dys*function, rather than the pure Latinism, *dis*function. I do so, of course, since I am borrowing the Greek-and-Latin hybrid, *dysfunction,* from the fields of medicine and biology where it has long since become thoroughly domesticated. I note that W. J. H. ("Sebastian") Sprott, the friend and lover of Maynard Keynes and E. M. Forster, who was surely the only psychologist and autodidactic sociologist in the Bloomsbury set, welcomed not only the SS&A paradigm but the concept of social *dysfunction* as well, but faithful to his Oxbridge education and associates, quietly transcribed it as *disfunction.* See his *Science and Social Action* (London: Watts and Co., 1954), 60, 113–16.

pattern of behavior, belief, or organization that interferes with a *designated* functional requirement of a *designated* social system. Otherwise, the term becomes little more than an epithet of disparagement or a vacuous expression of attitude. To say, for example, that a high rate of social mobility is "dysfunctional" (or, for that matter, "functional") without specifying its consequences for a social system is to say little. But it is quite another thing to propose, as such ideologically contrasting theorists as Karl Marx and Vilfredo Pareto did propose, that a high rate of upward social mobility from the working class is dysfunctional for maintaining its solidarity and attaining its goals, since such mobility exports talent from the class and depletes its potential leadership.[2] More recently, a related ambivalence toward upward mobility has been observed among "the masses of lower-class Negroes [who] regard this movement up the ladder with mixed feelings, both proud and resentful of the success of 'one of their own.'"[3] In the same ambivalent fashion, the collective efforts to have many more black scholars appointed to the faculties of major universities and colleges has been described as "the black brain-drain to white colleges."

Composite Functions-and-Dysfunctions: Different Groups

The same social pattern can be dysfunctional for some segments of a social system and functional for others. This arises from a basic characteristic of social structure: *in a differentiated society, the consequences of social patterns tend to differ for individuals, groups, and social strata variously located in the structure.*

If a social pattern persists, it is unlikely that it is uniformly dysfunctional for all groups. Thus, comparatively free access to higher education, irrespective of racial, ethnic, or other status, is dysfunctional for maintaining a relatively fixed system of caste. At the same time, it is functional for the attainment of higher education, a culturally induced goal, by people formerly excluded from it.

Various groups and strata in the structure of a society have *conflicting* interests and values as well as *shared* interests and values, and this

2. For essentially the same hypothesis about the dysfunctions of rapid, large-scale mobility see Karl Marx, *Capital* (1867; reprint, Chicago: Kerr, 1906), 648–49 and Vilfredo Pareto, *The Mind and Society* (1916; reprint, New York: Harcourt Brace, 1935), 3:1419–32; 4:1836–46. For analysis of the pattern of "cognitive agreement and value disagreement" evident in this case, see Robert K. Merton, *The Sociology of Science: Theoretical and Empirical Investigations*, ed. N. W. Storer (Chicago: University of Chicago Press, 1973) 65–66.

3. Kenneth B. Clark, *Dark Ghetto: Dilemmas of Social Power* (New York: Harper and Row, 1965), 57–58.

means that one group's problem sometimes becomes another group's solution. This structural condition is one reason why the periodically popular notion of a society in which everything works together for good is literally utopian. But abandoning this image of a perfect society does not mean that nothing can be done to reduce social disorganization. Quite the contrary: it is by discovering and disclosing dysfunctional social formations that sociology links up with critical morality as opposed to conventional reality.[4]

Composite Functions-and-Dysfunctions: Same Group

Not only is the same pattern sometimes functional for some groups and dysfunctional for others, it can also serve some and defeat other functional requirements of the *same* group. The reason for this resembles the reasons for a pattern having different consequences for different groups. A group has diverse functional requirements.

One example of composite function-and-dysfunction aptly illustrates the general idea. A group requires enough social cohesion to provide a sense of group identity, but prime attention to this need can conflict with the need to work effectively toward collective goals. Activities functional for one requirement can become dysfunctional for the other.[5] Sociologists have found that up to a certain point, social cohesion facilitates the productivity of a group. People feel at one with each other and so are more willing to work together for joint objectives. But this mutually reinforcing relation between the two sets of activities holds only within certain limits. Beyond those limits, a dysfunctional imbalance develops between activity that serves chiefly to maintain social cohesion and activity that results chiefly in getting work done. There can be too much of a good thing. Members of an exceedingly cohesive group become reciprocally indulgent and fail to hold one another to effective standards of performance; or they devote too much of their social interaction to expressing solidarity, at the expense of time and energy for getting the job done. Correlatively, a group may become so exacting in its demands for instrumental activity that it fails to maintain a sufficient degree of cohesion.

This is a prototype of the functional decisions that must be made in social systems of all kinds. Morale and productivity, compassion and efficiency, personal ties and impersonal tasks—these are familiar

4. Ralph Ross, *Obligation: A Social Theory* (Ann Arbor: University of Michigan Press, 1970), esp. chap. 5, "Critical Morality," and chaps. 8 and 9.

5. The classic experimental work on instrumental and expressive interaction in groups is Robert F. Bales, *Interaction Process Analysis* (Reading, Mass.: Addison-Wesley, 1951).

enough pairs of values not simultaneously realizable to the fullest extent. This way of thinking is a sociological equivalent to the economist's concept of opportunity costs, which means in effect, that under certain conditions one commitment reduces the opportunities for making other commitments. By recognizing the composite pattern of function and dysfunction, we guard ourselves against the utopian thinking which neglects the social constraints that result from prior commitment to other objectives. Neglecting these constraints leads to the false assumption that all values can be totally fulfilled simultaneously in society. But cost-free social action is a sociological fantasy.

A Sociological, not Moral, Concept

Above all else, it must be emphasized that the concept of social dysfunction does not harbor a concealed moral judgment.[6] Social dysfunction is not equivalent to immorality, unethical practice, or social disrepute. It is a concept referring to an objective state of affairs. Whether one judges a particular social dysfunction as ethically good or bad, as desirable or undesirable, depends upon an entirely individual judgment of the moral worth of that system, not upon sociological analysis. When we observed, for example, that extending opportunities for higher education to all is dysfunctional for the maintenance of a caste system, we surely were not suggesting that this dysfunction was undesirable or evil. Or when it is observed that the extremely authoritarian character of the Nazi bureaucracy was dysfunctional for its effective operation by excessively restricting lines of communication among its several echelons, this is surely not to deplore the breakdown of Nazism. Correlatively, when sociologists specify the functions of social conflict in general and of racial conflict in particular, they are engaged in sociological analysis, not in making moral judgments.[7] Sociological analyses of function and dysfunction are in a different universe of discourse from that of moral judgments; they are not merely different expressions of the same thing.

All this would not require emphasis except for the widespread assumption that nonconforming behavior is necessarily dysfunctional to

6. Dorothy Emmet, *Function, Purpose, and Powers* (London: Macmillan, 1958), 78–82.

7. Lewis A. Coser, *The Functions of Social Conflict* (Glencoe, Ill.: Free Press, 1956) and *Continuities in the Study of Social Conflict* (New York: Free Press, 1967); Jonathan H. Turner, *The Structure of Sociological Theory* (Homewood, Ill.: Dorsey Press, 1974), esp. chap. 6, "Dialectical Conflict Theory: Ralf Dahrendorf" and chap. 7, "Conflict Functionalism: Lewis A. Coser;" Robert A. Dentler and Kai T. Erikson, "The Functions of Racial Conflict," *Social Problems* 7 (1959): 98–107.

a social system and that social dysfunctions, in turn, necessarily violate ethical codes. Yet frequently the nonconforming minority in a society represents its ultimate values and interests more fully than the conforming majority. This is not a moral but a functional judgment, not a statement in ethical theory but a statement in sociological theory. In the history of many societies, one supposes, some of its culture heroes became heroes partly because they had the courage and vision to challenge the beliefs and routines of their society. The rebel, revolutionary, nonconformist, heretic, or renegade of an earlier day is often the culture hero of today. The distinction between nonconforming and aberrant behavior is designed to capture basic *functional* differences in forms of deviant behavior. For the accumulation of dysfunctions in a social system is often the prelude to concerted social change toward a system that better serves the ultimate values of the society.

9

Paradigm for a Structural Analysis in Sociology (1975)

[. . .] Structural analysis in sociology has generated a problematics I find interesting and a way of thinking about problems I find more effective than any other I know. Moreover, it connects with other sociological paradigms that, the polemics notwithstanding, are anything but contradictory in much of what they suppose or assert. This is no doubt an unbecoming pacifist position to adopt at a time when the arena of sociology echoes with the claims of gladiators championing rival doctrines. Still, recent work in structural analysis leads me to spheres of agreement and of complementarity rather than to the alleged basic contradictions between various sociological paradigms. This is nothing strange. For it is not easy to achieve even mildly plausible sociological doctrines (paradigms, theories, conceptual schemes, models) that contradict one another in basic assumptions, concepts, and ideas. Many ideas in structural analysis and symbolic interactionism, for example, are opposed to one another in about the same sense as ham is opposed to eggs: they are perceptibly different but mutually enriching. [. . .]

The following paradigm sketches out basic components of a variant of structural analysis in the form of a series of stipulations. Although the term "stipulation" is taken from the adversary culture of the law, I use it here only to indicate provisional agreement on one kind of structural analysis. With such agreement, I can proceed to the rest of my subject: the place of that mode of theorizing in the cognitive and social structure of sociology.

Fourteen Stipulations for Structural Analysis

Here, then, are stipulations of this one variant of structural analysis. *It is stipulated*

From "Structural Analysis in Sociology," in *Sociological Ambivalence and Other Essays* (New York: Free Press, 1976), 119–29. Reprinted by permission of The Free Press, an imprint of Simon and Schuster.

1. *That* the evolving notion of "social structure" is polyphyletic and polymorphous (but not, one hopes, polymorphous perverse): that is, the notion has more than one ancestral line of sociological thought, and these differ partly in substance and partly in method.[1]

2. *That* the basic ideas of structural analysis in sociology long antedated the composite intellectual and social movement known as structuralism.[2] Spanning a variety of core disciplines, structuralism has lately become the focus of a popular, sometimes undiscriminating social movement that has exploited through undisciplined extension the intellectual authority of such iconic figures as Ferdinand de Saussure and Roman Jakobson in linguistics, Claude Lévi-Strauss in anthropology, Jean Piaget in psychology and, most recently, François Jacob in biology. In short, although structural analysis in sociology today has been affected by certain communalities of structuralism serving as cognitive context—for example, certain parallels between Saussure and Durkheim—it does not historically derive from these intellectual traditions any more than it does, say, from the input-output form of structural analysis developed by Wassily Leontief in economics.[3]

3. *That* structural analysis in sociology involves the confluence of ideas deriving principally from Durkheim and Marx. Far from being contradictory, as has sometimes been assumed, basic ideas drawn from their work have been found to be complementary in a long series of inquiries through the years, ranging from the social-structural sources of deviant behavior and the formation of bureaucratic personality to

1. Boudon adopts the image of a "polymorphism of sociology" in a related but different sense to refer to various forms of sociological work: a "brilliant essay," "an empirical descriptive study," a verifiable "analytical theory" or a "speculative theory" pointing to directions of inquiry. Raymond Boudon, *La crise de la sociologie* (Paris: Droz, 1971), 9–10.

2. The burgeoning literature on structuralism is practically inexhaustible, and no good purpose would be served by supplying a long list of titles here. The works of the masters are easily accessible and require no mention, except, perhaps, for the overview of Jean Piaget, *Structuralism* (New York: Basic Books, 1970), and the masterly history of heredity with its successive disclosure of structures by François Jacob, *The Logic of Life* (New York: Pantheon Books, 1973). Raymond Boudon has made a serious effort to differentiate and formalize the major conceptions of social structure in relation to notions of structure in other disciplines in *The Uses of Structuralism* (London: Heinemann, 1971). For other secondary works, see Jean Viet, *Les méthodes structuralistes dans les sciences sociales* (Paris: Mouton and Co., 1965); Oswald Ducrot et al., *Qu'est-ce que le structuralisme?* (Paris: Editions du Seuil, 1968); and *Structuralism: An Introduction,* ed. David Robey (Oxford: Clarendon Press, 1973).

3. Wassily Leonteif, "Some Basic Problems of Structural Analysis," *The Review of Economics and Statistics* 34 (1952): 1–9.

the growth and institutional structure of science.[4] In a more compact form, a paradigm proposed for functional analysis in the 1930s and published in 1949 drew attention to the overlaps, not identity, of these theoretical orientations. Examples are the basic concepts of "contradictions" in the one and of "dysfunctions" in the other; the fundamental place accorded "conditions" of society in Marx and of "structural context" or "structural constraint" in structural analysis and, in the domain of the sociology of knowledge, Marx's postulate that men's changing "social existence determines their consciousness" corresponding to Durkheim's conception that collective representations reflect a social reality.[5]

The intertwining of these strands of thought has not gone unnoticed. Stinchcombe's analyses[6] of the overlapping sets of theoretical ideas generated his term "Marxian functionalism," while Gouldner takes repeated note of my "emphasizing [the] affinities between them, concluding with the compact observation about the analysis in "Social Structure and Anomie" that "here, in effect, Merton uses Marx to pry open Durkheim."[7] Kalàb[8] describes Marx's method as "dialectically conceived structural analysis" and notes the interdependence of historical and structural analysis as did the exemplary historian Herbert Butterfield years ago when he described the major contribution of Marxism to historiography as one of having "taught us to make our history a structural piece of analysis."[9] In one instructive volume, Giddens[10] has recently analyzed congruities in the writings of Marx, Durkheim, and Weber, and in another, Sztompka[11] finds close congruities between functional and Marxian analysis, just as Pierre L. van den

4. Robert K. Merton, *Social Theory and Social Structure,* 3d ed., rev. and enl. (New York: Free Press, 1968).

5. Ibid., 93–5, 160–61, 516ff.

6. Arthur Stinchcombe, *Constructing Social Theories* (New York: Harcourt Brace Jovanovich, 1968), 80–101; and Arthur Stinchcombe, "Merton's Theory of Social Structure" in *The Idea of Social Structure,* ed. Lewis A. Coser (New York: Harcourt Brace Jovanovich, 1975), 11–33.

7. Alvin Gouldner, *The Coming Crisis of Western Sociology* (New York: Basic Books, 1970), 335, 402, 426, 448, and, for the aperçu, 477.

8. Miloš Kalàb, "The Marxist Conception of the Sociological Method," *Quality and Quantity* 3 (1969): 5–23.

9. Herbert Butterfield, History and Human Relations (London and Glasgow: Collins, 1951), 79–80.

10. Anthony Giddens, *Capitalism and Modern Social Theory* (Cambridge: Cambridge University Press, 1971), 79–80.

11. Piotr Sztompka, *System and Function* (New York: Academic Press, 1974).

Berghe did in short compass more than a decade ago (see also Malewski).[12] Van den Berghe's conclusion states the case pointedly:

> Our central contention is that the two major approaches which have dominated much of social science present partial but complementary views of reality. Each body of theory raises difficulties which can be resolved, either by rejecting certain unnecessary postulates, or by introducing concepts borrowed from the other approach. As functionalism and the dialectic show, besides important differences, some points of convergence and overlap, there is hope of transcending ad hoc eclecticism and of reaching a balanced theoretical synthesis.[13]

4. *That* since the confluence of elements of Durkheim and Marx has been evident from at least the 1930s, it cannot be taken, as Gouldner[14] proposes it should be taken, as another sign of the crisis he ascribes to both functional and Marxist sociology in the 1960s.[15] Put more generally, it is being stipulated here that far from necessarily constituting a sign of theoretical crisis or decline, the convergence of separate lines of thought can, and in this case, does involve a process of consolidation of concepts, ideas, and propositions that result in more general paradigms.[16]

5. *That,* like theoretical orientations in the other social sciences, to say nothing of the physical and life sciences, structural analysis in sociology must deal with successively micro- and macro-level phenomena. Like them, it therefore confronts the formidable problem, lately taken up anew by Peter Blau[17] and many another, of developing concepts, methods, and data for linking micro- and macro-analysis.[18]

12. Andrzej Malewski, "Der empirische Gehalt der Theorie des historischen Materialismus," *Kölner Zeitschrift für Soziologie und Sozialpsychologie* 11 (1959): 281–305; and *Verhalten und Interaktion* (Tübingen: J. C. B. Mohr, 1967).

13. Pierre van den Berghe, "Dialectic and Functionalism: Toward a Theoretical Synthesis," *American Sociological Review* 28 (1963): 695–705.

14. Gouldner, *The Coming Crisis,* 341 ff.

15. In this connection, I must disown Gouldner's avowed conjecture that, in the 1930s and 1940s, I "sought to make peace between Marxism and Functionalism precisely by emphasizing their affinities, and thus make it easier for Marxist students to become Functionalist professors." Gouldner, *The Coming Crisis,* 335. Here, Gouldner surely does me too much honor. I had neither the far-seeing intent nor the wit and powers thus to transmogrify my students.

16. This stipulation is of long standing. I have been arguing for the importance of theoretical consolidation in sociology since the 1940s (Merton, *Social Theory and Social Structure,* chap. 2, esp. 49–53).

17. Peter Blau, *Exchange and Power in Social Life* (New York: John Wiley and Sons, 1964).

18. It seems safe to stipulate rather than to discuss this conception at length now that it has found its way into that depository of "established knowledge," the textbook. (On

6. *That,* to adopt Stinchcombe's important and compact formulation on the micro-level,

> the core process conceived as central to social structure is *the choice between socially structured alternatives.* This differs from the choice process of economic theory, in which the alternatives are conceived to have inherent utilities. It differs from the choice process of learning theory, in which the alternatives are conceived to emit reinforcing or extinguishing stimuli. It differs from both of these in that . . . the utility or reinforcement of a particular alternative choice is thought of as socially established, as part of the institutional order.[19]

7. *That,* on the macro-level, the social distributions (i.e., the concentration and dispersion) of authority, power, influence, and prestige comprise structures of social control that change historically, partly through processes of "accumulation of advantage and disadvantage" accruing to people occupying diverse stratified positions in that structure (subject to processes of feedback under conditions still poorly understood).[20]

---

the significance of the textbook in different disciplines, see Thomas S. Kuhn, *The Structure of Scientific Revolutions* [(Chicago: University of Chicago Press, 1962), 163–65] and Ludwik Fleck, *Genesis and Development of a Scientific Fact* [(1935; reprint, Chicago: University of Chicago Press, 1979), 118–24]). Thus, in discussing Blau's "exchange structuralism," Jonathan Turner writes, "Bridging the Micro-Macro Gap. One of the most important analytical problems facing sociological theorizing revolves around the question: To what extent are structures and processes at micro *and* macro levels of social organization subject to analysis by the same concepts and to description by the same sociological laws? At what levels of sociological [*sic*] organization do emergent properties require use of additional concepts and description in terms of their own social laws?" Jonathan Turner, *The Structure of Sociological Theory* (Homewood, Ill.: Dorsey Press, 1974), 292.

And without indulging in easy and misplaced analogizing, sociologists must take a degree of interest in the reminder by the polymathic physicist, Richard Feynman, that in connection with the laws of physics, "we have found that the behaviour of matter on a small scale obeys laws very different from things on a large scale. So the question is, how does gravity look on a small scale? That is called the Quantum Theory of Gravity. There is no Quantum Theory of Gravity today. People have not succeeded completely in making a theory which is consistent with the uncertainty principles and the quantum mechanical principles." Richard Feynman, *The Character of Physical Law* (London: Cox and Wyman, 1965), 32–33.

19. Stinchcombe, *The Idea of Social Structure,* 12.

20. Since appearing in 1942, the idea of "accumulation of advantage" (which relates to the notions of "the self-fulfilling prophecy" and "the Matthew effect") in systems of social stratification has been developed in a series of investigations: Merton, *The Sociology of Science,* 273, 416, 439–59; Harriet Zuckerman, *Scientific Elite* (New York: Free Press, 1977), chap. 3; Jonathan R. Cole and Stephen Cole, *Social Stratification in Science* (Chicago: University of Chicago Press, 1973), 237–47; passim; Paul D. Allison and

8. *That* it is fundamental, not incidental, to the paradigm of structural analysis that *social structures generate social conflict* by being differentiated, in historically differing extent and kind, into interlocking arrays of social statuses, strata, organizations, and communities that have their own and therefore potentially conflicting as well as common interests and values.[21] (I shall have more to say about this presently.)

9. *That* normative structures do not have unified norm-sets. Instead, that *sociological ambivalence* is built into normative structures in the form of incompatible patterned expectations and a "dynamic alternation of norms and counternorms" in social roles, as this sociological ambivalence has been identified, for example, in the spheres of bureaucracy, medicine and science.[22]

10. *That social structures generate differing rates of deviant behavior,* variously so defined by structurally identifiable members of the society. The behavior defined as deviant results, in significant degree, from socially patterned discrepancies between culturally induced personal aspirations and patterned differentials in access to the opportunity structure for moving toward those aspirations by institutional means.[23]

11. *That,* in addition to exogenous events, *social structures generate both change within the structure and change of the structure,* and that these types of change come about through cumulatively patterned choices in behavior and the amplification of dysfunctional consequences resulting from certain kinds of strains, conflicts, and contradictions in the differentiated social structure.[24]

John A. Stewart, "Productivity Differences among Scientists: Evidence for Accumulative Advantage," *American Sociological Review* 39 (1974): 596–606; Harriet Zuckerman and Jonathan R. Cole, "Women in American Science," *Minerva* 13 (1975): 82–102; Harriet Zuckerman, "Accumulation of Advantage and Disadvantage: The Theory and Its Intellectual Biography," in *L'Opera di R. K. Merton e la sociologia contemporanea,* ed. Carlo Mongardini and Simonetta Tabboni (Genoa: ECIG, 1989), 153–76.

21. Robert K. Merton, "Social Problems and Sociological Theory," in *Contemporary Social Problems,* 3d ed., ed. Robert K. Merton and Robert A. Nisbet (New York: Harcourt Brace Jovanovich, 1971), 796; and Merton, *Social Theory and Social Structure,* 424–25.

22. Robert K. Merton and Elinor Barber, "Sociological Ambivalence," in *Sociological Theory, Values, and Sociocultural Change* (Glencoe, Ill.: Free Press, 1963), 91–120; Merton, *The Sociology of Science,* chap. 18; and Ian Mitroff, "Norms and Counter-Norms in a Selected Group of the Apollo Moon Scientists: A Case Study in the Ambivalence of Scientists," *American Sociological Review* 39 (1974): 579–95.

23. Merton, *Social Theory and Social Structure,* 185–248; Merton, "Social Problems," 793–846; and chap. 13 in this volume.

24. Merton, *Social Theory and Social Structure,* 176–77. This is stipulated in spite of the recent critiques by Runciman and Nisbet. Both agree that it is badly misplaced to

12. *That,* in accord with preceding stipulations, every new cohort born into a social structure it never made proceeds differentially, along with other age cohorts, to modify that structure, both unwittingly and by design, through its responses to the objective social consequences, also both unanticipated and designed, of previous organized and collective action.[25]

13. *That* it is analytically useful to distinguish between manifest and latent levels of social structure as of social function (with the aside that structuralism as set forth in other disciplines—for example, by Jakobson, Lévi-Strauss, and Chomsky—finds it essential to distinguish "surface" from "deep" structures).

14. And, finally, it is stipulated as a matter of theoretical principle (rather than as a stab at conspicuous modesty) *that,* like other theoretical orientations in sociology, structural analysis can lay no claim to being able to account exhaustively for the entire range of social and cultural phenomena.

From these severely condensed stipulations, it must be plain that this variant of structural analysis owes much to the classic mode of structural–functional analysis developed by my teacher, friend, and colleague-at-a-distance, Talcott Parsons. But the variant differs from the standard form in, for me, two major respects—substantive and metatheoretical.

Structural Sources of Conflict and Deviant Behavior

Substantively, the variant doctrine makes a large place for the structural sources and differential consequences of conflict, dysfunctions, and contradictions in the social structure, thus representing, as has been noted, an intertwining of central strands of thought in Marx and Durkheim. I find it significant that Ralf Dahrendorf, long tagged as a

charge functional or structural analysis with not having any "theory of social change," and they make their case in the best way: by stating that theory and criticizing it. In a series of works, Nisbet strongly criticizes the idea of structurally or immanently generated social change as being theoretically untenable. I remain unpersuaded. His analysis only shows that sources exogenous to social structure also operate to produce change, a position altogether congenial, as he evidently recognizes, to those of us who do not take structural analysis to exhaust all aspects of social phenomena. Robert A. Nisbet, "Developmentalism: A Critical Analysis," in *Theoretical Sociology,* ed. John C. McKinney and Edward A. Tiryakian (New York: Appleton-Century-Crofts, 1970), 167–294 at 178, 194–96; Nisbet, *Social Change and History* (New York: Oxford University Press, 1969); Nisbet, ed., *Social Change* (New York: Harper and Row, 1972); W. G. Runciman, *Social Science and Political Theory* (Cambridge: Cambridge University Press, 1963), 43.

25. See chaps. 15 and 16 in this volume.

"conflict theorist" in the sometimes demimythic classifications of theoretical sociology, noted this basic point years ago. In his chapter, signally entitled "Die Funktionen sozialer Konflikte," Dahrendorf observed that this mode of structural analysis

> enables Merton, in contrast to Mayo, to accept the idea that conflicts may be *systematically produced by social structures.* There are for him circumstances where the structures of roles, reference-groups, and institutions to some degree *necessarily generate conflict.* But where do these conflicts arise, and what is their significance? It is at this point that he introduces the concept of "dysfunction" which has been used so much since. . . . This step forward [in the development of functional analysis] lay above all in its indication of the possibility *of a systematic explanation of conflict* ("on the structural level").[26]

Much the same observation was made independently by Hans Goddijn in noting that this mode of structural analysis finds "the origins of social conflict within the social structure itself, namely in the antithesis of social positions. For that reason, this analysis can be seen within the context of a sociology of conflict."[27]

Gouldner has made the same kind of historical and analytical observation about the structural analysis of deviant behavior. Easily breaking through the make-believe barricades that would obstruct even restricted passage between theoretical orientations stemming from Marx and Durkheim, he notes the overlap between them. As I cannot improve upon Gouldner's own formulation, I borrow it. He observes that certain theorizing on deviant behavior

> should be seen *historically,* in terms of what it meant when it first appeared and made the rounds. In this context, it needs emphasis that Merton's work on *anomie* as well as Mills's work on "social pathology" was a *liberative* work, for those who lived with it as part of a *living* culture as distinct from how it may now appear as part of the mere *record* of that once-lived culture.
>
> There are several reasons for this. One is that both Merton and Mills kept open an avenue of access to Marxist theory. Indeed both of them had a kind of tacit *Marxism.* Mills's Marxism was always much more tacit than his own radical position made it seem, while Merton was always much more Marxist than his silences on that question may make it seem. . . . Merton always knew his Marx and knew thoroughly the nuances of controversy in living Marxist culture. Merton developed his generalized analysis of the various forms of deviant behaviour by locating them within a systematic

26. Ralf Dahrendorf, *Pfade aus Utopia* (Munich: R. Piper and Co., 1967), 268–69 [italics added].

27. H. P. M. Goddijn, *Het Funktionalisme in de Sociologie* (Assen: Van Gorcum, 1963), chap. 4.

> formalization of Durkheim's theory of *anomie*, from which he gained analytic distance by tacitly grounding himself in a Marxian ontology of social contradiction. It is perhaps this Hegelian dimension of Marxism that has had the most enduring effect on Merton's *analytic* rules, and which dispose him to view *anomie* as the unanticipated outcome of social institutions that thwarted men in their effort to acquire the very goods and values that these same institutions had encouraged them to pursue.[28]

These observations on deviant behavior like those on social conflict are sharply at odds with the hackneyed and immutable notion, current in some sociological quarters, which holds that a theoretical orientation called "conflict sociology" is inescapably opposed to the mode of structural analysis under discussion here. In a way, Dahrendorf, Goddijn, Gouldner, and not a few others had falsified that claim before it became current. The fixed claim, made out of whole cloth, imputes to this kind of structural analysis the undisclosed assumption that societies or groups have a *total consensus* of values, norms, and interests. This imputed (rather than documented) assumption presumably contrasts with the assumption that social conflict is somehow indigenous to human society. But, of course, social conflict cannot occur without a clash of values, norms, or interests variously shared by each of the social formations that are in conflict. As we noted in the eighth stipulation, it is precisely that kind of socially patterned differentiation of interests and values that leads structural analysis to hold that social conflict is not mere happenstance but is rooted in social structure.

All apart from the Dahrendorf–Goddijn–Gouldner observations and my own reiterations to the same effect in the development of structural analysis, there is ample evidence to negate the stereotype that describes it as "consensual sociology." After all, it is no accident (as one says) that Lewis Coser, a continuing exponent of this variant tradition of structural analysis, adopted for investigation the twin foci registered in the title of his early book, *The Functions of Social Conflict*,[29] then went on to develop *Continuities in the Study of Social Conflict*,[30] and most recently, focused on structural sources of social conflict in his *Greedy Institutions: Patterns of Undivided Commitment*.[31] [. . .]

28. Alvin W. Gouldner, "Foreword," in *The New Criminology: For a Social Theory of Deviance*, Ian Taylor, Paul Walton, and Jock Young (London: Routledge and Kegan Paul, 1973), ix–xiv at x–xi.

29. Glencoe, Ill.: Free Press, 1956.

30. New York: Free Press, 1967.

31. New York: Free Press, 1974.

II

SOCIAL STRUCTURE AND ITS VICISSITUDES

Structural Complexes and Contradictions

10

The Role-Set (1957)

[. . .] I begin with the premise that each social status involves not a single associated role, but an array of roles. This basic feature of social structure can be registered by the distinctive but not formidable term *role-set*. By role-set I mean that complement of role-relationships in which persons are involved by virtue of occupying a particular social status. Thus, in our current studies of medical schools,[1] we have begun with the view that the status of medical student entails not only the role of a student vis-à-vis teachers, but also an array of other roles relating the student diversely to other students, physicians, nurses, social workers, medical technicians, and the like. Again, the status of school teacher in the United States has its distinctive role-set, in which are found pupils, colleagues, the school principal and superintendent, the Board of Education, professional associations, and, on occasion, local patriotic organizations.

It should be made plain that the role-set differs from what sociologists have long described as "multiple roles." By established usage, the term multiple role refers not to the complex of roles associated with a single social status, but with the various social statuses (often, in differing institutional spheres) in which people find themselves—for illustration, the statuses of physician, husband, father, professor, church elder, Conservative Party member, and army captain. (This complement of distinct statuses of a person, each of these in turn having its own role-set, I would designate as a status-set. This concept gives rise to its own range of analytical problems which cannot be considered here.)[2]

From "The Role-Set: Problems in Sociological Theory," in *British Journal of Sociology* 8 (1957): 110–20. Reprinted by permission of Routledge.

1. Robert K. Merton, P. L. Kendall, and G. G. Reader, eds., *The Student-Physician: Introductory Studies in the Sociology of Medical Education* (Cambridge: Harvard University Press, 1957).

2. The analysis of status-sets is carried forward in Merton, *Social Theory and Social Structure* (New York: Free Press, 1968), 422–38; on role-set theory as a theory of the middle range, see chapter 3 in this volume.—*Ed.*

The notion of the role-set reminds us, in the unlikely event that we need to be reminded of this obstinate fact, that even the seemingly simple social structure is fairly complex. All societies face the functional problem of articulating the components of numerous role-sets, the functional problem of managing somehow to organize these so that an appreciable degree of social regularity obtains, sufficient to enable most people most of the time to go about their business of social life, without encountering extreme conflict in their role-sets as the normal, rather than the exceptional, state of affairs.

If this relatively simple idea of role-set has any theoretical worth, it should at the least generate distinctive problems for sociological theory, which come to our attention only from the perspective afforded by this idea, or by one like it. This the notion of role-set does. It raises the general problem of identifying the social mechanisms which serve to articulate the expectations of those in the role-set so that the occupant of a status is confronted with less conflict than would obtain if these mechanisms were not at work. It is to these social mechanisms that I would devote the rest of this discussion.

Before doing so, I should like to recapitulate the argument thus far. We depart from the simple idea, unlike that which has been rather widely assumed, that a single status in society involves, not a single role, but an array of associated roles, relating the status-occupant to diverse others. Second, we note that this structural fact, expressed in the term role-set, gives rise to distinctive analytical problems and to corresponding questions for empirical inquiry. The basic problem, which I deal with here, is that of identifying social mechanisms, that is, processes having designated effects for designated parts of the social structure, which serve to articulate the role-set more nearly than would be the case if these mechanisms did not operate. Third, unlike the problems centered upon the notion of multiple roles, this one is concerned with social arrangements integrating the expectations of those in the role-set; it is not primarily concerned with the familiar problem of how the occupants of a status manage to cope with the many, and sometimes conflicting, demands made of them. It is thus a problem of social structure, not an exercise in the no doubt important but different problem of how individuals happen to deal with the complex structures of relations in which they find themselves. Finally, by way of setting the analytical problem, the logic of analysis exhibited in this case is developed wholly in terms of the elements of social structure, rather than in terms of providing concrete historical description of a social system.

All this presupposes, of course, that there is always a potential for differing and sometimes conflicting expectations of the conduct appropriate to a status-occupant among those in the role-set. The basic source of this potential for conflict, I suggest—and here we are at one with theorists as disparate as Marx and Spencer, Simmel and Parsons—is that the members of a role-set are, to some degree, apt to hold social positions differing from that of the occupants of the status in question. To the extent that they are diversely located in the social structure, they are apt to have interests and sentiments, values and moral expectations differing from those of the status-occupants themselves. This, after all, is one of the principal assumptions of Marxist theory, as it is of all sociological theory: social differentiation generates distinct interests among those variously located in the structure of the society. To continue with one of our examples: the members of a school board are often in social and economic strata which differ greatly from that of the school teachers, and their interests, values, and expectations are consequently apt to differ, to some extent, from those of the teacher. The teacher may thus become subject to conflicting role-expectations among such members of the role-set as professional colleagues, influential members of the school board, and, say, the Americanism Committee of the American Legion. What is an educational essential for the one may be judged as an education frill, or as downright subversion, by the other. These disparate and contradictory evaluations by members of the role-set greatly complicate the task of coping with them all. The familiar case of the teacher may be taken as paradigmatic. What holds conspicuously for this one status holds, in varying degree, for the occupants of all other statuses who are structurally related, through their role-sets, to others, who themselves occupy diverse positions in society.

This, then, is the basic structural basis for potential disturbance of a role-set. And it gives rise, in turn, to a double question: which social mechanisms, if any, operate to counteract such instability of role-sets and, correlatively, under which circumstances do these social mechanisms fail to operate, with resulting confusion and conflict. This is not to say that role-sets invariably operate with substantial efficiency. We are concerned here, not with a broad historical generalization to the effect that social order prevails, but with an analytical problem of identifying social mechanisms which produce a greater degree of order than would obtain if these mechanisms were not called into play. Otherwise put, it is theoretical sociology, not history, that is of interest here.

Social Mechanisms Articulating Role-Sets

1. *Relative importance of various statuses.* The first of these mechanisms derives from the sociological circumstance that social structures designate certain statuses as having greater importance than others. Family and job obligations, for example, are defined in American society as having priority over membership in voluntary associations.[3] As a result, a particular role-relationship may be of peripheral concern for some; for others it may be central. Our hypothetical teacher, for whom this status holds primary significance, may by this circumstance be better able to withstand the demands for conformity with the differing expectations of those comprising the teacher's role-set. For at least some of these others, the relationship has only peripheral significance. This does not mean, of course, that teachers are not vulnerable to demands which are at odds with their own professional commitments. It means only that when powerful members of their role-set are only little concerned with this particular relationship, teachers are less vulnerable than they would otherwise be (or sometimes are). Were all those involved in the role-set *equally* concerned with this relationship, the plight of the teacher would be considerably more sorrowful than it often is. What holds for the particular case of the teacher presumably holds for the occupants of other statuses: the impact upon them of diverse expectations among those in their role-set is mitigated by the basic structural fact of differentials of involvement in the relationship among those comprising their role-set.

2. *Differences of power of those in the role-set.* A second potential mechanism for stabilizing the role-set is found in the distribution of power and authority. By power, in this connection, is meant the observed and predictable capacity to impose one's will in a social action, even against the opposition of others taking part in that action; by authority, the culturally legitimized organization of power.

As a consequence of social stratification, the members of a role-set are not apt to be equally powerful in shaping the behaviour of status-occupants. However, it does not follow that the individuals, group, or stratum in the role-set which are *separately* most powerful uniformly succeed in imposing their demands upon the status-occupant, say, the teacher. This would be so only in the circumstance that the one member of the role-set has either a monopoly of power in the situation or out-

3. Bernard Barber has drawn out the implications of this structural fact in his study of voluntary associations; see his "Participation and Mass Apathy in Associations," in *Studies in Leadership*, ed. A. W. Gouldner (New York: Harper and Brothers, 1950), 477–504, especially at 486 ff.

weighs the combined power of the others. Failing this special but, of course, not infrequent situation, there may develop coalitions of power among some members of the role-set which enable the status-occupants to go their own way. The familiar pattern of a balance of power is not confined to the conventionally defined political realm. In less easily visible form, it can be found in the workings of role-sets generally, as the boy who succeeds in having his father's decision offset his mother's opposed decision has ample occasion to know. To the extent that conflicting powers in the role-set neutralize one another, status-occupants have relative freedom to proceed as they intended in the first place.

Thus, even in those potentially unstable structures in which the members of a role-set hold contrasting expectations of what the status-occupant should do, the latter is not wholly at the mercy of the most powerful among them. Moreover, the structural variations of engagement in the role-structure, which I have mentioned, can serve to reinforce the relative power of the status-occupant. For to the extent that powerful members of the role-set are not centrally concerned with this particular relationship, they will be the less motivated to exercise their potential power to the full. Within varying margins of activity, status-occupants will then be free to act as they would.

Once again, to reiterate that which lends itself to misunderstanding, I do not say that status-occupants subject to conflicting expectations among members of role-sets are in fact immune to control by them. I suggest only that the power and authority-structure of role-sets is often such that the status-occupants have a larger measure of autonomy than they would have had if this structure of competing power did not obtain.

3. *Insulation of role-activities from observability by members of the role-set.* People do not engage in continuous interaction with all those in their role-sets. This is not an incidental fact, to be ignored because familiar, but one integral to the operation of social structure. Interaction with each member of a role-set tends to be variously intermittent. This fundamental fact allows for role-behaviour that is at odds with the expectations of some in the role-set to proceed without undue stress. For, as I suggest elsewhere at some length,[4] effective social control presupposes social arrangements making for the observability of behavior. (By observability, a conception which I have borrowed from

4. Robert K. Merton, *Social Theory and Social Structure,* 3d ed., rev. and enl. (New York: Free Press, 1968), 336–56. This discussion of role-set draws upon one part of chapter 9, "Continuities in the Theory of Reference Groups and Social Structures," 368–84.

Simmel and tried to develop, I mean the extent to which social norms and role-performances can readily become known to others in the social system. This is, I believe, a variable crucial to structural analysis, a belief which, unhappily, I cannot undertake to defend here.)

To the extent that the social structure insulates individuals from having their activities known to members of their role-sets, they are the less subject to competing pressures. It should be emphasized that we are dealing here with structural arrangements for such insulation, not with the fact that people *happen* to conceal part of their role-behavior from others. The structural fact is that social statuses differ in the extent to which the conduct of those in them is regularly insulated from observability by members of the role-set. Some have a functionally significant insulation of this kind, as for example, the status of the university teacher, insofar as norms hold that what is said in the classroom is privileged. In this familiar type of case, the norm clearly has the function of maintaining some degree of autonomy for the teacher. For if they were forever subject to observation by all those in the role-set, with their often differing expectations, teachers might be driven to teach not what they know or what the evidence leads them to believe, but to teach what will placate the numerous and diverse people who are ostensibly concerned with "the education of youth." That this sometimes occurs is evident. But it would presumably be more frequent, were it not for the relative exemption from observability by all and sundry who may wish to impose their will upon the instructor.

More broadly, the concept of privileged information and confidential communication in the professions has this same function of insulating clients from observability of their behavior and beliefs by others in their role-set. Were physicians or priests free to tell all they have learned about the private lives of their clients, the needed information would not be forthcoming and they could not adequately discharge their functions. More generally, if all the facts of one's conduct and beliefs were freely available to anyone, social structures could not operate. What is often described as the need for privacy—that is, insulation of actions and beliefs from surveillance by others—is the individual counterpart to the functional requirement of social structure that some measure of exemption from full observability be provided. Privacy is not only a personal predilection, though it may be that, too. It is also a requirement of social systems which must provide for a measure, as they say in France, of *quant-à-soi,* a portion of the self that is kept apart, immune from observation by others.

Like other social mechanisms, this one of insulation from full observability can, of course, miscarry. Were the activities of politicians fully removed from the public spotlight, social control of their behavior would be correspondingly reduced. And as we all know, anonymous power anonymously exercised does not make for a stable social structure meeting the values of a society. So, too, teachers or physicians who are largely insulated from observability may fail to live up to the minimum requirements of their statuses. All this means only that some measure of observability of role-performance by members of the role-set is required, if the indispensable social requirement of accountability is to be met. This statement does not contradict an earlier statement to the effect that some measure of insulation from observability is also required for the effective operation of social structures. Instead, the two statements, taken in conjunction, imply that there is an optimum zone of observability, difficult to identify in precise terms and doubtless varying for different social statuses, which will simultaneously make both for accountability and for substantial autonomy, rather than for a frightened acquiescence with the distribution of power which happens, at a particular moment, to obtain in the role-set.

4. *Observability of conflicting demands by members of a role-set.* This mechanism is implied by what has been said and therefore needs only passing comment here. As long as members of the role-set are happily ignorant that their demands upon the occupants of a status are incompatible, members may press their own cases. The pattern is then many against one. But when it becomes plain that the demands of some are in full contradiction with the demands of others, it becomes, in part, the task of members of the role-set, rather than that of the status-occupant, to resolve these contradictions, either by a struggle for overriding power or by some degree of compromise.

In such circumstances, the status-occupant subjected to conflicting demands often becomes cast in the role of the *tertius gaudens,* the third (or more often, the nth) party who draws advantage from the conflict of the others. Originally at the focus of the conflict, the status-occupant can virtually become a bystander whose function it is to highlight the conflicting demands being made by members of the role-set. It becomes a problem for them, rather than for the status-occupant, to resolve their contradictory demands. At the least, this serves to make evident that it is not wilful misfesance on the status-occupant's part which keeps that person from conforming to all the contradictory expectations imposed by the role-set.[5] When most effective, this serves to articulate the expec-

5. See the observations by William G. Carr, the executive secretary of the National Education Association, who has summarized some of the conflicting pressures exerted upon school curricula by voluntary organizations, such as the American Legion, the Association for the United Nations, the National Safety Council, the Better Business Bu-

tations of those in the role-set beyond a degree which would occur, if this mechanism of making contradictory expectations manifest were not at work.

5. *Mutual social support among status-occupants.* Whatever they may believe to the contrary, occupants of a social status are not alone. The very fact that they are placed in a social position means that there are others more or less like-circumstanced. To this extent, the actual or potential experience of facing a conflict of expectations among members of the role-set is variously common to all occupants of the status. The particular persons subject to these conflicts need not, therefore, meet them as wholly private problems that must be coped with in wholly private fashion.

It is this familiar and fundamental fact of social structure that is the basis for those in the same social status forming the associations intermediate to the individual and the larger society in a pluralistic system. These organizations constitute a structural response to the problems of coping with the (potentially or actually) conflicting demands by those

---

reau, the American Federation of Labor, and the Daughters of the American Revolution. His summary may serve through concrete example to indicate the extent of competing expectations among those in the complex role-set of school superintendents and local school boards in as differentiated a society as our own. Sometimes, Mr. Carr reports, these voluntary organizations "speak their collective opinions temperately, sometimes scurrilously, but always insistently. They organize contests, drives, collections, exhibits, special days, special weeks, and anniversaries that run all year long.

"They demand that the public schools give more attention to Little League baseball, first aid, mental hygiene, speech correction, Spanish in the first grade, military preparedness, international understanding, modern music, world history, American history, and local history, geography and homemaking, Canada and South America, the Arabs and the Israelis, the Turks and the Greeks, Christopher Columbus and Leif Ericsson, Robert E. Lee and Woodrow Wilson, nutrition, care of the teeth, free enterprise, labor relations, cancer prevention, human relationships, atomic energy, the use of firearms, the Constitution, tobacco, temperance, kindness to animals, Esperanto, the 3 Rs, the 3 Cs, and the 4 Fs, use of the typewriter and legible penmanship, moral values, physical fitness, ethical concepts, civil defense, religious literacy, thrift, law observance, consumer education, narcotics, mathematics, dramatics, physics, ceramics, and (that latest of all educational discoveries) phonics.

"Each of these groups is anxious to avoid overloading the curriculum. All any of them ask is that the nonessentials be dropped in order to get their material in. Most of them insist that they do not want a special course—they just want their ideas to permeate the entire daily program. Every one of them proclaims a firm belief in local control of education and an apprehensive hatred of national control.

"Nevertheless, if their national organization program in education is not adopted forthwith, many of them use the pressure of the press, the radiance of the radio, and all the props of propaganda to bypass their elected school board." An address at the inauguration of Hollis Leland Caswell, Teachers College, Columbia University, 21–22 November 1955, 10.

in the role-set of the status.[6] Whatever the intent, these constitute social formations serving to counter the power of the role-set; of being not merely amenable to its demands, but of helping to shape them. Such organizations—so familiar a part of the social landscape of differentiated societies—also develop normative systems designed to anticipate and thereby to mitigate such conflicting expectations. They provide social support to the individuals in the status under attack. They minimize the need for their improvising personal adjustments to patterned types of conflicting expectations. Emerging codes which state in advance what the socially supported conduct of the status-occupant should be also serve this social function. This function becomes all the more significant in the structural circumstances when status-occupants are highly vulnerable to pressures from their role-sets because they are relatively isolated from one another. Thus, thousands of librarians sparsely distributed among the towns and villages of America and not infrequently subject to censorial pressures received strong support from the code on censorship developed by the American Library Association.[7] This only illustrates the general mechanisms whereby status-peers curb the pressures exerted upon them individually by drawing upon the organizational and normative support of their peers.

6. *Abridging the role-set.* There is, of course, a limiting case in the modes of coping with incompatible demands by the role-set. Role-relations are broken off, leaving a greater consensus of role-expectations among those who remain. But this mode of adaptation by amputating the role-set is possible only under special and limited conditions. It can be effectively utilized only in those circumstances where it is still possible for status-occupants to perform their other roles without the support of those with whom they have discontinued relations. It presupposes that the social structure provides this option. By and large, however, this option is infrequent and limited, since the composition of the role-set is ordinarily not a matter of personal choice but a matter of the social organization in which the status is embedded. More typically, the individual goes, and the social structure remains.

Residual Conflict in the Role-Set

Doubtless, these are only some of the mechanisms which serve to articulate the expectations of those in the role-set. Further inquiry will

6. In this context, see the acute analysis of the formation of the National Union of Teachers by Asher Tropp, *The School Teachers* (London: Heinemann, 1957).

7. See Richard P. McKeon, Robert K. Merton, and Walter Gellhorn, *Freedom to Read* (New York: R. R. Bowker Co., 1957).

uncover others, just as it will probably modify the preceding account of those we have provisionally identified. But, however much the substance may change, I believe that the logic of the analysis will remain largely intact. This can be briefly recapitulated.

First, it is assumed that each social status has its organized complement of role-relationships which can be thought of as comprising a role-set. Second, relationships hold not only between the occupant of the particular status and each member of the role-set, but always potentially and often actually, between members of the role-set itself. Third, to the extent that members of the role-set themselves hold substantially differing statuses, they will tend to have some differing expectations (moral and actuarial) of the conduct appropriate for the status-occupant. Fourth, this gives rise to the sociological problem of how their diverse expectations become sufficiently articulated for the status-structure and the role-structure to operate with a modicum of effectiveness. Fifth, inadequate articulation of these role-expectations tends to call into play one or more social mechanisms, which serve to reduce the extent of patterned conflict below the level which would be involved if these mechanisms were not at work.

And now, sixth, finally and importantly, even when these (and probably other) mechanisms are operating, they may not, in particular cases, prove sufficient to reduce the conflict of expectations below the level required for the social structure to operate with substantial effectiveness. This residual conflict within the role-set may be enough to interfere materially with the effective performance of roles by the occupants of the status in question. Indeed, it may well turn out that this condition is the most frequent one—role-systems operating at considerably less than full efficiency. Without trying to draw tempting analogies with other types of systems, I suggest only that this is not unlike the case of engines which cannot fully utilize heat energy. If the analogy lacks force, it may nevertheless have the merit of excluding the utopian figment of a perfectly effective social system.

We do not yet know some of the requirements for fuller articulation of the relations between the occupant of a status and members of the corresponding role-set, on the one hand, and for fuller articulation of the values and expectations among those comprising the role-set, on the other. As we have seen, even those requirements which can now be identified are not readily satisfied, without fault, in social systems. To the extent that they are not, social systems are forced to limp alone with that measure of ineffectiveness and inefficiency which is often accepted because the realistic prospect of decided improvement seems so remote as sometimes not to be visible at all.

11

Sociological Ambivalence (1963)

The Concept of Sociological Ambivalence

The concept of ambivalence in psychology refers to the experienced tendency of individuals to be pulled in psychologically opposed directions, as love and hate for the same person, acceptance and rejection, affirmation and denial. The concept leads directly to distinctive problems: How is it that these opposed pressures persist? Why doesn't one or the other prevail? What psychic mechanisms are triggered by ambivalence, as, for example, the separation of the conflicting components with one of them—say, hate—being repressed while the overt reaction to the repressed hate takes the form of a marked expression of loving care? Such problems are not our concern here. For our purpose, the essential point is that whatever the psychological theory takes as the sources of ambivalence, it centers on how this or that type of personality develops a particular ambivalence and copes with it.

The sociological theory of ambivalence is directed to quite other problems. It refers to the social structure, not to the personality. *In its most extended sense,* sociological ambivalence refers to incompatible normative expectations of attitudes, beliefs, and behavior assigned to a status (i.e., a social position) or to a set of statuses in a society. *In its most restricted sense,* sociological ambivalence refers to incompatible normative expectations incorporated in a *single* role of a *single* social status (for example, the therapist role of the physician as distinct from other roles of his or her status as researcher, administrator, professional colleague, participant in the professional association, etc.). In both the most extended and the most restricted sense, the ambivalence is located in the social definition of roles and statuses, not in the feeling-state of one or another type of personality. To be sure, as we would expect and as we shall find, sociological ambivalence is one major source of psy-

Coauthored with Elinor Barber. From "Sociological Ambivalence," in *Sociological Ambivalence and Other Essays* (New York: Free Press, 1976), 6–12, 17–19. Reprinted by permission of The Free Press, an imprint of Simon and Schuster.

chological ambivalence. Individuals in a status or status-set that has a large measure of incompatibility in its social definition will tend to develop personal tendencies toward contradictory feelings, beliefs, and behavior. Although the sociological and psychological kinds of ambivalence are empirically connected, they are theoretically distinct. They are on different planes of phenomenal reality, on different planes of conceptualization, on different planes of causation and consequences.

The sociological theory deals with the processes through which social structures generate the circumstances in which ambivalence is embedded in particular statuses and status-sets together with their associated social roles. Anticipating our later discussion a bit, we suggest that one source of ambivalence is to be found in the structural context of a particular status. Another source is found in the multiple types of functions assigned to a status—for example, expressive and instrumental functions. These two sources have been identified in a research program of sociological analyses of ambivalence during the last twenty years or so: ambivalence in the role of the bureaucrat when individualized and personal attention is wanted by the client while the bureaucracy requires generalized and impersonal treatment[1]; the role of the intellectual expert in bureaucracy, embracing values derived from his profession and values derived from the organization[2]; the ambivalence about accomplishments by people not warranted by their social status (involving a positive response to the achievement and a negative attitude toward the devalued status)[3]; the ambivalent ex-member of a group[4]; the ambivalence of scientists about priority that results from their role incorporating potentially incompatible values ("the value of originality, which leads them to want their priority to be recognized, and the value of humility, which leads them to insist on how little they have been able to accomplish")[5]; the ambivalence in the role of physicians which requires them to try "to blend incompatible or potentially incompatible norms into a functionally consistent whole[6]; and the ambivalence inherent in a wide variety of roles that must deal with both maintenance of the pattern of behavior and with instrumental results,

1. Robert K. Merton, *Social Theory and Social Structure,* 3d ed., rev. and enl. (New York: Free Press, 1968), 256–59.

2. Merton, *Social Theory and Social Structure,* 266–73, 277.

3. Ibid., 483–86.

4. Ibid., 349.

5. Robert K. Merton, "Priorities in Scientific Discovery: A Chapter in the Sociology of Science," *American Sociological Review* 22 (1957): 647–49.

6. Robert K. Merton, "Some Preliminaries to a Sociology of Medical Education," in *The Student-Physician,* ed. Robert K. Merton, George Reader, M.D., and Patricia L. Kendall (Cambridge: Harvard University Press, 1957), 72–76.

with activity that serves chiefly to maintain social cohesion and activity that serves to get things done.[7]

In all these and kindred instances, people occupying the statuses are exposed to ambivalence. They are exposed to it not because of their idiosyncratic history or their distinctive personality but because the ambivalence is inherent in the social positions they occupy.[8] This is what we mean by saying that sociological ambivalence is a concept focused on social structure.

In its broadest sense of inconsistent normative expectations embodied in a social status or status-set, sociological ambivalence has been substantially investigated. But very little attention has been accorded ambivalence in its most *restricted, core sense of conflicting normative expectations socially defined for a particular social role associated with a single social status.* It is this special case which holds greatest sociological interest. We shall analyze the ways in which social function and social structure make for a socially prescribed ambivalence in a particular role, as, for example, in the therapist role of the physician which calls for *both* a degree of affective detachment from, and a degree of compassionate concern about the patient. This core type of sociological ambivalence puts contradictory demands upon the occupants of a status in a particular social relation. And *since these norms cannot be simultaneously expressed in behavior, they come to be expressed in an oscillation of behaviors: of detachment and compassion, of discipline and permissiveness, of personal and impersonal treatment.*

Before examining the core type of sociological ambivalence, we sketch out other, related types which have been the object of inquiry. The second type of ambivalence is perhaps the most thoroughly investigated: ambivalence involved in a conflict of statuses within a status-set (i.e., the set of social positions occupied by each individual). There are, for example, the familiar cases of conflict between men's or women's statuses in the occupational sphere and in the family sphere; in their religious status and their secular ones; in their public and private statuses, as say, the judge and friend or the school monitor and student

7. Robert K. Merton, "Social Problems and Sociological Theory," in *Contemporary Social Problems,* ed. Robert K. Merton and Robert A. Nisbet (New York: Harcourt Brace Jovanovich, 1961), 734–35.

8. For analyses bearing on this conception, see Melvin Seeman, "Role Conflict and Ambivalence in Leadership," *American Sociological Review* 18 (1953): 373–80; Lewis Coser, *The Functions of Social Conflict* (Glencoe, Ill.: Free Press, 1956), 61–65; W. C. Mitchell, "The Ambivalent Social Status of the American Politician," *Western Political Quarterly* 12 (1959): 683–89; Werner Cahn, "Social Status and the Ambivalence Hypothesis: Some Critical Notes and a Suggestion," *American Sociological Review* 25 (1960), 508–13.

colleagues.[9] This type has been studied particularly in the case of voting behavior under the concept of cross-pressures.[10]

This second kind of sociological ambivalence is essentially a pattern of a "conflict of interests or of values" incorporated in *different* statuses occupied by the same person which result in mixed feelings and compromise behavior. Put more compactly, this involves conflicting interests and values in the individual's status-set.[11] Since the pattern is induced by the social structure, it can be regarded as a form of sociological ambivalence. But this form of ambivalence differs from the first, core type in a basic respect: its frequency and dynamics differ according to the number of people who happen to have a particular combination of statuses. The more parents at work in the labor market, the more subject to competing obligations. But this is not inherent in their occupying a *single* status and performing a *single* role. That is part of what we intend by describing this type as a derivative rather than core type of sociological ambivalence. It differs from the core type also in that the conflicting demands of different statuses ordinarily involve *different* people in the role-sets of the conflicting statuses (the demands of an employer, for example, and of a spouse). But in the core type, the ambivalence is built into the social relation with the *same* people. It involves a structurally induced ambivalence in a single relation—say, of lawyer and client—and not a conflict between relations—say, of lawyer and family and lawyer and client. Since conflicts in the status-set have been examined repeatedly, we shall not deal with these except as they bear upon the ambivalence that is found in a particular role associated with a particular status.

9. The literature on this is so voluminous (and still uncollated) that a few citations must suffice. Everett C. Hughes, "Dilemmas and Contradictions of Status," *American Journal of Sociology* 50 (1945): 353–59; Samuel A. Stouffer, *Social Research to Test Ideas* (Glencoe, Ill.: Free Press, 1962), 39–67; Mirra Komarovsky, "Cultural Contradictions and Sex Roles," *American Journal of Sociology* 52 (1946): 184–89.

10. As originally set forth in Paul F. Lazarsfeld, Bernard Berelson, and Hazel Gaudet, *The People's Choice* (New York: Duell, Sloan and Pearce, 1944), 53–72, and in a variety of investigations since.

11. On the conflicting demands of several statuses in an individual's status-set, much germane and relatively unexplored data for the sociologist can be found in recurrent situations, subject to legal definition, involving a "conflict of interest." For one excellent account of this pattern from the legal standpoint, see Association of the Bar of the City of New York, *Conflict of Interest and Federal Service* (Cambridge: Harvard University Press, 1960.) The social arrangements expressly devised to reduce the frequency of such patterned conflict of interest provide a rich source for sociological analysis of this type of ambivalence. On the lack of integration of the status-set, see Robert K. Merton, *Social Theory and Social Structure,* 422–24, 434–38.

A third kind, comparable to the preceding one, is found in the conflict between several roles associated with a particular status. This too is a familiar type of sociological ambivalence. The one position of university professor or scientist in a research organization has variously multiple roles associated with it: the roles of teaching or training, of research, administration, and so on. And as some of the readers of this paper have ample reason to know from their own experience, the demands of these several roles in the one status can be at odds. Not only do they make competing demands for time, energy, and interest upon the occupants of the one status, but the kinds of attitudes, values, and activities required by each of these roles may also be incompatible[12] with the others.

A fourth kind of sociological ambivalence is found in the form of contradictory cultural values held by members of a society. These values are not ascribed to particular statuses, but are normatively expected of all in the society (e.g., patriotism and honesty). Thus, Robert Lynd listed twenty paired assumptions by which Americans live, noting that these run at once "into a large measure of contradiction and resulting ambivalence."[13] Examples follow:

Everyone should try to be successful.
But: The kind of person you are is more important than how successful you are.

The family is our basic institution and the sacred core of our national life.
But: Business is our most important institution, and, since national welfare depends upon it, other institutions must conform to its needs.

Honesty is the best policy.
But: Business is business, and a businessman would be a fool if he didn't cover his hand.

As long as these value premises are widely held and not organized into sets of norms for one or another role in particular, they can be regarded as cases of cultural conflict. When they are so organized, they result in the core type of sociological ambivalence, in which incompatible normative demands are built into a particular role of a particular status. The phenomena of cultural conflict have been investigated at length,

12. See, for example, Logan Wilson, *The Academic Man* (New York: Oxford University Press, 1942), chaps. 4, 5, 10, 11; William Kornhauser, *Scientists in Industry: Conflict and Accommodation* (Berkeley and Los Angeles: University of California Press, 1962), chap. 7.

13. Robert S. Lynd, *Knowledge for What?* (Princeton: Princeton University Press, 1939), chap. 3.

and although there are gaps in our understanding of the processes involved in such conflict, we shall not deal with them here.[14]

A fifth type of sociological ambivalence is found in the disjunction between culturally prescribed aspirations and socially structured avenues for realizing these aspirations (what one of us has long described as the opportunity structure).[15] It is neither cultural conflict nor social conflict, but a contradiction between the cultural structure and the social structure. It turns up when cultural values are internalized by those whose position in the social structure does not give them access to act in accord with the values they have been taught to prize. This type, examined in studies of "social structure and anomie,"[16] will also be largely ignored in this discussion which focuses on the core type of ambivalence built into a single role of a particular social status.

A sixth type of sociological ambivalence develops among people who have lived in two or more societies and so have become oriented to differing sets of cultural values. Best exemplified by immigrants, this special pattern has been intensively investigated at least since the concept of the "marginal man" was introduced by Robert E. Park and effectively developed by Everett V. Stonequist.[17] In a connected but somewhat different vein, reference-group theory has dealt with the ambivalence of people who accept certain values held by groups of which they are *not* members.[18] This type instructively combines elements of the fourth type of ambivalence (cultural conflict) and the second type (conflict within the status-set). Although in this orientation to a nonmembership group individuals do not "belong" to the group whose values they accept and so do not, in social fact, occupy conflicting statuses, their identification with that group, if only in aspiration or fan-

14. See ibid.; Karen Horney, *The Neurotic Personality of Our Time* (New York: W. W. Norton, 1937); and for an emphatic, literary version of this theme, Paul Goodman, *Growing Up Absurd* (New York: Random House, 1960).

15. See Robert K. Merton, "Social Conformity, Deviation, and Opportunity Structures," *American Sociological Review* 24 (1959): 177–89.

16. Merton, *Social Theory and Social Structure,* 185–248; Elinor G. Barber, *The Bourgeoisie in 18th Century France* (Princeton: Princeton University Press, 1955), 56 ff.; Leo Srole, "Social Integration and Certain Corollaries," *American Sociological Review* 21 (1956): 709–16; papers by Robert Dubin, Richard A. Cloward, *American Sociological Review* 24 (1959): 147–89; E. H. Mizruchi, "Social Structure and Anomia in a Small City," *American Sociological Review* 25 (1960): 645–54; and in these see citations to other studies of the subject. See chap. 12 in this volume.—*Ed.*

17. Robert E. Park, "Preface" in Everett V. Stonequist, *The Marginal Man* (New York: Charles Scribner's Sons, 1937), and of course the book itself.

18. Merton, *Social Theory and Social Structure,* 279–440.

tasy, subjects them to the conflicting demands of their own groups and of the group to which they aspire. This type of ambivalence is presumably most characteristic of the socially mobile. [. . .]

Norms and Counter-Norms

From the perspective of sociological ambivalence, we see a social role as a dynamic organization of norms and counter-norms, not as a combination of dominant attributes (such as affective neutrality or functional specificity). We propose that *the major norms and the minor counter-norms alternatively govern role-behavior to produce the core type of ambivalence.*

This line of inquiry differs from that indicated by the caveats of both Sorokin and Parsons to the effect that social relations cannot be exhaustively analyzed in terms of their dominant attributes. Sorokin warns that role-relations are only predominantly of one kind or another; they are seldom purely familistic *or* compulsory.[19] And Parsons, writing of the pattern-variables, refers to the primacy of one or another, so that they do not necessarily apply "to every specific act within the role."[20] The role of public officials, for example, which is primarily defined as collectivity-oriented, nevertheless allows officials to be self-oriented in choosing among jobs, although they are expected to be collectivity-oriented in taking stands on public policies.

Important as such caveats are in their own right, they do not direct us to the structure of social roles caught up in the notion of sociological ambivalence. It is true that the practice of characterizing social roles only in terms of their dominant attributes does not exhaust the normative complexity of these roles. This is plainly the case, for example, when in the Parsonsian scheme the role of the physician in relation to the patient is represented in the formula of affective neutrality, functional specificity, universalism, performance-orientation, and collectivity-orientation. Or when, in the Sorokinian scheme, it is characterized as having narrow extensivity (confined to matters pertinent to the health problem), variously intensive (depending on the acuteness of the problem), predominantly direct, of contingent duration and largely asymmetrical (with the physician largely governing the character of the interaction). Such formulae in terms of dominant attributes alone give no reason to

19. Pitirim A. Sorokin, *Society, Culture, and Personality* (New York: Harper and Row, 1947) 89 n.

20. Talcott Parsons, *The Social System* (Glencoe, Ill.: Free Press, 1953), 61.

suppose that sociological ambivalence is built into the relation of physician and patient. Since the attributes are not at odds, the connected social roles would appear integrated and stable.

From the standpoint of sociological ambivalence, however, the structure of the physician's role differs from these characterizations, consisting of *a dynamic alternation of norms and counter-norms.* These norms call for potentially contradictory attitudes and behaviors. For the social definitions of this role, as of social roles generally, in terms of dominant attributes alone would not be flexible enough to provide for the endlessly varying contingencies of social relations. Behavior oriented wholly to the dominant norms would defeat the functional objectives of the role. Instead, role-behavior is alternatively oriented to dominant norms and to subsidiary counter-norms in the role. This alternation of subroles *evolves* as a social device for helping people in designated statuses to cope with the contingencies they face in trying to fulfill their functions.[21] This is lost to view when social roles are analyzed only in terms of their major atttributes.

To continue with the example of physicians, it is only partly true that their role requires them to be affectively neutral in their professional relations with patients. Rather, this aspect of the role (and, we repeat, not merely the concrete behavior of this or that physician) is more complex than that. As the Columbia studies of the medical student have found, the physician is taught to be oriented toward both the dominant norm of affective neutrality (detachment) and the subsidiary norm of affectivity (the expression of compassion and concern for the patient). That is why, in these studies, we have treated this part of the physician's role not as one of affective neutrality (with only idiosyncratic departures from this norm) but as one involving "detached concern," calling for alternation between the instrumental impersonality of detachment and the functional expression of compassionate concern.[22] As physician and patient interact, different and abstractly contradictory norms are activated to meet the dynamically changing needs of the relation. *Only through such structures of norms and counter-norms,* we suggest, *can the various functions of a role be effectively discharged.* This is not merely a matter of social psychology but of role-structure. Po-

21. Henry L. Lennard and Arnold Bernstein, *The Anatomy of Psychotherapy* (New York: Columbia University Press, 1960), chaps. 7 and 8, passim.

22. The concept of detached concern calls attention to the major point that each of these abstractly opposed norms may be functionally necessary for the task in hand at differing times in the interaction between physician and patient. On the concept, see Merton, Reader, and Kendall, eds., *The Student-Physician,* 74; Renée C. Fox, *Experiment Perilous* (Glencoe, Ill.: Free Press, 1959).

tentially conflicting norms are built into the social definition of roles that provide for normatively acceptable alternations of behavior as the state of a social relation changes. This is a major basis for that oscillation between differing role-requirements that makes for sociological ambivalence. [. . .]

12

Social Structure and Anomie (1938)

[. . .] The conceptual framework set out in this essay is designed to provide a systematic approach to the analysis of social and cultural sources of deviant behavior. Our primary aim is to discover how some *social structures exert a definite pressure upon certain persons in the society to engage in nonconformist rather than conformist conduct.* If we can locate groups peculiarly subject to such pressures, we should expect to find fairly high rates of deviant behavior in these groups, not because the human beings comprising them are compounded of distinctive biological tendencies but because they are responding normally to the social situation in which they find themselves. Our perspective is sociological. We look at variations in the *rates* of deviant behavior, not at its specific incidence. Should our quest be at all successful, some forms of deviant behavior will be found to be as psychologically normal as conformist behavior, and the equation of deviation and abnormality will be put in question.

Patterns of Cultural Goals and Institutional Norms

Among the several elements of social and cultural structures, two are of immediate importance. These are analytically separable although they merge in concrete situations. The first consists of culturally defined goals, purposes, and interests, held out as legitimate objectives for all or for diversely located members of the society. The goals are more or less integrated—the degree is a question of empirical fact—and roughly ordered in some hierarchy of value. Involving various degrees of sentiment and significance, the prevailing goals comprise a frame of aspirational reference. They are the things "worth striving for." They are a basic, though not the exclusive, component of what Linton has called "designs for group living." And though some, not all, of these

From "Social Structure and Anomie," in *American Sociological Review* 3 (1938): 672–82. Reprinted by permission of the American Sociological Association.

cultural goals are related to the biological drives of man, they are not determined by them.

A second element of the cultural structure defines, regulates, and controls the acceptable modes of reaching out for these goals. Every social group invariably couples its cultural objectives with regulations, rooted in the mores or institutions, of allowable procedures for moving toward culturally defined objectives. These regulatory norms are not necessarily identical with technical or efficiency norms. Many procedures which from the standpoint of particular individuals would be most efficient in securing desired values—the exercise of force, fraud, power—are ruled out of the institutional area of permitted conduct. At times, the disallowed procedures include some which would be efficient for the group itself—e.g., historic taboos on vivisection, on medical experimentation, on the sociological analysis of "sacred" norms—since the criterion of acceptability is not technical efficiency but value-laden sentiments (supported by most members of the group or by those able to promote these sentiments through the composite use of power and propaganda). In all instances, the choice of expedients for striving toward cultural goals is limited by institutionalized norms.

Sociologists often speak of these controls as being "in the mores" or as operating through social institutions. Such elliptical statements are true enough, but they obscure the fact that culturally standardized practices are not all of a piece. They are subject to a wide gamut of control. They may represent definitely prescribed or preferential or permissive or proscribed patterns of behavior. In assessing the operation of social controls, these variations—roughly indicated by the terms *prescription, preference, permission,* and *proscription*—must of course be taken into account.

To say, moreover, that cultural goals and institutionalized norms operate jointly to shape prevailing practices is not to say that they bear a constant relation to one another. The cultural emphasis placed upon certain goals varies independently of the degree of emphasis upon institutionalized means. There may develop a very heavy, at times a virtually exclusive, stress upon the value of certain goals, involving comparatively little concern with the institutionally prescribed means of striving toward these goals. The limiting case of this type is reached when the range of alternative procedures is governed only by technical rather than institutional norms. Any and all procedures which promise attainment of the all-important goal would be permitted in this hypothetical polar case. This constitutes one type of malintegrated culture. A second polar type is found in groups where activities originally conceived as instrumental are transmuted into self-contained practices lacking further objectives. The original purposes are forgotten and close adherence to institutionally prescribed conduct becomes a matter of ritual.[1] Sheer conformity becomes a central value. For a time, social stability is ensured—at the expense of flexibility. Since the range of alternative behaviors permitted by the culture is severely limited, there is little basis for adapting to new conditions. There develops a tradition-bound, "sacred" society marked by neophobia. Between these extreme types are societies which maintain a rough balance between emphases upon cultural goals and institutionalized practices, and these constitute the integrated and relatively stable, though changing, societies.

An effective equilibrium between these two phases of the social structure is maintained so long as satisfactions accrue to individuals conforming to both cultural constraints, viz., satisfactions from the achievement of goals and satisfactions emerging directly from the institutionally canalized modes of striving to attain them. It is reckoned in terms of the product and in terms of the process, in terms of the outcome and in terms of the activities. Thus continuing satisfactions must derive from sheer participation in a competitive order as well as from eclipsing one's competitors if the order itself is to be sustained. If concern shifts exclusively to the outcome of competition, then those who perennially suffer defeat will, understandably enough, work for a change in the rules of the game. The sacrifices occasionally—not, as Freud assumed, invariably—entailed by conformity to institutional norms must be compensated by socialized rewards. The distribution of statuses through competition must be so organized that positive incentives for adherence to status obligations are provided *for every position* within the distributive order. Otherwise, as will soon become plain, aberrant behavior ensues. It is, indeed, my central hypothesis that aberrant behavior may be regarded sociologically as a symptom of dissociation between culturally prescribed aspirations and socially structured avenues for realizing these aspirations.

Of the types of societies that result from independent variation of cultural goals and institutionalized means, we shall be primarily concerned with the first—a society in which there is an exceptionally strong emphasis upon specific goals without a corresponding emphasis upon institutional procedures. If it is not to be misunderstood, this statement must be elaborated on. No society lacks norms governing conduct. But societies do differ in the degree to which the folkways,

1. This ritualism may be associated with a mythology which rationalizes these practices so that they appear to retain their status as means, but the dominant pressure is toward strict ritualistic conformity, irrespective of the mythology. Ritualism is thus most complete when such rationalizations are not even called forth.

mores and institutional controls are effectively integrated with the goals that stand high in the hierarchy of cultural values. The culture may be such as to lead individuals to center their emotional convictions on the complex of culturally acclaimed ends, with far less emotional support for prescribed methods of reaching out for these ends. With such differential emphases upon goals and institutional procedures, the latter may be so vitiated by the stress on goals as to have the behavior of many individuals limited only by considerations of technical expediency. In this context, the sole significant question becomes: Which of the available procedures is most efficient in netting the culturally approved value? The technically most effective procedure, whether culturally legitimate or not, becomes typically preferred to institutionally prescribed conduct. As this process of attenuation continues, the society becomes unstable and there develops what Durkheim called "anomie" (or normlessness).

The working of this process eventuating in anomie can be easily glimpsed in a series of familiar and instructive, though perhaps trivial, episodes. Thus, in competitive athletics, when the aim of victory is shorn of its institutional trappings and success becomes construed as "winning the game" rather than "winning under the rules of the game," a premium is implicitly set upon the use of illegitimate but technically efficient means. The star of the opposing football team is surreptitiously slugged; the wrestler incapacitates an opponent through ingenious but illicit techniques; university alumni covertly subsidize "students" whose talents are confined to the athletic field. The emphasis on the goal has so attenuated the satisfactions deriving from sheer participation in the competitive activity that only a successful outcome provides gratification. Through the same process, tension generated by the desire to win in a poker game is relieved by successfully dealing oneself four aces or, when the cult of success has truly flowered, by sagaciously shuffling the cards in a game of solitaire. The faint twinge of uneasiness in the last instance and the surreptitious nature of public delicts indicate clearly that the institutional rules of the game are *known* to those who evade them. But cultural (or idiosyncratic) exaggeration of the success-goal leads men to withdraw emotional support from the rules.[2]

2. It appears unlikely that cultural norms, once interiorized, are wholly eliminated. Whatever residuum persists will induce personality tensions and conflict, with some measure of ambivalence. A manifest rejection of the once-incorporated institutional norms will be coupled with some latent retention of their emotional correlates. "Guilt feelings," "a sense of sin," "pangs of conscience" are diverse terms referring to this unrelieved tension. Symbolic adherence to the nominally repudiated values or rationalizations for the rejection of these values constitute a more subtle expression of these tensions.

This process is not, of course, restricted to the realm of competitive sport, which has simply provided us with microcosmic images of the social macrocosm. The process whereby exaltation of the end generates a literal *demoralization,* i.e., a deinstitutionalization, of the means occurs in many[3] groups where the two components of the social structure are not highly integrated.

Contemporary American culture appears to approximate the polar type in which great emphasis upon certain success-goals occurs without equivalent emphasis upon institutional means. It would of course be fanciful to assert that accumulated wealth stands alone as a symbol of success just as it would be fanciful to deny that Americans assign it a place high in their scale of values. In some large measure, money has been consecrated as a value in itself, over and above its expenditure for articles of consumption or its use for the enhancement of power. Money is peculiarly well adapted to become a symbol of prestige. As Simmel emphasized, money is highly abstract and impersonal. However acquired, fraudulently or institutionally, it can be used to purchase the same goods and services. The anonymity of an urban society, in conjunction with these peculiarities of money, permits wealth, the sources of which may be unknown to the community in which the plutocrat lives or, if known, to become purified in the course of time, to serve as a symbol of high status. Moreover, in the American dream there is no final stopping point. The measure of monetary success is conveniently indefinite and relative. At each income level, as H. F. Clark has found, Americans want just about twenty-five per cent more (but this "just a bit more" continues to operate once it is obtained). In this flux of shifting standards, there is no stable resting point, or rather, it is the point which manages always to be "just ahead." . . . To say that the goal of monetary success is entrenched in American culture is only to say that Americans are bombarded on every side by precepts which affirm the right or, often, the duty of retaining the goal even in the face of repeated frustration. Prestigeful representatives of the society reinforce the cultural emphasis. The family, the school, and the workplace—the major agencies shaping the personality structure and goal formation of Americans—join to provide the intensive disciplining required if individuals are to retain intact a goal that remains elusively beyond reach, if they are to be motivated by the promise of a gratification which is not redeemed. Parents serve as a transmission belt for the

3. "Many," not all, unintegrated groups, for the reason mentioned earlier. In groups where the primary emphasis shifts to institutional means, the outcome is normally a type of ritualism rather than anomie.

values and goals of the groups of which they are a part—above all, of their social class or of the class with which they identify themselves. And the schools are of course the official agency for the passing on of prevailing values, with a large proportion of the textbooks used in city schools implying or stating explicitly "that education leads to intelligence and consequently to job and money success."[4] Central to this process of discipling people to maintain their unfulfilled aspirations are the cultural prototypes of success, the living documents testifying that the American dream can be realized if one has the requisite abilities. [. . .]

The symbolism of a commoner rising to the estate of economic royalty is woven deep in the texture of the American culture pattern, finding what is perhaps its ultimate expression in the words of one who knew whereof he spoke, Andrew Carnegie: "Be a king in your dreams. Say to yourself, 'My place is at the top.'"[5]

Coupled with an emphasis on the obligation to maintain lofty goals is a correlative emphasis on the penalizing of those who draw in their ambitions. Americans are admonished "not to be a quitter" for in the dictionary of American culture, as in the lexicon of youth, "there is no such word as 'fail.'" The cultural manifesto is clear: one must not quit, must not cease striving, must not lessen one's goals, for "not failure, but low aim, is crime."

Thus the culture enjoins the acceptance of three cultural axioms: First, all should strive for the same lofty goals since these are open to all; second, present seeming failure is but a way station to ultimate success; and third, genuine failure consists only in the lessening or withdrawal of ambition.

In rough psychological paraphrase, these axioms represent, first, a symbolic "secondary reinforcement" of incentive; second, curbing the threatened extinction of a response through an associated stimulus; third, increasing the motive-strength to evoke continued responses despite the continued absence of reward.

In sociological paraphrase, these axioms represent, first, the deflection of criticism of the social structure onto one's self among those so situated in the society that they do not have full and equal access to opportunity; second, the preservation of a given structure of social power by having individuals in the lower strata identify themselves, not

4. Malcolm S. MacLean, *Scholars, Workers and Gentlemen* (Cambridge: Harvard University Press, 1938), 29.

5. Cf. A. W. Griswold, "The American Cult of Success" (Ph.D. diss., Yale University, 1933).

with their compeers, but with those at the top (whom they will ultimately join); and third, providing pressures for conformity with the cultural dictates of unslackened ambition by the threat of less than full membership in the society for those who fail to conform.

It is in these terms and through these processes that contemporary American culture continues to be characterized by a heavy emphasis on wealth as a basic symbol of success, without a corresponding emphasis upon the legitimate avenues on which to march toward this goal. How do individuals living in this cultural context respond? And how do our observations bear upon the doctrine that deviant behavior typically derives from biological impulses breaking through the restraints imposed by culture? What, in short, are the consequences for the behavior of people variously situated in the social structure of a culture in which the emphasis on dominant success-goals has become increasingly separated from an equivalent emphasis on institutionalized procedures for seeking these goals?

Types of Individual Adaptation

Turning from these *culture patterns,* we now examine types of adaptation by individuals within the culture-bearing society. Though our focus is still the cultural and social genesis of varying rates and types of deviant behavior, our perspective shifts from the plane of patterns of cultural values to the plane of types of adaptation to these values among those occupying different positions in the social structure.

We here consider five types of adaptation, as these are schematically set out in table 12.1, where (+) signifies "acceptance," (−) signifies "rejection," and (±) signifies "rejection of prevailing values and substitution of new values."

Examination of how the social structure operates to exert pressure upon individuals for one or another of these alternative modes of behavior must be prefaced by the observation that individuals may shift from one alternative to another as they engage in different spheres of social activities. These categories refer to role behavior in specific types of situations, not to personality. They are types of more or less enduring response, not types of personality organization. To consider these types of adaptation in several spheres of conduct would introduce a complexity unmanageable within the confines of this paper. For this reason, we shall be primarily concerned with economic activity in the broad sense of "the production, exchange, distribution and consumption of goods and services" in our competitive society, where wealth has taken on a highly symbolic cast.

TABLE 12.1 A TYPOPLOGY OF MODES OF INDIVIDUAL ADAPTATION

Modes of Adaptation	*Culture Goals*	*Institutionalized Means*
I. Conformity	+	+
II. Innovation	+	−
III. Ritualism	−	+
IV. Retreatism	−	−
V. Rebellion[a]	±	±

[a]This fifth alternative is on a plane clearly different from that of the others. It represents a transitional response seeking to *institutionalize* new goals and new procedures to be shared by other members of the society. It thus refers to efforts to *change* the existing cultural and social structure rather than to accommodate efforts *within* this structure.

I. Conformity

To the extent that a society is stable, adaptation type I—conformity to both cultural goals and institutionalized means—is the most common and widespread. Were this not so, the stability and continuity of the society could not be maintained. The mesh of expectancies constituting every social order is sustained by the modal behavior of its members representing conformity to the established, though secularly changing, culture patterns. It is, in fact, only because behavior is typically oriented toward the basic values of the society that we can speak of a human aggregate as comprising a society. Unless there is a deposit of values shared by interacting individuals, there exist social relations, if the disorderly interactions may be so called, but no society. [. . .]

Since our primary interest centers on the sources of *deviant* behavior, and since we have briefly examined the mechanisms making for conformity as the modal response in American society, little more need be said regarding this type of adaptation at this point.

II. Innovation

Great cultural emphasis upon the success-goal invites this mode of adaptation through the use of institutionally proscribed but often effective means of attaining at least the simulacrum of success—wealth and power. This response occurs when the individual has assimilated the cultural emphasis upon the goal without equally internalizing the institutional norms governing ways and means for its attainment.

From the standpoint of psychology, great emotional investment in an objective may be expected to produce a readiness to take risks, and this attitude may be adopted by people in all social strata. From the standpoint of sociology, the question arises, which features of our social structure predispose toward this type of adaptation, thus producing greater frequencies of deviant behavior in one social stratum than in another?

On the top economic levels, the pressure toward innovation not infrequently erases the distinction between businesslike strivings this side of the mores and sharp practices beyond the mores. As Veblen observed, "It is not easy in any given case—indeed it is at times impossible until the courts have spoken—to say whether it is an instance of praiseworthy salesmanship or a penitentiary offense." The history of the great American fortunes is threaded with strains toward institutionally dubious innovation as is attested by many tributes to the Robber Barons. The reluctant admiration often expressed privately, and not seldom publicly, of these "shrewd, smart and successful" men is a product of a cultural structure in which the sacrosanct goal virtually consecrates the means. This is no new phenomenon.

[. . .] Not all large departures from institutional norms in the top economic strata are known, and possibly fewer deviations among the lesser middle classes come to light. Sutherland has repeatedly documented the prevalence of "white-collar criminality" among businessmen. He notes, further, that many of these crimes were not prosecuted because they were not detected or, if detected, because of "the status of the businessman, the trend away from punishment, and the relatively unorganized resentment of the public against white-collar criminals."[6] A study of some 1,700 prevalently middle-class individuals found that "off the record crimes" were common among wholly "respectable" members of society. Ninety-nine percent of those questioned confessed to having committed one or more of forty-nine offenses under the penal law of the State of New York, each of these offenses being sufficiently serious to draw a maximum sentence of not less than one year. The mean number of offenses in adult years—this excludes all offenses committed before the age of sixteen—was eighteen for men and eleven for women. Fully 64 percent of the men and 29 percent of the women acknowledged their guilt on one or more counts of felony which, under the laws of New York, would be ground for depriving them of all rights of citizenship. [. . .] On the basis of these results, the authors modestly

6. Edwin H. Sutherland, "White Collar Criminality," *American Sociological Review* 5 (1940): 1–12; "Crime and Business," *Annals, American Academy of Political and Social Science*, 1941, 217, 112–18; "Is 'White Collar Crime' Crime?" *American Sociological Review* 10 (1945): 132–39.

conclude that "the number of acts legally constituting crimes are far in excess of those officially reported. Unlawful behavior, far from being an abnormal social or psychological manifestation, is in truth a very common phenomenon."[7]

But whatever the differential rates of deviant behavior in the several social strata, and we know from many sources that the official crime statistics uniformly showing higher rates in the lower strata are far from complete or reliable, it appears from our analysis that the greatest pressures toward deviation are exerted upon the lower strata. Cases in point permit us to detect the social mechanisms involved in producing these pressures. Several researches have shown that specialized areas of vice and crime constitute a "normal" response to a situation where the cultural emphasis upon pecuniary success has been absorbed but where there is little access to conventional and legitimate means for becoming successful. The occupational opportunities of people in these areas are largely confined to manual labor and the lesser white-collar jobs. Given the American stigmatization of manual labor, which has been found to hold rather uniformly in all social classes,[8] and the absence of realistic opportunities for advancement beyond this level, the result is a distinct tendency toward deviant behavior. The status of unskilled labor and the consequent low income cannot readily compete, *in terms of established standards of worth,* with the promises of power and high income from organized vice, rackets, and crime.[9]

For our purposes, these situations exhibit two salient features. First, incentives for success are provided by the established values of the culture; and second, the avenues available for moving toward this goal are largely limited by the class structure to those of deviant behavior. It is the *combination* of the cultural emphasis and the social structure that produces intense pressure for deviation. Recourse to legitimate chan-

7. James S. Wallerstein and Clement J. Wyle, "Our Law-Abiding Law-Breakers," *Probation,* April, 1947.

8. National Opinion Research Center, *National Opinion on Occupations,* April, 1947.

9. See Joseph D. Lohman, "The Participant Observer in Community Studies," *American Sociological Review* 2 (1937): 890–98; and William F. Whyte, *Street Corner Society* (Chicago: University of Chicago Press, 1943). Note Whyte's conclusions, pages 273–74; "It is difficult for the Cornerville man to get onto the ladder [of success], even on the bottom rung. . . . He is an Italian, and the Italians are looked upon by upper-class people as among the least desirable of the immigrant peoples . . . the society holds out attractive rewards in terms of money and material possessions to the 'successful' man. For most Cornerville people these rewards are available only through advancement in the world of rackets and politics."

nels for "getting in the money" is limited by a class structure that is not fully open at each level to men of good capacity.[10] Despite our persisting open-class ideology,[11] advance toward the success-goal is relatively rare and notably difficult for those armed with little formal education and few economic resources. The dominant pressure leads toward the gradual attenuation of legitimate, but by and large ineffectual, strivings and the increasing use of illegitimate, but more or less effective, expedients.

Of those located in the lower reaches of the social structure, the culture makes incompatible demands. On the one hand, they are asked to orient their conduct toward the prospect of large wealth—"Every man a king," said Marden and Carnegie and Long—and on the other, they are largely denied effective opportunities to do so institutionally. The consequence of this structural inconsistency is a comparatively high rate of deviant behavior. The equilibrium between culturally designated ends and means becomes highly unstable with progressive emphasis on attaining the prestige-laden ends by any means whatsoever. Within this context, Al Capone represents the triumph of amoral intelligence over morally prescribed "failure," when the channels of vertical mobility are closed or narrowed *in a society which places a high premium on economic affluence and social ascent for* all *its members.*[12]

This last qualification is of central importance. It implies that other aspects of the social structure, besides the extreme emphasis on pecuniary success, must be considered if we are to understand the social sources of deviant behavior. A high frequency of deviant behavior is

10. Numerous studies have found that the educational pyramid operates to keep a large proportion of unquestionably able but economically disadvantaged youth from obtaining higher formal education.

11. The shifting historical role of this ideology would be a profitable subject for exploration. The "office-worker-to-president" imagery was once in approximate accord with the facts, in the loose sense that vertical mobility was probably more common then than now. The ideology persists however, possibly because it still performs an important function for motivating members of the society to work within the social framework. It probably operates to increase the probability of Adaptation I and to lessen the probability of Adaptation V. In short, the role of this doctrine has changed from that of a roughly valid theorem to that of an ideology.

12. The role of Blacks in this connection raises almost as many theoretical as practical questions. It has been reported that large segments of the Black population have assimilated the dominant caste's values of pecuniary success and social advancement, but have "realistically adjusted" themselves to the "fact" that social ascent is presently confined almost entirely to movement within the caste. See John Dollard, *Caste and Class in a Southern Town* (New Haven: Yale University Press, 1936), 66 ff.; Donald Young, *American Minority Peoples* (New York: Harper, 1932), 581. See also the further discussion in the present chapter.

not generated merely by lack of opportunity or by this exaggerated pecuniary emphasis. A comparatively rigidified class structure, a feudalistic caste order, may limit opportunities far beyond the point which obtains in American society today. It is only when a system of cultural values extols, virtually above all else, certain common success-goals *for the population at large* while the social structure rigorously restricts or completely closes access to approved modes of reaching these goals *for a considerable part of the same population,* that deviant behavior ensues on a large scale. Otherwise said, our egalitarian ideology denies by implication the existence of non-competing individuals and groups in the pursuit of pecuniary success. Instead, the same body of success-symbols is held to apply for all. Goals are held to transcend class lines, not to be bounded by them, yet the actual social organization is such that there exist class differentials in accessibility of the goals. In this setting, a cardinal American virtue, "ambition," promotes a cardinal American vice, "deviant behavior."

This theoretical analysis may help explain the varying correlations between crime and poverty. "Poverty" is not an isolated variable which operates in precisely the same fashion wherever found; it is only one in a complex of identifiably interdependent social and cultural variables. Poverty as such and consequent limitation of opportunity are not enough to produce a conspicuously high rate of criminal behavior. Even the notorious "poverty in the midst of plenty" will not necessarily lead to this result. But when poverty and associated disadvantages in competing for the culture-values approved for *all* members of the society are linked with a cultural emphasis on pecuniary success as a dominant goal, high rates of criminal behavior are the "normal" outcome. Thus, crude (and not necessarily reliable) crime statistics suggest that poverty is less highly correlated with crime in southeastern Europe than in the United States. The economic life-chances of the poor in these European areas would seem to be even less promising than in this country, so that neither poverty nor its association with limited opportunity is sufficient to account for the varying correlations. However, when we consider the full configuration—poverty, limited opportunity, and the assignment of cultural goals—there appears some basis for explaining the higher correlation between poverty and crime in our society than in others where rigidified class structure is coupled with *differential class symbols of success.*

The victims of this contradiction between the cultural emphasis on pecuniary ambition and the social bars to full opportunity are not always aware of the structural sources of their thwarted aspirations. To be sure, they are typically aware of a discrepancy between individual worth and social rewards. But they do not necessarily see how this comes about. Those who do find its source in the social structure may become alienated from that structure and become ready candidates for Adaptation V (rebellion). But others, and this appears to include the great majority, may attribute their difficulties to more mystical and less sociological sources. For as the distinguished classicist and sociologist-in-spite-of-himself, Gilbert Murray, has remarked in this general connection, "The best seed-ground for superstition is a society in which the fortunes of men seem to bear practically no relation to their merits and efforts. A stable and well-governed society does tend, speaking roughly, to ensure that the Virtuous and Industrious Apprentice shall succeed in life, while the Wicked and Idle Apprentice fails. And in such a society people tend to lay stress on the reasonable or visible chains of causation. But in [a society suffering from anomie] . . . , the ordinary virtues of diligence, honesty, and kindliness seem to be of little avail."[13] And in such a society people tend to put stress on mysticism: the workings of Fortune, Chance, Luck.

Actually, both the eminently "successful" and the eminently "unsuccessful" in our society often attribute the outcome to luck. Thus, the prosperous man of business, Julius Rosenwald, declared that 95 percent of the great fortunes were "due to luck."[14] And a leading business journal, in an editorial explaining the social benefits of great individual wealth, finds it necessary to supplement wisdom with luck as the factors accounting for great fortunes: "When one man through wise investments—aided, we'll grant, by good luck in many cases—accumulates a few millions, he doesn't thereby take something from the rest of us."[15] In much the same fashion, the worker often explains economic status in terms of chance. "The worker sees all about him experienced and skilled men with no work to do. If he is in work, he feels lucky. If he is out of work, he is the victim of hard luck. *He can see little relation between worth and consequences.*"[16]

But these references to the workings of chance and luck serve distinctive functions according to whether they are made by those who

13. Gilbert Murray, *Five Stages of Greek Religion* (New York: Columbia University Press, 1925), 164–65. Professor Murray's chapter, "The Failure of Nerve," from which I have taken this excerpt, must surely be ranked among the most civilized and perceptive sociological analyses in our time.

14. See the quotation from an interview cited in Gustavus Meyers, *History of the Great American Fortunes* (New York: Modern Library, 1937), 706.

15. *Nation's Business* 27 no. 9 (n.d.): 8–9.

16. E. W. Bakke, *The Unemployed Man* (New York: E. P. Dutton, 1934), 14.

have reached or those who have not reached the culturally emphasized goals. For the successful it is, in psychological terms, a disarming expression of modesty. It is far removed from any semblance of conceit to say, in effect, that one was lucky rather than altogether deserving of one's good fortune. In sociological terms, the doctrine of luck as expounded by the successful serves the dual function of explaining the frequent discrepancy between merit and reward while keeping immune from criticism a social structure which allows this discrepancy to become frequent. For if success is primarily a matter of luck, if it is just in the blind nature of things, if it bloweth where it listeth and thou canst not tell whence it cometh or whither it goeth, then surely it is beyond control and will occur in the same measure *whatever the social structure.*

For the unsuccessful, and particularly for those among the unsuccessful who find little reward for their merit and their effort, the doctrine of luck serves the psychological function of enabling them to preserve their self-esteem in the face of failure. It often entails the dysfunction of curbing motivation for sustained endeavor.[17] Sociologically, as implied by Bakke,[18] the doctrine may reflect a failure to comprehend the workings of the social and economic system, and may be dysfunctional inasmuch as it eliminates the rationale of working for structural changes making for greater equities in opportunity and reward.

[. . .] Among those who do not apply the doctrine of luck to the gulf between merit, effort, and reward there may develop an individuated and cynical attitude toward the social structure, best exemplified in the cultural cliché that "it's not what you know, but who you know, that counts."

In societies such as our own, then, the great cultural emphasis on pecuniary success for all and a social structure which unduly limits practical recourse to approved means for many has set up a tension toward innovative practices which depart from institutional norms. But this form of adaptation presupposes that individuals have been imperfectly socialized so that they abandon institutional means while retaining the success-aspiration. Among those who have fully internalized the institutional values, however, a comparable situation is more likely to

17. At its extreme, it invites resignation and routinized activity (Adaptation III) or a fatalistic passivism (Adaptation IV), of which more presently.

18. Bakke, *The Unemployed Man,* 14, where he suggests that "the worker knows less about the processes which cause him to succeed or have no chance to succeed than business or professional people. There are more points, therefore, at which events appear to have their incidence in good or ill luck."

lead to an alternative response in which the goal is abandoned but conformity to the mores persists. This type of response calls for further examination.

III. Ritualism

The ritualistic type of adaptation can be readily identified. It involves the abandoning or scaling down of the lofty cultural goals of pecuniary success and social mobility to the point where one's aspirations can be satisfied. But though one rejects the cultural obligation to attempt "to get ahead in the world," though one draws in one's horizons, one continues to abide almost compulsively by institutional norms.

It is something of a terminological quibble to ask whether this represents "genuinely deviant behavior." Since the adaptation is, in effect, an internal decision and since the overt behavior is institutionally permissive, though not culturally preferred, it is not generally considered to represent a "social problem." Intimates of individuals making this adaptation may pass judgment in terms of prevailing cultural emphases and may "feel sorry for them," they may, in the individual case, feel that "old Jonesy is certainly in a rut." Whether this is described as deviant behavior or no, it clearly represents a departure from the cultural model in which Americans are obliged to strive actively, through institutionalized procedures, to move onward and upward in the social hierarchy.

We should expect this type of adaptation to be fairly frequent in a society which makes one's social status largely dependent upon one's achievements. For, as has often been observed,[19] this ceaseless competitive struggle produces acute status anxiety. One device for allaying these anxieties is to lower one's level of aspiration—permanently. Fear produces inaction, or more accurately, routinized action.[20]

The syndrome of the social ritualist is both familiar and instructive. His implicit life-philosophy finds expression in a series of cultural clichés: "I'm not sticking *my* neck out"; "I'm playing safe"; "I'm satisfied with what I've got"; "Don't aim high and you won't be disappointed." The theme threaded through these attitudes is that high ambitions invite frustration and danger whereas lower aspirations produce satisfac-

19. See, for example, Harry Stack Sullivan, "Modern Conceptions of Psychiatry," *Psychiatry* 3 (1940): 111–12; and Robert K. Merton, Marjorie Fiske and Alberta Curtis, *Mass Persuasion* (New York: Harper, 1946), 159–60.

20. Pierre Janet, "The Fear of Action," *Journal of Abnormal Psychology* 16 (1921): 150–60, and the discussion by F. L. Wells, "Social Maladjustments: Adaptive Regression," in *Handbook of Social Psychology,* ed. Clark Murchison (n.p.: 1935), which bears closely on the type of adaptation examined here.

tion and security. It is a response to a situation that appears threatening and excites distrust. It is the attitude implicit among workers who carefully regulate their output to a constant quota in an industrial organization where they have occasion to fear that they will "be noticed" by managerial personnel and "something will happen" if their output rises and falls.[21] It is the perspective of the frightened employee, the zealously conformist bureaucrat in the teller's cage of the private banking enterprise or in the front office of the public works enterprise.[22] It is, in short, the mode of adaptation of individually seeking a *private* escape from the dangers and frustrations that seem inherent in the competition for major cultural goals by abandoning these goals and clinging all the more closely to the safe routines and the institutional norms.

If we should expect *lower-class* Americans to exhibit Adaptation II—"innovation"—to the frustrations enjoined by the prevailing emphasis on large cultural goals and the fact of small social opportunities, we should expect *lower-middle-class* Americans to be more often represented among those making Adaptation III, "ritualism." For it is in the lower-middle class that parents typically exert continuous pressure upon children to abide by the moral mandates of the society, and where the social climb upward is less likely to meet with success than in the upper-middle class. The strong disciplining for conformity with mores reduces the likelihood of Adaptation II and promotes the likelihood of Adaptation III. The socialization patterns of the lower-middle class thus promote the very character structure most predisposed toward ritualism,[23] and it is in this stratum, accordingly, that the adaptative pattern III should most often occur.

But we should note again, as at the outset of this paper, that we are here examining *modes of adaptation* to contradictions in the cultural and social structure; we are not focusing on character or personality types. Individuals caught up in these contradictions can and do move from one type of adaptation to another. Thus it may be conjectured that some ritualists, conforming meticulously to the institutional rules,

21. Fritz J. Roethlisberger and W. J. Dickson, Management and the Worker (Cambridge: Harvard University Press, 1939), chap. 18 and 531 ff.; and on the more general theme, the typically perspicacious remarks of Gilbert Murray, *Five Stages of Greek Religion*, 138–39.

22. Robert K. Merton, "Bureaucratic Structure and Personality," *Social Forces* 18 (1940): 560–68; and "Role of the Intellectual in Public Bureaucracy," *Social Forces* 23 (1945): 405–15.

23. See, for example, Allison Davis and John Dollard, *Children of Bondage* (Washington: 1940), chap. 12, "Child Training and Class," which, though it deals with the lower- and lower-middle-class patterns of socialization among Blacks in the far South, appears applicable, with slight modification, to the white population as well.

are so steeped in the regulations that they become bureaucratic virtuosos, that they over-conform precisely because they are subject to guilt engendered by previous nonconformity with the rules (i.e., Adaptation II). The occasional passage from ritualistic adaptation to dramatic kinds of illicit adaptation is well documented in clinical case histories and often set forth in insightful fiction. Defiant outbreaks not infrequently follow upon prolonged periods of overcompliance. But though the psychodynamic mechanisms of this type of adaptation have been fairly well identified and linked with patterns of discipline and socialization in the family, much sociological research is still required to account for their different frequencies in different groups and social strata. Our own discussion has merely set out one analytical framework for sociological research focused on this problem.

IV. Retreatism

Just as Adaptation I (conformity) remains the most frequent, Adaptation IV (the rejection of cultural goals and institutional means) is probably the least common. People who "adapt" (or maladapt) in this fashion are, strictly speaking, *in* the society but not *of* it. Sociologically, these constitute the true "aliens." Not sharing the common frame of values, they can be included as members of the *society* (in distinction from the *population*) only in a fictional sense.

In this category fall some of the adaptive activities of psychotics, autists, pariahs, outcasts, vagrants, tramps, chronic drunkards and drug addicts. They have relinquished culturally prescribed goals and their behavior does not accord with institutional norms. This is not to say that in some cases the source of their mode of adaptation is not the very social structure which they have in effect repudiated or that their very existence within an area does not constitute a problem for members of the society.

From the standpoint of its sources in the social structure, this mode of adaptation is most likely to occur when *both* the culture goals and the institutional practices have been thoroughly assimilated by the individual and imbued with affect and high value, but accessible institutional avenues do not prove effective. There results a twofold conflict: the interiorized moral obligation for adopting institutional means conflicts with pressures to resort to illicit means (which might attain the goal) and the individual is shut off from means which are both legitimate and effective. The competitive order is maintained but frustrated and handicapped individuals who cannot cope with this order drop out. Defeatism, quietism and resignation are manifested in escape mechanisms which ultimately lead them to "escape" from the requirements of the society. It is thus an expedient that arises from continued failure to near the goal by legitimate measures and from an inability to use the illegitimate route because of internalized prohibitions, *this process occurring while the supreme value of the success-goal has not yet been renounced.* The conflict is resolved by abandoning *both* precipitating elements, the goals and the means. The escape is complete, the conflict is eliminated and the individual is asocialized.

In public and ceremonial life, this type of deviant behavior is most heartily condemned by conventional representatives of the society. In contrast to the conformist, who keeps the wheels of society running, this type of deviant is a nonproductive liability; in contrast to the innovator who is at least "smart" and actively striving, he sees no value in the success-goal which the culture prizes so highly; in contrast to the ritualist who conforms at least to the mores, he pays scant attention to institutional practices.

Nor does the society lightly accept these repudiations of its values. To do so would be to put these values into question. Those who have abandoned the quest for success are relentlessly pursued to their haunts by a society insistent upon having its members orient themselves to success-striving. [. . .]

This fourth mode of adaptation, then, is that of the socially disinherited who, if they have none of the rewards held out by society, also have few of the frustrations attendant upon continuing to seek these rewards. It is, moreover, a "privatized" rather than a collective mode of adaptation. Although people exhibiting this deviant behavior gravitate toward centers where they come into contact with other deviants and although they may come to share in the subculture of these deviant groups, their adaptations are largely private and isolated rather than unified under the aegis of a new cultural code. The type of collective adaptation remains to be considered.

V. Rebellion

This adaptation leads people to envisage and seek to bring into being a new, that is to say, a greatly modified social structure. It presupposes alienation from reigning goals and standards. These come to be regarded as purely arbitrary. And the arbitrary is precisely that which can neither exact allegiance nor possess legitimacy, for it might as well be otherwise. In our society, organized movements for rebellion aim to introduce a social structure in which the cultural standards of success would be sharply modified and provision would be made for a closer correspondence between merit, effort, and reward.

But before examining rebellion as a mode of adaptation, we must distinguish it from a superficially similar but essentially different type, *ressentiment*. Introduced in a special technical sense by Nietzsche, the concept of *ressentiment* was taken up and developed sociologically by Max Scheler.[24] This complex sentiment has three interlocking elements. First, diffuse feelings of hate, envy, and hostility; second, a sense of being powerless to express these feelings actively against the person or social stratum evoking them; and third, a continual reexperiencing of this impotent hostility.[25] The essential point distinguishing *ressentiment* from rebellion is that the former does not involve a genuine change in values. *Ressentiment* involves a sour-grapes pattern which asserts that desired but unattainable objectives do not actually embody the prized values—after all, the fox in the fable does not say that he abandons all taste for sweet grapes; he says only that these particular grapes are not sweet. Rebellion, in contrast, involves a genuine transvaluation, where the direct or vicarious experience of frustation leads to denunciation of previously prized values—the rebellious fox renounces the prevailing taste for sweet grapes. In *ressentiment,* one condemns what one secretly craves; in rebellion, one condemns the craving itself. But though the two are distinct, organized rebellion may draw upon a vast reservoir of the resentful and discontented as institutional dislocations become acute.

When the institutional system is regarded as the barrier to the satisfaction of legitimized goals, the stage is set for rebellion as an adaptive response. To pass into organized political action, allegiance must not only be withdrawn from the prevailing social structure but must be transferred to new groups possessed of a new myth.[26] The dual function of the myth is to locate the source of large-scale frustrations in the social structure and to portray an alternative structure which would not, presumably, give rise to frustration of the deserving. It is a charter for action. In this context, the functions of the counter-myth of the conservatives—briefly sketched in an earlier section of this chapter—become further clarified: whatever the source of mass frustration, it is not to be found in the basic structure of the society. The conservative myth may thus assert that these frustrations are in the nature of things and

24. Max Scheler, *L'homme du ressentiment* (Paris: n.d.). This essay first appeared in 1912; revised and completed, it was included in Scheler's *Abhandlungen und Aufsätze,* appearing thereafter in his *Vom Umsturz der Werte* (n.p.: 1919).

25. Scheler, *L'homme du ressentiment,* 55–56. No English word fully reproduces the complex of elements implied by the word *ressentiment;* its nearest approximation in German would appear to be *Groll.*

26. George S. Pettee, *The Process of Revolution* (New York: Harper, 1937), 8–24; see particularly his account of "monopoly of the imagination."

would occur in *any* social system: "Periodic mass unemployment and business depressions can't be legislated out of existence; it's just like a person who feels good one day and bad the next."[27] Or, if not the doctrine of inevitability, then the doctrine of gradual and slight adjustment: "A few changes here and there, and we'll have things running as shipshape as they can possibly be." Or, the doctrine which deflects hostility from the social structure onto the individual who is a "failure" since "every man really gets what's coming to him in this country."

The myths of rebellion and of conservatism both work toward a "monopoly of the imagination" seeking to define the situation in such terms as to move the frustrated toward or away from Adaptation V. It is above all the "renegade" who, though himself "successful," renounces the prevailing values that becomes the target of hostility among those in rebellion. For he not only puts the values in question, as does the out-group, but he signifies that the unity of the group is broken.[28] Yet, as has often been noted, it is typically a rising class rather than the most depressed strata that organizes the resentful and the rebellious into a revolutionary group.

The Strain Toward Anomie

The social structure we have examined produces a strain toward anomie and deviant behavior. The pressure of such a social order is upon outdoing one's competitors. So long as the sentiments supporting this competitive system are distributed throughout the entire range of activities and are not confined to the final result of "success," the choice of means will remain largely within the ambit of institutional control. When, however, the cultural emphasis shifts from the satisfactions deriving from competition itself to exclusive concern with the outcome, the resulting stress makes for the breakdown of the regulatory structure. With this attenuation of institutional controls, there occurs an approximation to the situation erroneously held by the utilitarian philosophers to be typical of society, a situation in which calculations of personal advantage and fear of punishment are the only regulating forces.

This strain toward anomie does not operate evenly throughout the society. The present analysis suggests the strata most vulnerable to the pressures for deviant behavior and sets forth some of the mechanisms

27. Robert S. and Helen M. Lynd, *Middletown in Transition* (New York: Harcourt Brace, 1937), 408, for a series of cultural cliches exemplifying the conservative myth.

28. See the acute observations by Georg Simmel, *Soziologie* (Leipzig: 1908), 276–77.

operating to produce those pressures. To simplify the problem, monetary success was taken as the major cultural goal, although there are, of course, alternative goals in the repository of common values. The realms of intellectual and artistic achievement, for example, provide alternative career patterns that seldom entail large pecuniary rewards. To the extent that the cultural structure attaches prestige to these alternatives and the social structure permits access to them, the system is somewhat stabilized. Potential deviants may still conform in terms of auxiliary sets of values.

But the central tendencies toward anomie remain, and it is to these that the analytical scheme here set forth calls particular attention. [. . .]

Concluding Remarks

It should be apparent that the foregoing discussion is not pitched on a moralistic plane. Whatever the sentiments of the reader concerning the moral desirability of coordinating the goals- and means-phases of the social structure, it is clear that imperfect coordination of the two leads to anomie. Insofar as one of the most general functions of social structure is to provide a basis for predictability and regularity of social behavior, it becomes increasingly limited in effectiveness as these elements of the social structure become dissociated. At the extreme, predictability is minimized and what may be called anomie or cultural chaos supervenes.

This essay on the structural sources of deviant behavior remains a brief prelude. It has not included a detailed treatment of the structural elements which predispose towards one rather than another of the alternative responses open to individuals living in an ill-balanced social structure; it has largely neglected but not denied the relevance of the social-psychological processes determining the specific incidence of these responses; it has only briefly considered the social functions fulfilled by deviant behavior; it has not put the explanatory power of the analytical scheme to full empirical test by determining group variations in deviant and conformist behavior; it has only touched upon rebellious behavior that seeks to refashion the social framework.

It is suggested that these and related problems may be advantageously analyzed by use of this scheme.

13

Opportunity Structure (1995)

Emergence of the Concept of Opportunity Structure in the Columbia Micro-environment of the 1950s

A new phase of the evolving Social Structure and Anomie (SS&A) paradigm first found expression in the form of oral publication—this, in the Columbia "classes at that time" which James Coleman, self-described as "an informal course assistant," recollects for us in tranquillity. These, he recalls, were "the Merton lectures in Sociology 215–216 ["Analysis of Social Structure"] in 1951–52, and Sociology 213–214 [with the odd-sounding but resonating title of "Social Theory Applied to Social Research"] in 1952–53. It was in the latter course of lectures, he reports, that he and fellow students were introduced to the notions of "*reconceptualization*" and "*respecification*" of a lower-level empirical generalization as procedures in sociological theorizing.[1] It was in the prior course, evolving from year to year in the early 1950s, that the concept of socially structured "differential access to legitimate opportunity," set forth in the original 1938 SS&A, became expressly related to the correlative concept of the "legitimate opportunity structure."

[. . .] *Opportunity structure* designates the scale and distribution of conditions that provide various probabilities for individuals and groups to achieve specifiable outcomes. From time to time, the opportunity structure expands or contracts, as do segments of that structure. However, as indicated by the correlative concepts of socially structured "differential access to opportunity" in the original paradigm of SS&A and of "structural context" in the paradigm for functional analysis, location in the social structure strongly influences, though it does not wholly determine, the extent of access to the opportunity structure. By concept, then, an expanding or contracting opportunity structure does

From "Opportunity Structure," in *The Legacy of Anomie Theory,* ed. Freda Adler and William Laufer (New Brunswick, N.J.: Transaction Publishers, 1995), 24–33. Reprinted by permission of Transaction Publishers.

1. James S. Coleman, "Robert K. Merton as Teacher" in *Robert K. Merton: Consensus and Controversy,* ed. Jon Clark et al. (London and New York: Falmer Press, 1990), 28–29, 31.

not carry with it the uniform expansion or contraction of opportunities for all sectors of a socially stratified population, a familiar enough notion with diverse implications.

By way of providing an early example in the oral publication of the concept of opportunity structure as it was inflicted upon graduate students in the course, Sociology 215: Analysis of Social Structure, I retrieve an early set of sketchy notes that, I now notice, are dated in the distinctive fashion adopted by George Sarton, my mentor in the history of science. Thus, in typefaced facsimile:

> 5311.04
> place of "CONDITIONS" in "THEORY"/i.e., paradigm/of ACTION
> 1. PERTINENT ELEMENTS OF THE SITUATION WHICH ARE BEYOND CONTROL OF THE ACTING INDIVIDUAL
> —concrete e.g.'s: feminization of cities a very general pattern . . .
> at any particular age, therefore, fewer possibilities of woman finding marriage partner . . .
> (may be known or unknown
> experienced difficulty: reaction may be self-appraisal
> —e.g., higher morale (and conflict) among one sample in Craftown; less among matched sample . . . summer and winter interaction patterns general e.g.: same logic of analysis in Social Structure and Anomie
> condition: OPPORTUNITY STRUCTURE for individuals variously located in social structure
> STEEPNESS OR FLATNESS OF STRATIFICATION STRUCTURE
>
> 2. SALIENT THEORETICAL SIGNIFICANCE OF CONDITIONS
> may be regarded FUNCTIONALLY: as *facilitative* of means for accomplishing objectives or as frustrating ends of action

Not quite cryptic, these notes bring to mind a long-forgotten phase in the emergence of the concept of opportunity structure out of the focus on "differential access to opportunities" in the first SS&A paper. This was the evolving theoretical formulation of the interplay between structural context and individual action in which (1) the (variously changing) arrays of opportunities represent objective *conditions* confronting acting individuals—*agents,* as they are now often described—who (2) are constrained or aided in gaining access to diverse types of opportunities by their positions in the social structure, which (3) locate individuals in terms of their status-sets (i.e., the array of their social attributes such as class, sex, race, ethnicity, age, religion, etc.). This conceptual scheme thus allowed for individual variability of socially structured choices, both conscious and unwitting, among the many or few opportunities accessible to them as well as for individual variability in the extent to which those opportunities are realized in the actual event.

Furthermore, as first became evident in oral publications and, as we shall now see, was then signaled in the 1957 analysis of "Continuities in the Theory of Social Structure and Anomie," the concept of "differential access to opportunities" introduced in the first SS&A paper corresponds to what was described as the "still loosely utilized but important concept of what Weber called 'life-chances.'"[2] Whereas, as we have noted, the correlative concept of "opportunity structure" refers to the scale and distribution of opportunities for achieving designated outcomes rather than to structurally patterned differential access to those opportunities.

The still ongoing oral publication of the evolving paradigm was severely condensed when it was put in print. The analysis of continuities in the SS&A paradigm dealt with a variety of theoretical ideas and problems that had emerged in the near-decade since the prior extension of the paradigm but for the purpose of this truncated retrospective I note only the introduction into print of the term-and-concept *opportunity structure* (and of such variants as the "structure of opportunity" and "socially structured opportunities").[3]

In setting forth further problematics of the SS&A paradigm "schematically," the chapter on continuities observed that next steps call for systematic data on "*socially patterned differentials in . . . relative accessibility to the goal:* life-chances in the opportunity structure."[4] The term-and-concept of opportunity structure had been introduced first in lectures and then in print for a definite propaedeutic purpose. I had acquired the distinct impression that a growing proportion of articles and monographs in the SS&A tradition and representations of that theoretical tradition in textbooks had come to focus on the "strains" induced by disjunctions between culturally induced or reinforced aspirations and access to institutionally legitimate means for realizing those aspirations—leading that theoretical tradition to become known in criminology and sociology as "strain theory"—while devoting far less theoretical and empirical attention to the basic sociological ideas of socially structured differentials in access to such legitimate opportunities and of changes in the structure of those opportunities.

My impression back then in the early 1950s may have been mis-

2. Robert K. Merton, *Social Theory and Social Structure,* 3d ed., rev. and enl. (New York: Free Press, 1968), 230.

3. Ibid., 229–30, 232, 246, 247.

4. Ibid., 229.

placed. Or perhaps it was correct for the period prior to the appearance in print of the new term-and-evolving-concept in the 1957 *Social Theory and Social Structure* (ST&SS). At any rate, that impression evidently did not correspond to the later representation of SS&A in introductory textbooks of sociology in the 1960s and 1970s as these were reported in an empirical study by Hilbert and Wright.[5] They found that twenty-seven of twenty-nine such textbooks indicated that those strains were primarily generated by "variations in access to legitimate means or opportunity structures." However that may have been later on, it seemed to me in the 1950s that making the proto-concept of opportunity structure explicit might make for theoretical clarification of the twin sources of such social strains. It would also register the further theoretical idea that not only were there socially structured differentials in access to the opportunities that then and there did exist but that the scale, character, and distribution of those opportunities which formed objective conditions affecting the probability of successful outcomes of choices were subject to varying rates and degrees of structural change that differentially affect those variously located in the social structure. As was being emphasized back then and as I have been repeatedly noting here (for the point bears reiteration since it often continues to be lost to view), the sociological concepts of opportunity and opportunity structure are generic; they are not confined to economic opportunities or to opportunities for social mobility.

Once the concept of opportunity structure had become differentiated as a correlative of the concept of "differing access to opportunities," it became clear to me that the latter idea was equivalent to Weber's important idea of "life-chances." Writing in the 1970s, Gerhard Hufnagel and Ralf Dahrendorf[6] have each variously pointed out that "Weber's use of the notion of chance"—the deliberate phrase, "*notion* of chance" is Dahrendorf's—still requires clarification and elucidation. Dahrendorf's lucid exegesis is especially in point for us. He observes that

> while Weber in fact uses the word[s] "life chances," this for him has specific and, in our context, irrelevant connotations. . . . "Life" is an emphatic word here [in a quotation from Weber], nearer to survival than to the fullness of human opportunities. But in many other places, Weber speaks of

5. R. E. Hilbert and Charles W. Wright, "Representations of Merton's Theory of Anomie," *The American Sociologist* 14 (1979): 150–56, at 152.

6. Gerhard Hufnagel, *Kritik als Beruf: Der kritische Gehalt im Werk Max Webers* (Frankfurt am Main: Propyläen, 1972), 225, 273–74, 289, 294; Ralf Dahrendorf, *Life Chances: Approaches to Social and Political Theory* (Chicago: University of Chicago Press, 1979) esp. 28–30, 62–84.

> social chances in ways which are highly relevant to our discussion [of social theories of social change and the political theory of liberty]: "economic chances," such as "market chances," "chances of acquisition," "price chances," "supply chances," but also "social chances," such as "the typical chance offered by class position," and indeed "future chances including those of rebirth and religious justification." Max Weber never included chances, let alone life chances, in his catalogue of "basic concepts"; but he might well have done so. For him the concept signifies in the first instance a methodological reservation; he is trying to avoid dogmatism. Furthermore, it indicates a characteristic of social structure; while social action is not random, there can never be more than a probability that things will happen as prescribed in norms and guaranteed by sanctions. Above and beyond that, however, Weber uses the notion of chance to indicate opportunities provided by social structure, and it is this usage which we intend to develop.[7]

I have paused to quote the Dahrendorf interpretation at some length since it bears directly upon the distinction I had begun to draw between life chances and opportunity structure. That interpretation treats the idea of life chances as equivalent to the idea of "differential access to opportunities among those variously located in the social structure"—an idea that remains central in Dahrendorf's thought-provoking discussion of the SS&A paradigm[8] since, understandably enough, he focuses on its original use for the analysis of deviant behavior while the present retrospective examines an evolving focus on the *generic* concept of opportunity structure. That concept, I had begun to realize, should apply in principle to every kind of socially patterned choice. It should find empirical expression in aggregative social patterns of choices and outcomes while allowing for individuality by being coupled with *the concept of a distribution of choices among individuals similarly situated in the social structure.*

Thus, as an immediate case in point, the notes for that lecture in the

7. Dahrendorf, *Life Chances,* 28–29. Apropos of Dahrendorf's observations, note the rather limited contexts in which Weber makes use of the concepts of "life chances" (*Lebenschancen*) and "survival chances" (*Ueberlebenschancen*). Those concepts appear chiefly as special types of his much-deployed generic concept of "chances" (*Chancen*). See his foundational *Wirtschaft und Gesellschaft,* 2d ed. (Tübingen: J. C. B. Mohr, 1925), esp. 21–22. It is perhaps indicative of Weber's limited utilization of the term *Lebenschancen* that the excellent translation edited by Guenther Roth and Claus Wittich elects not to translate it as "life chances" but as "struggle for advantages" or "struggle of individuals for personal advantages;" cf. their complete edition of Max Weber, *Economy and Society: An Outline of Interpretive Sociology* (Berkeley: University of California Press, 1978), 38–42.

8. Dahrendorf, *Life Chances,* 45, 84–88.

early 1950s which referred to interaction patterns in the worker's community of Craftown evidently called for further research exemplification of the generic concept of opportunity structure by drawing upon the "housing study" of the 1940s.[9] We had found that social, spatial, and architectural configurations provided unintended and unrecognized opportunity structures affecting the probabilities of forming social ties (such as local friendships) with particular kinds of significant others leading to patterns of homophily (i.e., friendships among social similars) and under determinate conditions, patterns of heterophily (i.e., friendships among social dissimilars).

Just as it proved necessary to emphasize that the concepts of opportunity and opportunity structure were generic and not confined to opportunities for some sort of economic success or social mobility, so it had become apparent from many published representations that the SS&A paradigm was still mistakenly assumed to hold only for such cases. This, although the original, 1938, paper had explicitly noted that "to treat the development of this process [posited in the paradigm] in various spheres of conduct would introduce a complexity unmanageable within the confines of this paper" and that "for this reason, we shall be concerned primarily with economic activity. . . ."[10] Such "patterned misunderstandings" of the kind contributing to the "fallacy of the latest word"[11] led to this plaintive observation in the 1957 piece, "Continuities in the Theory of Social Structure and Anomie":

> It will be remembered that we have considered the emphasis on monetary success as *one* dominant theme in American culture, and have traced the strains which it differentially imposes upon those variously located in the social structure. That is not to say, of course—as was repeatedly indicated,—that the disjunction between cultural goals and institutionally legitimate means derives only from *this* extreme goal-emphasis. The theory holds that *any* extreme emphasis upon personal wealth or, by a small stretch of the imagination, the conquests of a Don Juan—will attenuate conformity to the institutional norms governing behavior designed to achieve the particular form of "success," especially among those who are socially

9. Robert K. Merton, Patricia S. West and Marie Jahoda, *Patterns of Social Life: Explorations in the Sociology of Housing* (1948; reprint, New York: Columbia University, Bureau of Applied Social Research, 1951), mimeographed.

10. Robert K. Merton, "Social Structure and Anomie," *American Sociological Review* 3:672–82; and chap. 12 in this volume.

11. For an attempt to identify the phenomenon of such patterned misunderstandings in science and scholarship, see esp. Robert K. Merton, "The Fallacy of the Latest Word," *American Journal of Sociology* 89 (1984): 1091–121, esp. at 1101–12.

> disadvantaged in the competitive race. It is the conflict between cultural goals and the availability of using institutional means—whatever the character of the goals—which produces a strain toward anomie.[12]

Once again, this retrospective leads to, at least plausible, new perceptions. I can now better understand how it was (or might have been) that I decided to focus on priority-conflicts in science as the subject of my presidential address to the American Sociological Society (not yet American Sociological Association) back in 1957.[13] For in effect that subject would provide a strategic research site for application of SS&A theory to the privileged domain of science rather than to the more familiar scenes of crime and delinquency narrowly conceived. Further, the analysis of that social phenomenon would provide emphatic evidence that this probabilistic theory of deviant behavior was not confined to the lower reaches of the social structure and perhaps might contribute to a further understanding of the culture and practice of science. The paper concluded that

> the culture of science is, in this [i.e., some] measure, pathogenic. It can lead scientists to develop an extreme concern with recognition which is in turn the validation by peers of the worth of their work. Contentiousness, self-assertive claims, secretiveness lest one be forestalled, reporting only the data that support an hypothesis, false charges of plagiarism, even the occasional theft of ideas and, in rare cases, the fabrication of data,—all these have appeared in the history of science and can be thought of as deviant behavior in response to a discrepancy between the enormous emphasis in the culture of science upon original discovery and the actual difficulty many scientists experience in making an original discovery. In this situation of stress, all manner of adaptive behaviors are called into play, some of these being far beyond the mores of science.[14]

In view of the attention recently lavished in the mass media on fraud in science,[15] it is theoretically interesting (and rather consoling) that we

12. Merton, *Social Theory and Social Structure,* 220, and on 257: "we must emphasize again that the general theory of social structure and anomie is *not* confined to the specific goal of monetary success."

13. Robert K. Merton, "Priorities in Scientific Discovery: A Chapter in the Sociology of Science," *American Sociological Review* 22 (1957): 635–59, esp. the section on "Type of Response to Cultural Emphasis on Originality: *Fraud in Science,*" 649–59.

14. Ibid., 659.

15. Initially, most notably emphasized by the science journalists, William Broad and Nicholas Wade (*Betrayers of the Truth: Fraud and Deceit in the Halls of Science* [New York: Simon and Schuster, 1982]), who draw extensively on sociological treatments of the subject.

sociologists began long before to identify the social and cultural pressures making for such deviant behavior.[16] It seems that once theoretical problems are generated, social phenomena previously regarded as peripheral or uninteresting become reassessed as providing, in effect, strategic research sites for sociological understanding.

So it was that I continued to explore the theoretical utility of extending the generic concept of legitimate opportunity structure to social phenomena far beyond the confines of deviant behavior. As one case in point, it led to recognition of a theoretical and derivatively empirical linkage of that concept with what I had assumed to be the independent theories of "the accumulation of advantage and disadvantage" and the "Matthew effect" as these were first applied to the processes making for inequalities in the social stratification of scientists.[17] But since that theoretical consolidation took place in the 1960s and 1970s and is therefore beyond the time-frame of this retrospective, I confine myself here only to two summaries of that consolidation, the first of these exemplified in a case study of the career of the philosopher-historian of science, Thomas S. Kuhn:

> Processes of individual self-selection and institutional social selection interact to affect *successive probabilities of access to the opportunity-structure in a given field of activity*. When the role-performance of individual[s] measures up to demanding institutional standards, and especially when it greatly exceeds them, this initiates *a process of cumulative advantage* in which the individual[s] acquire enlarged opportunities to advance [their] work (and the rewards that go with it) even further. Since elite institutions have comparatively large resources for advancing work in their domains, talent that finds its way into those institutions has the heightened potential of acquiring *differentially accumulating advantages*. The systems of reward, allocation of resources, and social selection thus operate to create and maintain a class structure in science by providing a stratified distribution of chances among scientists for enlarging their role as investigators. Differentially accumulating advantages work in such a way that, in the words of Matthew, Mark, and Luke, unto every one that hath shall be given, and he shall have abundance, but from him that hath not shall be taken away even that which he hath.

16. For analytical overviews of that sociological literature, see Harriet Zuckerman, "Deviant Behavior and Social Control in Science" in *Deviance and Social Change,* ed. Edward Sagarin (Beverly Hills: Sage, 1977), 87–138; and her chapter, "The Sociology of Science" in *Handbook of Sociology,* ed. Neil Smelser (Newbury Park: Sage, 1988), 511–74, at 520–26.

17. Harriet Zuckerman, "Accumulation of Advantage and Disadvantage: The Theory and Its Intellectual Biography" in *L'Opera di R. K. Merton e la sociologia contemporanea,* ed. Carlo Mongardini and Simonetta Tabboni (Genoa: Edizioni Culturali Internazionali, 1989), 153–76.

> *Mutatis mutandis, cumulative advantages accrue for organizations and institutions as they do for individuals,* subject to countervailing forces that dampen exponential cumulation.[18]

That these theoretical linkages were readily evident to onetime Columbia students of the 1950s, at least in retrospect, can be seen from this concise summary by Lewis Coser and Charles Kadushin, which notes my having

> suggested that scientific recognition exemplifies the "Matthew effect"—to him that hath shall be given. Merton argues that there is a continuing interplay between the status system, based on honor and esteem, and the class system, based on differential life chances; this interplay locates scientists in differing positions within the opportunity structure of science. Without deliberate intent, the Matthew Effect operates to penalize the young and unknown and, in the process, reinforces the already unequal distribution of rewards.[19]

Thus, the 1950s saw the explicit introduction and elucidation of both the concept and the term of *opportunity structure* as a correlative to the concept of differential access to opportunities (which could be taken to correspond to Weber's concept-and-term of *life chances*). Linked with that theoretical development was the renewed and extended *application* of the concept of opportunity structure to socially patterned choices and their consequences, an application far removed from the initial focus on *deviant* behavior. As the concept was no longer confined to serving as an element in sociological accounts of crime and delinquency, its initially generic rather than specific character barely noted in the original, 1938, SS&A paper began to be elucidated and applied to such social phenomena as the formation of social ties and, later, to the social process of the accumulation of advantage and disadvantage. [. . .]

18. See Robert K. Merton, *The Sociology of Science: An Episodic Memoir* (Carbondale, Ill.: Southern Illinois University Press, 1979) for a fine-grained account of the workings of the opportunity structure in relation to life chances. (The brackets inserted in the quotation register self-conscious changes from the singular to the plural in the original text; this, in order to eliminate unwittingly gendered terminology while avoiding clumsy, "he-or-she" and "her-or-his" constructions; the added italics are there to highlight theoretical linkages between the concepts of the opportunity structure, the Matthew effect, and the accumulation of advantage and disadvantages.)

19. That succinct summary is from a collaborative work not only by these two authors but by a third as well (who, as it happens, was in turn a one-time student of Coser); see Lewis A. Coser, Charles Kadushin, and Walter W. Powell, *Books: The Culture and Commerce of Publishing* (New York: Basic Books, 1982), 235.

14

Socially Expected Durations (1984)

SED: The Concept in Course of Formation

Socially expected durations—hereafter, SEDs for short—are socially prescribed or culturally patterned expectations about temporal durations imbedded in social structures of various kinds. For example, the length of time that individuals are institutionally permitted to occupy particular statuses (such as an office in an organization or membership in a group); the assumed probable durations of diverse kinds of social relationships (such as friendship or a professional–client relationship); and the patterned and therefore anticipated longevity of individual occupants of statuses, of groups, and of organizations.

SEDs are ubiquitous in the social life of complex societies. They encompass social phenomena as seemingly unconnected as "the lame duck" pattern in political and other organizational structures; determinate and indeterminate prison sentences; prolonged and terminal illnesses; the statute of limitations; tenure of various kinds, academic and other; deadlines; residential mobility; master–apprentice relations in science; status sequences and succession in organizations; time-bounded and unbounded sports, and so on and on, through a great array of social structures and temporally patterned events.[1] [. . .]

Along with their omnipresence, SEDs are consequential. They constitute a class of social expectations that significantly affect the current behavior of groups and individuals, and of those in their role-sets and organization-sets. Since purposive social action involves varying anti-

From "Socially Expected Durations: A Case Study of Concept Formation in Sociology," in *Conflict and Consensus: A Festschrift for Lewis A. Coser,* ed. W. W. Powell and R. Robbins (New York: Free Press, 1984), 265–67, 274–75, 277–81. Reprinted by permission of the Free Press, an imprint of Simon and Schuster.

1. As witness the current (1995) debates in the U.S. Congress over term limits for senators and representatives and the recent study by Bryan R. Roberts, "Socially Expected Durations and the Economic Adjustment of Immigrants" in the *Economic Sociology of Immigration,* ed. Alejandro Portes (New York: Russell Sage Foundation, 1995), 42–86.—*Ed.*

cipations of what are taken to be relevant futures and since social structures involve intermeshed networks of socially supported normative expectancies in the forms of statuses and roles, SEDs constitute a fundamental class of patterned expectations linking social structures and individual action. SEDs affect *anticipatory social behavior,* ongoing behavior that is significantly shaped by socially prescribed and otherwise patterned anticipations; in this case, the anticipated durations of occupancy of statuses, of membership in groups, of interpersonal relations, and of groups and organizations.

But if the concept of SED refers to a ubiquitous and consequential element of social structure, it has scarcely been given concerted systematic attention by the sociological community. Occasional investigators have importantly focused on particular SEDs—in monographic studies of such total institutions as prisons and hospitals, for example—but in the aggregate, social scientists have not undertaken to identify and elucidate SEDs as a general and fundamental concept of social structure.

Socially *expected* durations are to be strongly distinguished from *actual* durations.[2] Indeed, the problematics of SEDs have to do in part with the behavior, structural dynamics, and consequences of interacting actual and expected durations. As we shall now see in a prime example, the expected duration of residence in a community has been found to have social consequences, such as the degree of involvement in the community, that are independent of the actually elapsed duration of residence.

Actual and Expected Durations: Craftown

Put in terms of the structural analysis of SED, an early study of mass persuasion had focused on aggregated consequences of *indeterminate duration*, actual and prospective, of a radio war-bond marathon for the behaviors of the audience, not on the consequences for the agent or focal actor (in this instance, the then-popular songbird of the air waves,

2. As emphasized, far too briefly, in Robert K. Merton, *Social Theory and Social Structure,* 3d ed., rev. and enl. (New York: Free Press, 1968), 365–66. The empirical study of actual durations was evidently inaugurated about half a century ago by Strumilin in Moscow (as cited by Alexander Szalai et al., eds., *The Use of Time: Daily Activities of Urban and Suburban Populations in Twelve Countries* [The Hague and Paris: Mouton and Co., 1972], 7) and in the United States by George A. Lundberg, Mirra Komarovsky, and Mary A. McInery (*Leisure: A Suburban Study* [New York: Columbia University Press, 1934]) and by Pitirim A. Sorokin and Clarence Q. Berger (*Time-Budgets of Human Behavior* [Cambridge: Harvard University Press, 1939]). The most comprehensive comparative study of time-utilization is of more recent vintage: Szalai et al., eds., *The Use of Time.*

Kate Smith).[3] It therefore represented a new conceptual step in the slowly evolving notion of SED when, in another study conducted during World War II, the already elapsed and still expected durations of residence in a community were adopted as central variables to help account for the agents' own current behaviors, with their second-order consequences, for the formation of community structure. [. . .]

This was the study of Craftown, a community of some 750 families built from scratch in 1941 in the great industrial flatlands south and west of New York City.[4] The study centered on the processes shaping the emergence of the social structure of this workers' community (with four-fifths of its wage earners being blue-collar workers in shipyards and aircraft plants, and the rest at work in lesser white-collar jobs). From the start it became plain that many residents—especially an idealistic subset of them concerned to develop a truly working-class community—were strongly oriented toward the future of this new community and of their own place in it.

Members of the community were classified in one of two broad categories—as relative "permanents" or as relative "transients"—based on analysis of their detailed responses to what seems, in long retrospect, a decidedly fragile indicator-question: "How long do you expect to stay in Craftown?" Yet this seemingly slight question about expected duration of residence evidently tapped a significant social and psychological reality. Craftowners who expected "to settle down and stay permanently" differed from the transients in a range of social behaviors: in the extent and intensity of their participation in local social life, in interpersonal relationships (close friendships), membership in local organizations, and in depth of community morale. More in point for our conceptual odyssey, it was found that the *expected* duration of residence worked independently of the *actual* duration to affect these behaviors.

Further evidence that the apparently fragile indicator-question about expected duration of stay was fairly robust appeared some twenty years later in a limited follow-up study of Craftown. Philip Sidel found that 61 percent of the Craftowners who had long before announced their intention of settling down were still there, in comparison with 30 percent of those who had intended to leave at some contingent time. Moreover, the minority of permanently oriented residents who *had* left dur-

3. Robert K. Merton, Marjorie Fiske, and Alberta Curtis, *Mass Persuasion: The Social Psychology of a War-Bond Drive* (New York: Harper and Brothers, 1946).

4. Robert K. Merton, Patricia S. West, and Marie Jahoda, *Patterns of Social Life: Explorations in the Sociology of Housing* (1948; reprint, New York: Columbia University Bureau of Applied Social Research, 1951), mimeographed.

ing this two-decade interval had stayed for a longer period than the more transiently minded who had declared their intention to stay only "as long as their jobs lasted, or until they found a better place, or until they owned their own home."[5]

The follow-up study held a certain degree of methodological interest in checking up on the adequacy of that primitive indicator-question about expected durations of residence. Apart from such matters of research procedure, the Craftown study proved to be something of a watershed in the conceptual evolution of SED. To begin with, it distinguished sharply between expected durations and actual durations, together with their separate and joint correlates. Second, it treated individually and collectively expected durations as strategic, consequential variables that interacted with their behavioral and social consequences for the developing local community. Put in another vocabulary, it adopted the hypothesis that individual and collective expected durations operated as both independent and dependent variables. They affected and in turn were affected by modes of participation in the organized social life and the interpersonal relations in the community. By this point the notion of individual and family expected durations was becoming somewhat less a tacit proto-concept and somewhat more a fledgling concept, beginning to be explicated and variously employed in empirical research. But it still focused on individual expected durations without explicitly attending to structurally defined durations as a major type of SED. That was to come later, when the focus of attention shifted to identifying significant properties of social structure. [. . .]

SED as Structural Property

To this point, the social researches that deployed a tacit or explicit notion of expected durations had focused on individual and collective expectations. These, to be sure, were seen as variously shaped by social and other contextual influences and, to that degree, could be construed as socially caused. But these expected durations were not structural in character, in the strict sense of being socially prescribed.

It was only in the latter 1950s, when I embarked on an effort to identify major properties of social structure rather than engage in the empirical study of a particular sociological problem, that I found myself centering on socially expected durations as a generic concept. It was proposed that SEDs could be thought of as a consequential property of social structures of every kind: of groups and organizations, of

5. Philip S. Sidel, "Residential Mobility in Craftown" (master's thesis, Columbia University, 1966).

social statuses and role-sets, and of social relationships.[6] Such structurally defined durations were not expectations that individuals or aggregates of individuals *happened* to have; they were, rather, normatively prescribed and, in varying degree, authoritatively enforced durations. Thus, to take a conspicuous kind of structural case, it was noted that organizations often provide for "a fixed term of duration" for occupancy of certain positions in it, while leaving other positions with unspecified and therefore open-ended durations.

It was further noted that socially prescribed and enforced durations were also structural properties of groups and organizations as a whole, not only properties of memberships and statuses in them. Exceedingly condensed statements alluded to such tacit parameters of SEDs as "fixed duration" and "unlimited duration" and to certain kinds of behavioral and structural consequences of differing kinds of SEDs. More specifically, it was said,

> The actual duration of a group should be distinguished from patterned expectations of [its] probable duration . . . : whether it is an association established "temporarily" to meet a need which, once met, involves self-liquidation or whether it is established with the expectation of unlimited duration for the indefinitely prolonged future. Variations in expected duration [will] presumably affect the self-selection of members, the kind and degree of involvement of members, the internal structure of the organization, its power, and other properties still to be considered.[7]

With this proposal that socially expected durations constitute a structural property that relates both to individual behaviors and to other structural properties and processes, some of the principal ingredients of SED were there for transformation from a particularized proto-concept into a generalized, analytically developed concept that consolidates first our recognition and then our systematic utilization in research of this temporal component of diverse kinds of social structure. But these theoretical ingredients were there only as loose conceptual particles in suspension, still needing to be systematically articulated in terms of methodically identified parameters and an explicit sociological problematics.

The change of SED from proto-concept to a (still largely unelucidated) concept came about, it seems, in accord with a corollary of the Burke theorem. Just as "a way of seeing is also a way of not seeing—a focus on object A involves a neglect of object B," so a changed angle of vision entails a new way of seeing that brings part of the formerly ne-

6. Robert K. Merton, *Social Theory and Social Structure,* 365–66.
7. Ibid., 366.

glected observation into focus. By the mid-1950s my own research program, as distinct from that of the local thought collective of which I was a member, had shifted from a monographic focus on particular sets of empirical sociological questions to a renewed focus on problems in structural sociology. This change in intellectual objectives led to a renewed emphasis on identifying uncodified elements of social structure, such as role-sets, status-sets, and status-sequences, and various properties of social structure, such as observability, degrees of tolerated nonconformity, and expected durations.[8]

In long retrospect, the emergence of an explicit notion of SED seems to be yet another instance of a pattern of intellectual development in which

> sustained emphasis on one range of problems evokes, after a time, corrective emphasis upon inadvertently neglected problems. On occasion, this calls for the revision of "analytical models" that have led sociologists to concentrate on a restricted range of problems at the expense of other problems that the model neglects. . . . Periodically, investigation of a range of problems is found to have gone about as far as it can with the use of existing concepts. Useful for a time, the concepts now prove to be insufficiently differentiated, thus typically introducing the problem of devising appropriate classifications.[9]

What intuitively seems an inordinately long time for the emergence of an explicit, albeit still much to be developed, concept of SED—"seems" since there are, of course, no standards for gauging rates of conceptual development—may in part have resulted from the very ubiquity of the aspects of social reality that we now know as SEDs. As we have seen, we experience socially expected durations in every department of social life and in most varied forms. [. . .] That ubiquity of phenomenal SEDs may lead them to blend, conceptually unnoticed, into the taken-for-granted social background rather than to be differentiated into a possibly illuminating concept directing us to their underlying similarities. As Wittgenstein once observed with italicized feeling: "How hard I find it to see what *is right in front of my eyes!*"[10]

That this excessively compact identification of expected durations as a sociostructural property failed to attract the notice of the sociological community presents no puzzle. Those two dense paragraphs on actual

8. Ibid., 364–440.

9. Robert K. Merton, "Notes on Problem-Finding in Sociology," in *Sociology Today,* ed. Robert K. Merton, Leonard Broom, and Leonard S. Cottrell (New York: Basic Books, 1959), xxx–xxxi.

10. Ludwig Wittgenstein, *Culture and Value,* ed. G. H. von Wright (1977; reprint, Chicago: University of Chicago Press, 1980), 39.

and expected durations failed to meet basic criteria of consequential problem finding: (1) They did not formulate the originating questions indicating what one wants to know; (2) they failed to state the rationale for the concept; and (3) they failed to show how the concept helped specify our ignorance about the workings of this proposed temporal property of social structure. [. . .]

The Concept and Problematics of SED

Along with *individual* expectations of durations, we have identified three major kinds of SEDs:

1. Structural or institutionalized durations, being formally prescribed and supported by the authority and power of the structures in which they are imbedded, are highly visible and confidently anticipatable. This familiar type appears in such structures as prisons, schools, and armies in the forms of determinate sentences, educational requirements for students and tenure for faculty, and term enlistments. The visibility and relative predictability of durations of this type shape current anticipatory behavior of members of the organizations and of nonmembers oriented to these organizations.

2. Unlike socially prescribed durations, collectively expected durations are relatively *un*certain, as we found in the war-bond drive. Such expected durations enter into millenarian movements, crazes, panics, and the many other forms of collective behavior[11] to affect the behavior of participants in these events as well as the behavior of those socially connected with them through their status-sets and role-sets. Even when formal organizations engage in collective phenomena, as in the case of war and the periods between wars, their durations remain far less predictable[12] than structurally prescribed durations. That is why, I suppose, we find rather improbable the story which has that gallant young man of the seventeenth century bravely announcing, "I'm off to fight in the Thirty Years' War."

3. Social life in its concrete complexity involves a third type of SED along with socially prescribed and collectively expected durations. These are the patterned temporal expectations found in the various kinds of interpersonal and social relations. Thus, predetermined and therefore relatively predictable durations are functionally appropriate to *gesellschaftliche* relations of a strictly contractual character, as with

11. Neil Smelser, *The Theory of Collective Behavior* (Glencoe, Ill.: Free Press, 1963).

12. Samuel Stouffer et al., *The American Soldier,* 2 vols. (Princeton: Princeton University Press, 1949).

business transactions, just as precisely defined durations are at odds with *gemeinschaftliche* relations, such as friendship and marriage.

The further analytical and interpretative questions about SEDs are legion. How do they vary in structure, operation, and consequences? What, in particular, are the attributes and parameters of SEDs—parameters such as socially expected durations relatively bounded or unbounded, flexible or fixed, relatively long or short, relatively certain or variously probable, determinate or indeterminate, valued or disvalued; durations symmetrically or asymmetrically expected by the parties to them, self-imposed or socially imposed, and so with other attributes that have been identified among the great array of substantively differing SEDs? Which formal attributes such as these combine to generate the conceptual types of SEDs, each of them populated by empirical specimens? Further, what are the *anticipatory* behavioral adaptations of individuals, status-occupants, groups, organizations, and collectivities to socially expected durations of the several kinds? In short, how do SEDs come to take certain forms rather than others; what are the parameters of the several forms; and finally, how do they operate as temporal components of social structure to affect patterns of organizational and individual behavior? Plainly, there is still much theoretical and empirical work to be done. [. . .]

Paradoxes of Social Process

15

The Unanticipated Consequences of Social Action (1936)

[. . .] Although the phrase "unanticipated consequences of purposive social action" is in a measure self-explanatory, the setting of the problem demands further specification. In the first place, the greater part of this paper deals with isolated purposive acts rather than with their integration into a coherent system of action (though some reference will be made to the latter). This limitation is prescribed by expediency, for a treatment of systems of action would introduce further complications. Furthermore, *unforeseen* consequences should not be identified with consequences which are necessarily undesirable (from the standpoint of the actor). For though these results are unintended, they are not upon their occurrence always deemed axiologically negative. In short, undesired effects are not always undesirable effects. The intended and anticipated outcomes of purposive action, however, are always, in the very nature of the case, *relatively* desirable to the actor, though they may seem axiologically negative to an outside observer. This is true even in the polar instance where the intended result is "the lesser of two evils" or in such cases as suicide, ascetic mortification and self-torture which, in given situations, are deemed desirable relative to other possible alternatives.

Rigorously speaking, the *consequences* of purposive action are limited to those elements in the resulting situation that are exclusively the outcome of the action, i.e., those elements that would not have occurred had the action not taken place. Concretely, however, the consequences result from the interplay of the action and the objective situation, the conditions of action.[1] We will be primarily concerned with

From "The Unanticipated Consequences of Purposive Social Action," in *American Sociological Review* 1 (1936): 894–96, 898–904. Reprinted by permission of the American Sociological Association.

1. Cf. Frank H. Knight, *Risk, Uncertainty and Profit* (Boston and New York: Houghton Mifflin Co., 1921), 201–2. Professor Knight's doctoral dissertation represents by far the most searching treatment of certain phases of this problem that I have yet seen.

the sum total results of action under certain conditions. This still involves the problem of causal imputation (of which more later), though to a less pressing degree than consequences in the rigorous sense. These sum-total or concrete consequences may be differentiated into (a) consequences to the actor(s), (b) consequences to other persons mediated through (1) the social structure, (2) the culture, and (3) the civilization.[2]

In considering *purposive* action, we are concerned with "conduct" as distinct from "behavior," that is, with action which involves motives and consequently a choice between various alternatives.[3] For the time being, we shall take purposes as given, so that any theories which reduce purpose to conditioned reflexes or tropisms, which assert that motives are simply compounded of instinctual drives and the experiential shaping of these drives, may be considered as irrelevant. Psychological considerations of the source or origin of motives, though they are undoubtedly important for a more complete understanding of the mechanisms involved in the development of unexpected consequences of conduct, will thus be ignored.

Moreover, it is not assumed that in fact social action always involves clear-cut, explicit purpose. It may well be that such awareness of purpose is unusual, that the aim of action is more often than not nebulous and hazy. This is certainly the case with habitual action which, though it may originally have been inducted by conscious purpose, is characteristically performed without such awareness. The significance of such habitual action will be discussed later.

Above all, it must not be inferred that purposive action implies rationality of human action, that persons always use the objectively most adequate means for the attainment of their end.[4] In fact, part of the present analysis is devoted to the determination of those elements that account for concrete deviations from rationality of action. Moreover, rationality and irrationality are not to be identified with the success and failure of action, respectively. For in a situation where the number of *possible* actions for attaining a given end is severely limited, one acts rationally by selecting the means which, on the basis of the available evidence, has the greatest probability of attaining this goal and yet the

2. For the distinction between society, culture, and civilization, see Alfred Weber, "Prinzipielles zur Kultursoziologie: Gesellschaftsprozess, Civilisationsprozess und Kulturbewegung," *Archiv für Sozialwissenschaft und Sozialpolitik* 47 (1920): 1–49; Robert K. Merton, "Civilization and Culture," *Sociology and Social Research* 21 (1936): 103–13.

3. Knight, *Risk, Uncertainty, and Profit,* 52.

4. Max Weber, *Wirtschaft und Gesellschaft* (Tübingen: J. C. B. Mohr, 1925), 3 ff.

goal may actually *not* be attained.[5] Contrariwise, an end may be attained by action which, on the basis of the knowledge available to the actor, is nonrational (as in the case of "hunches").

Turning now to *action,* we may differentiate this into two types: (a) unorganized and (b) formally organized. The first refers to actions of individuals considered distributively out of which grow the second when like-minded individuals form an association in order to achieve a common purpose. Unanticipated consequences may, of course, follow both types of action, though the second type would seem to afford a better opportunity for sociological analysis since the very process of formal organization ordinarily involves an explicit statement of purpose and procedure. [. . .]

Sources of Unanticipated Consequences

The most obvious limitation to a correct anticipation of consequences of action is provided by the existing state of knowledge. The extent of this limitation may be best appreciated by assuming the simplest case where this lack of adequate knowledge is the *sole* barrier to a correct anticipation.[6] Obviously, a large number of concrete reasons for inadequate knowledge may be found, but it is also possible to summarize several classes of factors that are most important.

The first class derives from the type of knowledge—usually, perhaps exclusively—attained in the sciences of human behavior. Properly speaking, the social scientist almost invariably finds stochastic associations and not, as in most fields of the physical sciences, functional associations.[7] This is to say, in the study of human behavior, there is

5. See Joseph Bertrand, *Calcul des probabilités* (Paris: Gautier, 1889), 90 ff.; J. M. Keynes, *A Treatise on Probability* (London: Macmillan Co., 1921), chap. 26.

6. Most previous discussions limit the explanation of unanticipated consequences to this one factor of ignorance. Such a view either reduces itself to a tautology or exaggerates the role of but one of many factors. In the first instance, the argument runs in this fashion: "if we had only known enough, we could have anticipated the consequences which, as it happens, were unforeseen." The apparent fallacy of this *ex post* argument rests in the word "enough," which is implicitly taken to mean "enough knowledge to foresee" the consequences of our action. It is then no difficult matter to uphold the contention which then reads in effect: "if we had known, we would have known." This viewpoint is basic to several schools of educational theory, just as it was to Comte's dictum, *savoir pour prevoir, prevoir pour pouvoir.* This intellectualist stand has gained credence partly because of its implicit optimism and because of the indubitable fact that ignorance does actually account for the occurrence of unforeseen consequences *in some cases.*

7. Cf. A. A. Tschuprow, *Grundbegriffe und Grundprobleme der Korrelationstheorie* (Leipzig: B. G. Teubnner, 1925), 20 ff., where he introduces the term "stochastic." It is apparent, of course, that stochastic associations are obtained because we have not ascer-

found a set of different values of one variable associated with each value of the other variable(s), or in less formal language, the set of consequences of any repeated act is not constant but there is a range of consequences, *any one of which may follow the act in any given case.* In some instances, we may have sufficient knowledge of the limits of the range of possible consequences, and even adequate knowledge for ascertaining the statistical (empirical) probabilities of the various possible sets of consequences, but it is impossible to predict with certainty the results in any particular case. Our classifications of acts and situations never involve completely homogeneous categories or even categories whose approximate degree of homogeneity is sufficient for the prediction of particular events.[8] We have here the paradox that whereas past experience is the sole guide to our expectations on the assumption that certain past, present, and future acts are sufficiently alike to be grouped in the same category, these experiences are in fact different. To the extent that these differences are pertinent to the outcome of the action and appropriate corrections for these differences are not adopted, the actual results will differ from the expected. As Poincaré has put it, "small differences in the initial conditions produce very great ones in the final phenomena; a slight error in the former would produce an enormous error in the latter. Prediction becomes impossible, and we have the fortuitous phenomenon."[9]

However, deviations from the usual consequences of an act may be anticipated by the actor who recognizes in the given situation some differences from previous similar situations. But, insofar as these differences can themselves not be subsumed under general rules, the direction and extent of these deviations cannot be anticipated.[10] It is clear, then, that the partial knowledge in the light of which action is commonly carried on permits a varying range of unexpected outcomes of conduct.

---

tained, or having ascertained, have not controlled the other variables in the situation which influence the final result. Thus, stochastic associations are not inherent in social knowledge but derive from our lack of experimental control.

8. A classification into completely homogeneous categories would, of course, lead to functional associations and would permit successful prediction, but the aspects of social action which are of practical importance are too varied and numerous to permit such homogeneous classification.

9. Henri Poincaré, *Calcul des probabilités* (Paris: G. Carré, 1912), 2.

10. The actor's awareness of his ignorance and its implications is perhaps most acute in the type of conduct which Thomas and Znaniecki attribute to the wish for "new experience." This is the case where unforeseen consequences actually constitute the purpose of action, but there is always the tacit assumption that these consequences will be desirable.

Although no formula for the exact *amount* of knowledge necessary for foreknowledge is presented, one may say in general that consequences are fortuitous when an exact knowledge of many details and facts (as distinct from general principles) is needed for even a highly approximate prediction. In other words, "chance consequences" are those which are occasioned by the interplay of forces and circumstances which are so complex and numerous that prediction of them is quite beyond our reach. This area of consequences should perhaps be distinguished from that of "ignorance," since it is related not to the knowledge actually in hand but to knowledge which can conceivably be obtained.[11]

The importance of ignorance as a factor is enhanced by the fact that the exigencies of practical life frequently compel us to act with some confidence even though it is manifest that the information on which we base our action is not complete. We usually act, as Knight has properly observed, not on the basis of scientific knowledge, but opinion and estimate. Thus, situations that demand (or what is for our purposes tantamount to the same thing, appear to the actor to demand) immediate action of some sort will usually involve ignorance of certain aspects of the situation and will bring about unexpected results.

Moreover, even when immediate action is not exacted, there is the *economic* problem of distributing our fundamental resources, time and energy. Time and energy are scarce means and economic behavior is concerned with the rational allocation of these means among alternative wants, only one of which is the anticipation of consequences of action.[12] It is manifestly uneconomic behavior to concern ourselves with attempts to obtain knowledge for predicting the outcomes of action to such an extent that we have practically no time or energy for other pursuits. An economy of social engineers is no more conceivable or practicable then an economy of laundrymen. It is the fault of the extreme antinoetic activists who promote the idea of action above all

11. Cf. Keynes, *Treatise on Probability,* 295. This distinction corresponds to that made by Keynes between "subjective chance" (broadly, ignorance) and "objective chance" (where even additional wide knowledge of general principles would not suffice to foresee the consequences of a particular act). Much the same distinction appears in the works of Poincaré and Venn, among others.

12. Cf. Knight, *Risk, Uncertainty, and Profit,* 348. The reasoning is also applicable to cases where the occupation of certain individuals (e.g., social engineers and scientists) is devoted solely to such efforts, since then it is simply a question of the distribution of the resources of society. Furthermore, there is the practical problem of the communicability of knowledge so obtained, since it may be of a very complex order and the effort of persons other than social engineers to assimilate such knowledge leads us back to the same problem of distribution of our resources.

else to exaggerate this limit and to claim (in effect) that virtually no time or energy be devoted to the acquisition of knowledge. On the other hand, the grain of truth in the anti-intellectualist position is, as was just observed, that there are not only economic limits to the advisability of not acting until all or as much as possible uncertainty is eliminated, but also psychological limits since excessive "forethought" of this kind precludes any action at all.

A second major factor of unexpected consequences of conduct, which is perhaps as pervasive as ignorance, is error. Error may intrude itself, of course, in any phase of purposive action: we may err in our appraisal of the present situation, in our inference from this to the future objective situation, in our selection of a course of action, or finally in the execution of the action chosen. A common fallacy is frequently involved in the too-ready assumption that actions which have in the past led to the desired outcome will continue to do so. This assumption is often fixed in the mechanism of habit and it there finds pragmatic justification, for habitual action does in fact often, even usually, meet with success. But precisely because habit is a mode of activity which has previously led to the attainment of certain ends, it tends to become automatic and undeliberative through continued repetition so that the actor fails to recognize that procedures which have been successful *in certain circumstances* need not be so *under any and all conditions.*[13] Just as rigidities in social organization often balk and block the satisfaction of new wants, so rigidities in individual behavior may block the satisfaction of old wants in a changing social environment.

Error may also be involved in instances where the actor attends to only one or some of the pertinent aspects of the situation which influence the outcome of the action. This may range from the case of simple neglect (lack of systematic thoroughness in examining the situation) to pathological obsession where there is a determined refusal or inability to consider certain elements of the problem. This last type has been extensively dealt with in the psychiatric literature. In cases of wish fulfillment, emotional involvements lead to a distortion of the objective situation and of the probable future course of events; such action predicated upon "imaginary" conditions must inevitably evoke unexpected consequences.

The third general type of factor, the "imperious immediacy of interest," refers to instances where the actor's paramount concern with the

13. Similar fallacies in the field of thought have been variously designated as "the philosophical fallacy" (Dewey), the "principle of limits" (Sorokin, Bridgman) and, with a somewhat different emphasis, "the fallacy of misplaced concreteness" (Whitehead).

foreseen immediate consequences excludes the consideration of further other consequences of the same act. The most prominent elements in such immediacy of interest may range from physiological needs to basic cultural values. Thus, Vico's imaginative example of the "origin of the family," which derived from the practice of men carrying their mates into caves to satisfy their sex drive out of the sight of God, might serve as a somewhat fantastic illustration of the first. The doctrine of classical economics according to which the individual endeavoring to employ his capital where most profitable to him and thus tending to render the annual revenue of society as great as possible is, to quote Adam Smith, led "by an invisible hand to promote an end which was no part of his intention," may serve as an example of economic interest leading to this sequence.

However, after the acute analysis by Max Weber, it goes without saying that action motivated by interest is not antithetical to an exhaustive investigation of the conditions and means of successful action. On the contrary, it would seem that interest, if it is to be satisfied, demands such objective analysis of situation and instrumentality, as is assumed to be characteristic of *hominis oeconomici.* But it is equally undeniable that intense interest does in fact often tend to preclude such analysis precisely because strong concern with the satisfaction of the immediate interest is a psychological generator of emotional bias, with consequent lopsidedness or failure to engage in the required calculations. It is as much a fallacious assumption to hold that interested action in fact necessarily entails a rational calculation of elements in the situation[14] as to deny rationality any and all influence over such conduct. Moreover, action in which this element of immediacy of interest is involved may be rational in terms of the values basic to that interest but irrational in terms of the life organization of the individual. Rational, in the sense that it is an action which may be expected to lead to the attainment of the specific goal; irrational, in the sense that it may defeat the pursuit or attainment of other values which are not, at the moment, paramount but which nonetheless form an integral part of the individual's scale of values. Thus, precisely because a particular action is not carried out in a psychological or social vacuum, its effects will ramify into other spheres of value and interest. For example, the practice of birth control for economic reasons influences the

14. This assumption is tenable only in a normative sense. It is indubitable that such calculation, within the limits specified in our previous discussion, should be made if the probability of satisfying the interest is to be at a maximum. The error lies in confusing norm with actuality.

age-composition and size of sibships with profound consequences of a psychological and social character.

Superficially similar to the factor of immediacy of interest, but differing from it in a highly significant theoretical sense, is that of basic values. This refers to instances where there is no consideration of further consequences because of the felt necessity of certain action enjoined by certain fundamental values. The classical analysis of the influence of this factor is Weber's study of the Protestant ethic and the spirit of capitalism. He has properly generalized this case, saying that active asceticism paradoxically leads to its own decline through the accumulation of wealth and possessions entailed by decreased consumption and intense productive activity.

The Self-Defeating Prophecy

This process contributes much to the dynamic of social and cultural change, as has been recognized with varying degrees of cogency by Hegel, Marx, Wundt, and others. The empirical observation is incontestable: activities oriented toward certain values release processes which so react as to change the very scale of values that precipitated them. These processes may in part be due to the fact that when a system of basic values enjoins certain *specific* actions, adherents are not concerned with the objective consequences of these actions but only with the subjective satisfaction of duty well performed. Or, action in accordance with a dominant set of values tends to be focussed upon that particular value-area. But with the complex interaction that constitutes society, action ramifies, its consequences are not restricted to the specific area in which they were initially intended to center, they occur in interrelated fields explicitly ignored at the time of the action. Yet it is because these fields are in fact interrelated that the further consequences in adjacent areas tend to *react* upon the fundamental value-system. It is this usually unlooked-for reaction which constitutes a most important element in the process of secularization, of the transformation or breakdown of basic value-systems. Here is the essential paradox of social action—the "realization" of values may lead to their renunciation. We may paraphrase Goethe and speak of "Die Kraft, die stets das Gute will, und stets das Böse schafft."

There is one other circumstance, peculiar to human conduct, that stands in the way of successful social prediction and planning. Public predictions of future social developments are frequently not sustained precisely because the prediction has become a new element in the concrete situation, thus tending to change the initial course of developments. This is not true of prediction in fields which do not pertain to human conduct. Thus, the prediction of the return of Halley's comet does not in any way influence the orbit of that comet; but, to take a concrete social example, Marx's prediction of the progressive concentration of wealth and increasing misery of the masses did influence the very process predicted. For at least one of the consequences of socialist preaching in the nineteenth century was the spread of organization of labor, which, made conscious of its unfavorable bargaining position in cases of individual contract, organized to enjoy the advantages of collective bargaining, thus slowing up, if not eliminating, the developments which Marx had predicted.[15]

Thus, to the extent that the predictions of social scientists are made public and action proceeds with full cognizance of these predictions, the "other-things-being-equal" condition tacitly assumed in all forecasting is not fulfilled. Other things will not be equal just because the scientists have introduced a new "other thing"—their predictions.[16] This contingency may often account for social movements developing in utterly unanticipated directions and it hence assumes considerable importance for social planning.

The foregoing discussion represents no more than the briefest exposition of the major elements involved in one fundamental social process. It would take us too far afield, and certainly beyond the compass of this paper, to examine exhaustively the implications of this analysis for social prediction, control and planning. We may maintain, however, even at this preliminary juncture, that no blanket statement categorically affirming or denying the practical feasibility of *all* social planning is warranted. Before we may indulge in such generalizations, we must examine and classify the *types* of social action and organization with reference to the elements here discussed and then refer our gener-

15. Corrado Gini, *Prime linee di patologia economica* (Milan: A. Giuffré, 1935), 72–75. John Venn uses the picturesque term "suicidal prophecies" to refer to this process and properly observes that it represents a class of considerations which have been much neglected by the various sciences of human conduct. See his *Logic of Chance* (London: Macmillan, 1888), 225–26.

16. For the correlative process, see chapter 16 in this volume, "The Self-Fulfilling Prophecy," which was first published a dozen years after the present chapter. For an account of later extensions of the sociological theme of unintended consequences, see Robert K. Merton, "Unanticipated Consequences and Kindred Sociological Ideas: A Personal Gloss," in *L'Opera di R. K. Merton e la sociologia contemporanea,* ed. Carlo Mongardini and Simonetta Tabboni (Genoa: Edizioni Culturali Internazionali, 1989), 307–29–*Ed.*

alizations to these essentially different types. If the present analysis has served to set the problem, if only in its most paramount aspects, and to direct attention toward the need for a systematic and objective study of the elements involved in the development of unanticipated consequences of purposive social action, the treatment of which has for much too long been consigned to the realm of theology and speculative philosophy, then it has achieved its avowed purpose.

16

The Self-Fulfilling Prophecy (1948)

It was W. I. Thomas, then the dean of American sociologists, who set forth a theorem basic to the social sciences: "If men define situations as real, they are real in their consequences."[1] Although that pregnant theorem appeared only twice in Thomas's voluminous writings and although it was each time elucidated in only a single paragraph, it has claimed the attention of many, for it possesses the gift of deep sociological relevance, being instructively applicable to a wide range of social processes.

There is all the more reason to assume that Thomas had hold of a crucial idea when we note that essentially the same theorem had been aperiodically stated by observant and disciplined minds long before. When we find such otherwise discrepant minds as the redoubtable Bishop Bossuet in his passionate seventeenth-century defense of Catholic orthodoxy; the ironic Mandeville in his eighteenth-century allegory, *The Fable of the Bees,* honeycombed with observations on the paradoxes of human society; the irascible genius of social science Marx in his revision of Hegel's theory of historical change; the seminal Freud in works which have perhaps gone further than any others of his day toward modifying our outlook on the psychology of men and women;

From "The Self-Fulfilling Prophecy," in *Social Research and the Practicing Professions,* ed. Aaron Rosenblatt and Thomas F. Gieryn (Cambridge, Mass.: Abt Books, 1982), 248–67. Reprinted by permission of Abt Books.

1. What can be described as the Thomas theorem appears on page 572 of the book he wrote with Dorothy Swaine Thomas in 1928: *The Child in America* (New York: Knopf). I ascribe the theorem to W. I. Thomas alone rather than to the Thomases jointly because Dorothy, who became Dorothy Thomas Thomas when they were married eight years after that book appeared, confirmed that the consequential sentence and the paragraph in which it was encased were written by him. Thus, nothing in this attribution smacks of "the Matthew effect," which operates in cases of collaboration between scholars of decidedly unequal reputation to ascribe credit to the eminent scholars and little or none to the collaborator(s).—Supplementary note, 1982 (See also Robert K. Merton, "The Thomas Theorem and the Matthew Effect" *Social Forces* 74 [1995], 379–422). —*Ed.*

and the erudite, dogmatic, and often sound Yale professor, William Graham Sumner, who lives on as spokesman for the middle classes—when we find this mixed company (and I select from a longer if less celebrated list) agreeing on the truth and the pertinence of what is substantially the Thomas theorem, we may conclude that perhaps it is worth our attention as well.

To what, then, are Thomas and Bossuet, Mandeville, Marx, Freud, and Sumner directing our attention?

The first part of the theorem provides an unceasing reminder that we respond not only to the objective features of a situation, but also, and at times primarily, to the meaning this situation has for us. And once we have assigned some meaning to the situation, our consequent behavior and some of the consequences of that behavior are determined by the ascribed meaning. But this is rather abstract, and sociological abstractions have a way of becoming unintelligible if they are not occasionally tied to concrete data. What is a case in point?

A Sociological Parable

It is the year 1932. The Last National Bank is a flourishing institution. A large part of its resources is liquid without being watered. Cartwright Millingville has ample reason to be proud of the banking institution over which he presides. Until Black Wednesday. As he enters his bank, he notices that business is unusually brisk. A little odd, that, since the workers at the A.M.O.K. steel plant and the K.O.M.A. mattress factory usually are not paid until Saturday. Yet here are two dozen workers, obviously from the factories, queued up in front of the tellers' cages. As he turns into his private office, Millingville muses rather compassionately: "Hope they haven't been laid off in midweek. They should be in the shop at this hour."

But speculations of this sort have never made for a thriving bank, and Millingville turns to the pile of documents upon his desk. His precise signature is affixed to fewer than a score of papers when he is disturbed by the absence of something familiar and the intrusion of something alien. The low discreet hum of bank business has given way to a strange and annoying stridency of many voices. A situation has been defined as real. And that is the beginning of what ends as Black Wednesday—the last Wednesday, it may be noted, of the Last National Bank.

Cartwright Millingville had never heard of the Thomas theorem. But he had no difficulty in the recognizing a special case of its workings. He knew that, despite the comparative liquidity of the bank's assets, a rumor of insolvency, once believed by enough depositors, would result in the insolvency of the bank. And by the close of Black Wednesday—and Blacker Thursday—when the long lines of anxious depositors, frantically seeking to salvage their own, grew to longer lines of even more anxious depositors, it turned out that he was right.

The stable financial structure of the bank had depended upon one set of definitions of the situation: belief in the validity of the interlocking system of economic promises people live by. Once depositors defined the situation otherwise, once they questioned the possibility of having these promises fulfilled, the consequences of this unreal definition were real enough.

A familiar case this, and one doesn't need the Thomas theorem to understand how it happened—not, at least, if one is old enough to have voted for Franklin Roosevelt in 1932. But perhaps with the aid of the theorem the tragic history of Millingville's bank can be converted into a sociological parable which may help us understand not only what happened to hundreds of banks in the 1930s, but also what still happens to relations between blacks and whites, between Protestants and Catholics and Jews in these days.

Varieties of Self-Fulfilling Prophecy

The parable tells us that certain kinds of definitions of a situation—we focus on the important class of *public* prophecies, beliefs, and expectations—become an integral part of the situation and thus affect subsequent developments. This is peculiar to human affairs. It is not found in the world of nature, untouched by human hands. Predictions of the return of Halley's comet do not influence its orbit. But the rumored insolvency of Millingville's bank did affect the actual outcome. The prophecy of collapse led to its fulfillment.

The self-fulfilling prophecy is, in the beginning, a *false* definition of the situation evoking behavior that makes the originally false conception come true. The specious validity of the self-fulfilling prophecy perpetuates a reign of error. For the prophet will cite the actual course of events as proof that he was right from the beginning. (Yet we know that Millingville's bank was solvent, that it would have survived for years had not the misleading rumor *created* the conditions of its own fulfillment.) Such are the perversities of social logic.[2]

2. The counterpart of the self-fulfilling prophecy is the self-defeating or "suicidal" prophecy which so alters human behavior from what would have been its course had it not been made, that it fails to be borne out. So it is that in occasional elections where large numbers of constituents come to believe that their candidate is bound to win, sizable

The self-fulfilling prophecy works its ways in every sphere of human experience, *social or public and individual or private*. The *social type* involves prophecy-dominated interactions between individuals and between collectivities. Thus, the political leaders of hostile nations become persuaded that war between them is inevitable. Actuated by this conviction, they become successively more alienated from one another, apprehensively countering each "offensive" move of the other with a "defensive" move of their own. Stockpiles of armaments, raw materials, and armed men and women grow ever larger, and in due course the anticipation of war helps create the actuality. The examination neurosis provides a familiar case of the *individual or private type* of self-fulfilling prophecy. Groundlessly convinced that he is destined to fail, the anxious student devotes more time and energy to worry than to study and then, not altogether surprisingly, turns in a poor examination. The initially fallacious expectation makes for a seemingly confirming outcome.

It is the social or public self-fulfilling prophecy that goes far toward explaining the dynamics of ethnic and racial conflict in the America of today. That this is the case, at least for relations between blacks and whites, may be gathered from the fifteen hundred pages which make up Gunnar Myrdal's *An American Dilemma*.[3] That the self-fulfilling prophecy has even more general bearing upon the relations between ethnic groups than Myrdal has indicated is the thesis of the considerably briefer discussion that follows.

Social Beliefs and Social Reality

As a result of their failure to comprehend the operation of the self-fulfilling prophecy, many Americans of goodwill (sometimes reluctantly) retain enduring ethnic and racial prejudices. They experience these beliefs not as prejudices, not as prejudgments, but as irresistible products of personal observation. "The facts of the case" permit them no other conclusion.

Thus, fair-minded white citizens strongly support a policy of excluding blacks from their labor unions. Their views are, of course, based not upon prejudice, but upon the cold, hard facts. And the facts seem

---

fractions fail to vote and the prophecy destroys itself. This important class of consequential prophecies is not considered here.

On the suicidal prophecy, see chapter 15 in this volume, "The Unanticipated Consequences of Social Action," especially the final selection on the "Self-Defeating Prophecy."

3. Gunnar Myrdal, *An American Dilemma: The Negro Problem and Modern Democracy* (New York: Harper and Row, 1944).

clear enough. Blacks, "lately from the nonindustrial South, are undisciplined in traditions of trade unionism and the art of collective bargaining." Black workers are strikebreakers. With their "low standard of living," the blacks rush in to take jobs at less than prevailing wages. The black workers are, in short, "traitors to the working class," and should manifestly be excluded from union organizations.[4] So run the facts of the case as seen by our tolerant but hard-headed union members, innocent of any understanding of the self-fulfilling prophecy as a basic process of society.

Our unionists fail to see, of course, that they have produced the very "facts" which they sometimes observe. For by defining the situation as one in which blacks are held to be incorrigibly at odds with principles of unionism and by excluding them from unions, the unionists then invite a series of consequences which indeed make it difficult if not impossible for many blacks to avoid the role of scab. Out of work after World War I, and kept out of unions, thousands of blacks could not resist strikebound employers who invitingly held a door open upon a world of jobs from which they were otherwise excluded.

History creates its own test of the theory of self-fulfilling prophecies. That some black workers were strikebreakers because they were excluded from unions (and from a wide range of jobs) rather than excluded because they were strikebreakers was shown in the last decades by their virtual disappearance as scabs in industries where they gained admission to unions.

The application of the Thomas theorem also suggests how the tragic, often vicious, circle of self-fulfilling prophecies can be broken. The initial definition of the situation which has set the circle in motion must be abandoned. Only when the originating assumption is questioned and a new definition of the situation introduced, does the consequent flow of events give the lie to the assumption. Only then does the belief no longer father the reality.

But to question these deep-rooted definitions of the situation is no simple act of the will. The will, or for that matter, goodwill, cannot be turned on and off like a faucet. Social intelligence and goodwill are themselves *products* of distinct social forces. They are not brought into being by mass propaganda and mass education, in the usual sense of these terms so dear to the sociological panaceans. No more in the social realm than in the psychological realm do false ideas quietly vanish when confronted with the truth. No one expects a paranoiac to aban-

4. This stands as written more than a generation ago. The far greater extent to which blacks now take part in unions is one measure of social change in this sphere.—*Ed.*

don hard-won distortions and delusions upon being informed that they are altogether groundless. If psychic ills could be cured merely by the dissemination of truth, psychiatrists would suffer from technological unemployment rather than overwork. Nor will a continuing "educational campaign" itself destroy racial prejudice and discrimination.

This is not a particularly popular position. The appeal to education as a cure-all for the most varied social problems is rooted deep in the mores of America. Yet it is nonetheless illusory for all that. For how would this program of racial education proceed? Who is to do the educating? The teachers in our communities? But, in some measure like many other Americans, the teachers share the same prejudices they are being urged to combat. And when they don't, aren't they being asked to serve as conscientious martyrs in the cause of educational utopianism? How long the tenure of elementary schoolteachers in racist communities who attempt meticulously to disabuse their young pupils of the racial beliefs they acquired at home? Education may serve as an operational adjunct but not as the chief basis for any except excruciatingly slow change in the prevailing patterns of race relations.

To understand further why educational campaigns cannot be counted on to eliminate prevailing ethnic hostilities, we must examine the operation of in-groups and out-groups in our society. Ethnic out-groups, to adopt Sumner's useful bit of sociological vernacular, consists of all those who are believed to differ significantly from "ourselves" in terms of nationality, race, or religion. The counterpart of the ethnic out-group is of course the ethnic in-group, constituted by those who "belong." Nothing is fixed or eternal about the lines separating the in-group from out-groups. As situations change, the lines of separation change. For a large number of white Americans, Joe Louis was a member of an out-group—when the situation was defined in racial terms. On another occasion, when Louis defeated the nazified Schmeling, many of these same white Americans acclaimed him as a member of the (national) in-group. National loyalty took precedence over racial separatism. These abrupt shifts in group boundaries sometimes prove embarrassing. Thus, when black Americans ran away with the honors in the Olympic games held in Berlin, the Nazis, pointing to the second-class citizenship assigned blacks in various regions of this country, denied that the United States had really won the games, since by our own admission the black athletes were "not full-fledged" Americans. And what could Bilbo or Rankin[5] say to that?

5. Two Southern congressmen in the 1940s known for their extreme racist views.—*Ed.*

Under the benevolent guidance of the dominant in-group, ethnic out-groups are continuously subjected to a lively process of prejudice which goes far toward vitiating mass education and mass propaganda for ethnic tolerance. This is the process whereby "in-group virtues become out-group vices," to paraphrase a remark by the sociologist Donald Young.[6] Or, more colloquially and perhaps more instructively, it may be called the "damned-if-you-do and damned-if-you-don't" pattern in ethnic and racial relations.

In-Group Virtues and Out-Group Vices

To discover that ethnic out-groups are damned if they do embrace the values of white Protestant society and damned if they don't, we turn to one of the in-group cultures heroes, examine the qualities with which he is endowed by biographers and popular belief, and thus distill the qualities of mind and action and character that are generally regarded as altogether admirable.

Periodic public opinion polls are not needed to justify the selection of Lincoln as the culture hero who most fully embodies the cardinal American virtues. As the Lynds point out in *Middletown,*[7] the people of that typical small city allow George Washington alone to join Lincoln as the greatest of Americans. He is claimed as their very own by almost as many well-to-do Republicans as less well-to-do Democrats.[8]

6. Donald Young, *American Minority Peoples* (New York: Harper, 1932).

7. Robert S. Lynd and Helen M. Lynd, *Middletown* (New York: Harcourt Brace, 1929).

8. On Lincoln as culture hero, see the perceptive essay by David Donald, "Getting Right with Lincoln" in *Lincoln Reconsidered* (New York: Alfred A. Knopf, 1956), 3–18.

Though Lincoln nominally remains, of course, the symbolic leader of the Republicans, this may be just another paradox of political history of the same kind which Lincoln noted in his day with regard to Jefferson and the Democrats: "Remembering, too, that the Jefferson party was formed upon its supposed superior devotion to the personal rights of men, holding the rights of property to be secondary only, and greatly inferior, and assuming that the so-called Democrats of to-day are the Jefferson, and their opponents the anti-Jefferson, party, it will be equally interesting to note how completely the two have changed hands as to the principle upon which they were originally supposed to be divided. The Democrats of to-day hold the liberty of one man to be absolutely nothing, when in conflict with another man's right of property: Republicans, on the contrary, are for both the man and the dollar, but in case of conflict the man before the dollar.

"I remember being once much amused at seeing two partially intoxicated men engaged in a fight with their great-coats on, which fight, after a long and rather harmless contest, ended in each having fought himself out of his own coat and into that of the other. If the two leading parties of this day are really identical with the two in the days of Jefferson and Adams, they have performed the same feat as the two drunken men" (Lin-

Even the inevitable schoolboy knows that Lincoln was thrifty, hard-working, ambitious, eager for knowledge, devoted to the rights of the average man, and eminently successful in climbing the ladder of opportunity from the lowermost rung of laborer to the respectable heights of merchant and lawyer. (We need follow his dizzying ascent no further.)

If one did not know that these attributes and achievements are numbered high among the values of middle-class America, one would soon discover it by glancing through the Lynds' account of "The Middletown Spirit." For there we find the image of the Great Emancipator fully reflected in the values in which Middletown believes. And since these are their values it is not surprising to find the Middletowns of America condemning and disparaging those individuals and groups who fail, presumably, to exhibit these virtues. If it appears to the white in-group that blacks are *not* educated in the same measure as themselves, that they have an "unduly" high proportion of unskilled workers and "unduly" low proportion of successful business and professional men and women, that they are thriftless, and so on through the catalogue of middle-class virtue and sin, it is not difficult to understand the charge that "the black" is "inferior" to "the white."

Sensitized to the workings of the self-fulfilling prophecy, we should be prepared to find that the antiblack charges which are not patently false are only speciously true. The allegations are true in the Pickwickian sense that we have found self-fulfilling prophecies in general to be true. Thus, if the dominant in-group believes that blacks are inferior, and sees to it that funds for education are not "wasted on these incompetents," and then proclaims as final evidence of this inferiority that blacks have proportionately "only" one-fifth as many college graduates as whites,[9] one can scarcely be amazed by this transparent bit of social legerdemain. Having seen the rabbit carefully though not too adroitly placed in the hat, we can only look askance at the triumphant air with which it is finally produced [. . .]

So, too, when Senator Bilbo, the gentleman from Mississippi (a state which then spent five times as much on the average white pupil as on the average black pupil), proclaimed the essential inferiority of blacks by pointing to the per capita ratio of physicians among them as less than one-fourth that of whites, we are impressed more by his scrambled logic than by his profound prejudices. So plain is the mechanism of the

coln to H. L. Pierce and others, 6 April 1859, *Complete Works of Abraham Lincoln,* vol. 5, ed. John G. Nicolay and John Hay [New York: Century Co., 1894], 125–26).

9. Again, this remains as written in 1948. Social change in this sphere continues.—*Ed.*

self-fulfilling prophecy in these instances that only those forever devoted to the victory of sentiment over fact can take such specious evidence seriously. Yet the specious evidence often creates a genuine belief. Self-hypnosis through one's own propaganda is a not infrequent phase of self-fulfilling prophecy.

So much for out-groups being damned if they don't (apparently) manifest in-group virtues. It is a tasteless bit of ethnocentrism, seasoned with self-interest. But what of the second phase of this process? Are out-groups also damned if they *do* possess these virtues? They are.

Through a faultlessly bisymmetrical prejudice, ethnic and racial out-groups get it coming and going. The systematic condemnation of out-groupers continues largely *irrespective of what they do*. More: through a freakish exercise of capricious judicial logic, the victim is punished for the crime.[10] Superficial appearances notwithstanding, prejudice and discrimination aimed at the out-group are not a result of what the out-group does, but are rooted deep in the structure of our society and the social psychology of its members.

To understand how this happens, we must examine the moral alchemy through which the in-group readily transmutes virtue into vice and vice into virtue, as the occasion may demand. Our studies will proceed by the case-method.

We begin with the engagingly simple formula of moral alchemy: the same behavior must be differently evaluated according to the person who exhibits it. For example, the proficient alchemist knows that the word "firm" is properly declined as follows:

I am firm,
Thou art obstinate,
He is pigheaded.

There are some, unversed in the skills of this science, who will tell you that one and the same term should be applied to all three instances of identical behavior. Such unalchemical nonsense should simply be ignored.

With this experiment in mind, we are prepared to observe how the very same behavior undergoes a complete change of evaluation in its transition from the in-group Abe Lincoln to the out-group Abe Cohen or Abe Kurokawa. We proceed systematically. Did Lincoln work far into the night? This testifies that he was industrious, resolute, perseverant, and eager to realize his capacities to the full. Do the out-group

10. See William Ryan, *Blaming the Victim* (New York: Pantheon Books, 1971).—*Ed.*

Jews or Japanese keep these same hours? This only bears witness to their sweatshop mentality, their ruthless undercutting of American standards, their unfair competitive practices. Is the in-group hero frugal, thrifty, and sparing? Then the out-group villain is stingy, miserly, and penny-pinching. All honor is due the in-group Abe for having been smart, shrewd, and intelligent, and, by the same token, all contempt is owing the out-group Abes for being sharp, cunning, crafty, and too clever by far. Did the indomitable Lincoln refuse to remain content with a life of work with his hands? Did he prefer to use his brain? Then all praise for his plucky climb up the shaky ladder of opportunity. But, of course, the shunning of manual work for brain work among the merchants and lawyers of the out-group deserves nothing but censure for their having chosen a parasitic way of life.

Was Abe Lincoln eager to learn the accumulated wisdom of the ages by unending study? The trouble with Jews is that they're greasy grinds, with their heads always in a book, while decent people are going to a show or a ball game. Was the resolute Lincoln unwilling to limit his standards to those of his provincial community? We should expect no less of a man of vision. But if the out-groupers criticize the vulnerable areas in our society, send 'em back where they came from. Did Lincoln, rising high above his origins, never forget the rights of the common man and applaud the right of workers to strike? This testifies only that, like all real Americans, this greatest of Americans was deathlessly devoted to the cause of freedom. But, as you examine the statistics on strikes, remember that this un-American practice is the result of out-groupers pursuing their evil agitation among otherwise contented workers.

Once stated, the classical formula of moral alchemy is clear enough. Through the adroit use of these rich vocabularies of encomium and opprobrium, the in-group readily transmutes its own virtues into others' vices. But why do so many in-groupers qualify as moral alchemists? Why are so many in the dominant in-group so fully devoted to this continuing experiment in moral transmutation?

An explanation may be found by putting ourselves at some distance from this country and following the anthropologist Malinowski to the Trobriand Islands.[11] For there we find an instructively similar pattern. Among the Trobrianders, to a degree which Americans, despite Hollywood and the confession magazines, have apparently not yet approximated, success with women confers honor and prestige on a man.

11. Bronislaw Malinowski, *Argonauts of the Western Pacific* (London: Routledge and Kegan Paul, 1922).

Sexual prowess is a positive value, a moral virtue. But if a rank-and-file Trobriander has "too much" sexual success, if he achieves "too many" triumphs of the heart, an achievement which should of course be limited to the elite, the chiefs, or men of power, then this glorious record becomes a scandal and an abomination. The chiefs are quick *to resent any personal achievement not warranted by social position.* The moral virtues remain virtues only so long as they are jealously confined to the proper in-group. The right activity by the wrong people becomes a thing of contempt, not of honor. For clearly, only in this way, by holding these virtues exclusively to themselves, can the men of power retain their distinction, their prestige, and their power. No wiser procedure could be devised to hold intact a system of social stratification and social power.

The Trobrianders could teach us more. For it seems clear that the chiefs have not calculatedly devised this program for entrenching their power. Their behavior is spontaneous, unthinking, and immediate. Their resentment of "too much" ambition or "too much" success for the ordinary Trobriander is not contrived; it is genuine. It just happens that this prompt emotional response to the "misplaced" manifestation of in-group virtues also serves the useful expedient of reinforcing the chiefs' special claims to the good things of Trobriand life. Nothing could be more remote from the truth and more distorted a reading of the facts than to assume that this conversion of in-group virtues into out-groups vices is part of a deliberate plot of Trobriand chiefs to keep Trobriand commoners in their place. It is merely that the chiefs have been indoctrinated with an appreciation of the proper order of things, and see it as their heavy burden to enforce the mediocrity of others.

Nor, in quick revulsion from the culpabilities of the moral alchemists, need we succumb to the equivalent error of simply upending the moral status of the in-group and the out-groups. It is not that Jews and blacks are one and all angelic while Gentiles and whites are one and all fiendish. It is not that individual virtue will now be found exclusively on the wrong side of the ethnic-racial tracks and individual viciousness on the right side. Conceivably there are as many corrupt and vicious men and women among Jews and blacks as among Gentile whites. It is only that the ugly fence enclosing the in-group happens to exclude the people who make up the out-groups from being treated with decency.

Social Functions and Dysfunctions

We have only to look at the consequences of this peculiar moral alchemy to see that there is no paradox at all in damning out-groupers when they do and when they don't exhibit in-group virtues. Condemnation on these two scores performs one and the same social function. Seeming opposites coalesce. When blacks are tagged as incorrigibly inferior because they (apparently) don't manifest these virtues, this confirms the natural rightness of their being assigned an inferior status in society. And when Jews or Japanese are seen as having too much of the fruits of in-group values, it becomes plain that they must be securely controlled by the high walls of discrimination. In both cases, the special status assigned the several out-groups can be seen to be eminently reasonable.

Yet this distinctly reasonable arrangement persists in having most unreasonable consequences, both logical and social. Consider only a few of these.

In some contexts, the limitations enforced upon the out-group—say, rationing the number of Jews permitted to enter colleges and professional schools—logically imply a fear of the alleged superiority of the out-group. Were it otherwise, no discrimination would need be practiced. The unyielding, impersonal forces of academic competition would soon trim down the number of Jewish (or Japanese or black) students to an "appropriate" size.

This implied belief in the superiority of the out-group seems premature. There is simply not enough scientific evidence to demonstrate Jewish or Japanese or black superiority. The effort of the in-group discriminator to supplant the myth of Aryan superiority with the myth of non-Aryan superiority is condemned to failure by science. Moreover, such myths are ill-advised. Eventually, a world of myth must collide with a world of reality. As a matter of simple self-interest and social therapy, therefore, the in-group might be wise to abandon the myth and cling to the reality.

The pattern of being damned-if-you-do and damned-if-you-don't has further consequences for out-group members. Their response to alleged deficiencies is as clear as it is predictable. If one is told repeatedly that one is inferior, that one lacks any positive accomplishments, it is all too human to seize upon every bit of evidence to the contrary. The in-group definitions force upon the allegedly inferior out-group a defensive tendency to magnify and exalt "race accomplishments." As the distinguished black sociologist E. Franklin Frazier has noted, the black newspapers are "intensely race conscious and exhibit considerable pride in the achievements of the Negro most of which are meager performances as measured by broader standards." Self-glorification, found in some measure among all groups, becomes a frequent counter-response to persistent belittlement from without.

It is the damnation of out-groups for excessive achievement, however, which gives rise to truly bizarre behavior. For, after a time and often as a matter of self-defense, these out-groups become persuaded that their virtues really are vices. And this provides the final episode in a tragicomedy of inverted values.

Let us try to follow the plot through its intricate maze of self-contradictions. Respectful admiration for the arduous climb from office boy to president is rooted deep in American culture. This long and strenuous ascent carries with it a twofold testimonial: first, that careers are abundantly open to genuine talent in American society; second, that individuals who have distinguished themselves by their heroic rise are praiseworthy. It would be invidious to choose among the many stalwart figures who fought their way up, against all odds, until they reached the summit, there to sit at the head of the long conference table in the longer conference room of The Board. Taken at random the saga of Frederick H. Ecker, once chairman of the board of one of the largest privately managed corporations in the world, the Metropolitan Life Insurance Company, will suffice as the prototype. From a menial and poorly paid job, he rose to a position of eminence. Understandably enough, an unceasing flow of honors came to this man of large power and large achievement. It so happens, though it is a matter personal to this eminent man of finance, that Mr. Ecker is a Presbyterian. Yet at last report, no elder of the Presbyterian church had risen publicly to announce that Mr. Ecker's successful career should not be taken too seriously, that, after all, relatively few Presbyterians have risen from rags to riches, and that Presbyterians do not actually "control" the world of life insurance or finance or investment housing. Rather, one would suppose, Presbyterian elders joined with other Americans imbued with middle-class standards of success to felicitate the eminently successful Mr. Ecker and to acclaim other sons and daughters of the faith who have risen to almost equal heights. Secure in their in-group status, they point the finger of pride rather than the finger of dismay at individual success.

Prompted by the practice of moral alchemy, noteworthy achievements by out-groupers can elicit other responses. Patently, if achievement is a vice, then achievements must be disclaimed—or at least, discounted. Under these conditions, an occasion for Presbyterian pride can become an occasion for Jewish dismay. If the Jew is condemned for educational or professional or scientific or economic success, then, understandably enough, many Jews will come to feel that these accomplishments must be minimized in simple self-defense. Thus is the circle of paradox closed by out-groupers busily engaged in assuring the powerful in-group that they have not, in fact, been guilty of inordinate contributions to science, the professions, the arts, the government, and the economy.

In a society, which ordinarily looks upon wealth as a warrant of one kind of ability, an out-group is compelled by the inverted attitudes of the dominant in-group to deny that many men and women of wealth are among its members. "Among the 200 largest non-banking corporations . . . only ten have a Jew as president or chairman of the board." Is this an observation of an anti-Semite, intent on proving the incapacity and inferiority of Jews who have done so little "to build the corporations which have build America?" No; it is a response of the Anti-Defamation League of B'nai B'rith to anti-Semitic propaganda.

In a society where, as a survey by the National Opinion Research Center has shown, the profession of medicine ranks highest in social prestige among ninety occupations, we find some Jewish spokesmen maneuvered by the attacking in-group into the position of announcing their "deep concern" over the number of Jews in medical practice, which is "disproportionate to the number of Jews in other occupations." The oversupply of Jewish doctors becomes a deplorable situation meriting deep concern, rather than a joyous accomplishment meriting applause for their hard-won acquisition of knowledge and skills and for their social utility. Only when the New York Yankees publicly announce deep concern over their numerous World Series triumphs, so disproportionate to the number achieved by other major league teams, will this self-abnegation seem part of the normal order of things.

In a society which consistently judges professionals as higher in social value than even the most skilled hewers of wood and drawers of water, the out-group finds itself in the anomalous position of pointing with defensive relief to the large number of Jewish painters and paperhangers, plasterers and electricians, plumbers and sheetmetal workers.

But the ultimate reversal of values is yet to be noted. Each succeeding census finds more Americans in the city and its suburbs. Americans have traveled the road to urbanization until fewer than one-fifth of the nation's population live on farms. Plainly, it is high time for the Methodist and the Catholic, the Baptist and the Episcopalian to recognize the iniquity of this trek of their coreligionists to the city. For, as is well known, one of the central accusations leveled against Jews is their heinous tendency to live in cities. Jewish leaders, therefore, find themselves maneuvered into the position of defensively urging their people to move into the very farm areas being hastily vacated by city-bound hordes of

Christians. Perhaps this is not altogether necessary. As the Jewish crime of urbanism becomes ever more popular among the in-group, it may be reshaped into transcendent virtue. But, admittedly, one can't be certain. For in this daft confusion of inverted values, it soon becomes impossible to determine when virtue is sin and sin, moral perfection.

Amid this confusion, one fact remains unambiguous. The Jews, like other peoples, have made distinguished contributions to world culture. Consider only an abbreviated catalogue. The field of creative literature (with acknowledgment of large variations in the caliber of achievement) includes Heine, Kraus, Borne, Hofmannsthal, Schnitzler, Kafka. In the realm of musical composition, there are Meyerbeer, Mendelssohn, Offenbach. Mahler, and Schonberg. Among the musical virtuosi, consider only Rosenthal, Schnabel, Godowsky, Pachmann, Kreisler, Hubermann, Milstein, Elman, Heifetz, Joachim, and Menuhin. And among scientists of a stature sufficient to merit the Nobel prize, examine the familiar list which includes Bárány, Mayerhof, Ehrlich, Michelson, Lippmann, Haber, Willstätter, and Einstein.[12] Or in the esoteric and imaginative universe of mathematical invention, take note only of Kronecker, the creator of the modern theory of numbers; Hermann Minkowski,[13] who supplied the mathematical foundations of the special theory of relativity; or Jacobi, with his basic work in the theory of elliptical functions. And so through each special province of cultural achievement, we are supplied with a list of preeminent men and women who happened to be Jews.

And who is thus busily engaged in singing the praises of the Jews? Who has so assiduously compiled the list of many hundreds of distinguished Jews who contributed notably to science, literature, and the arts—a list from which these few cases were excerpted? A philo-Semite, eager to demonstrate that the Jews have contributed their due share to world culture? No, by now we should know better than that. The complete list is found in the thirty-sixth edition of the anti-Semitic handbook by the Nazi racist Fritsch. In accord with the alchemical formula

12. A monograph on the Nobel Prize in science finds that, as of 1972, Jews constituted about a fifth of all 286 and about a fourth of the 89 laureates in the United States. Harriet Zuckerman, *Scientific Elite,* 2d ed. (1977; reprint, New York: Transaction Books, 1996), 68.—*Ed.*

13. Obviously, the forename must be explicitly mentioned, else Hermann Minkowski, the mathematician, may be confused with Eugen Minkowski (who contributed so notably to our knowledge of schizophrenia), or with Mieczyslaw Minkowski (high in the ranks of brain anatomists), or even with Oscar Minkowski (discoverer of pancreatic diabetes).

for transmuting in-group virtues into out-group vices, he presents this as a roll call of sinister spirits who have usurped the accomplishments properly owing the Aryan in-group.

Once we comprehend the predominant role of the in-group in defining the situation, the further paradox of the seemingly opposed behavior of the black out-group and the Jewish out-group falls away. The behavior of both minority groups is in response to the majority-group allegations.

If blacks are accused of inferiority, and their alleged failure to contribute proportionately to world culture is cited in support of this accusation, the human urge for self-respect and concern for security lead some of them *defensively* to magnify each and every achievement by members of the race. If Jews are accused of excessive achievements and excessive ambitions, and lists of preeminent Jews are compiled in support of this accusation, then the urge for security leads some of them *defensively* to minimize the actual achievements of members of the group. Apparently opposed types of behavior have much the same psychological and social functions. Self-assertion and self-effacement become devices for seeking to cope with condemnation for alleged group deficiencies and condemnation for alleged group excesses, respectively. And with a fine sense of moral superiority, the secure in-group looks on these curious performances by the out-groups with mingled derision and contempt.

Enacted Institutional Change

Will this desolate tragicomedy run on and on, marked only by minor changes in the cast? Not necessarily.

Were moral scruples and a sense of decency the only bases for bringing the play to an end, it would continue its run indefinitely. In and of themselves, moral sentiments are not much more effective in curing social ills than in curing physical ills. Moral sentiments no doubt help to motivate efforts for change, but they are no substitute for hardheaded institutional arrangements designed to achieve the objective, as the thickly populated graveyard of softheaded utopias bears witness.

There are ample indications that a deliberate and planned halt can be put to the workings of the self-fulfilling prophecy and the vicious circle in society. The sequel to our sociological parable of the Last National Bank provides one clue to the way in which this can be achieved. During the fabulous 1920s, when Coolidge undoubtedly caused a Republican era of lush prosperity, an annual average of 635 banks quietly suspended operations. And during the four years immediately before and after The Crash, when Hoover undoubtedly did not cause a Republican era of sluggish depression, this zoomed to the more spectacular average of 2,276 bank suspensions annually. But, interestingly enough, in the twelve years following the establishment of the Federal Deposit Insurance Corporation and the enactment of other banking legislation, while Roosevelt presided over Democratic depression and revival, recession and boom, bank suspensions dropped to a niggardly average of twenty-eight a year. Perhaps money panics have not been exorcised by legislation. Nevertheless, millions of depositors no longer have occasion to give way to panic-motivated runs on banks simply because deliberate institutional change has removed the grounds for panic. Occasions for racial hostility are no more inborn psychological constants than are occasions for panic. Despite the teachings of amateur psychologists, blind panic and racial aggression are not rooted in human nature. These patterns of human behavior are largely a product of the modifiable structure of society.

For a further clue, return to the instance of widespread hostility of white unionists toward black strikebreakers brought into industry by employers after the close of the very First World War. Once the initial definition of blacks as not deserving of union membership had largely broken down, black workers with a wider range of opportunities no longer found it necessary to enter industry through doors help open by strikebound employers. Again, appropriate institutional change broke through the tragic circle of the self-fulfilling prophecy. Deliberate social change gave the lie to the firm conviction that "it just ain't in the nature of the Nigra" to join trade unions.

A final instance is drawn from a study of a public biracial housing project, one of the first to be established in the country.[14] Located in Pittsburgh, the community of Hilltown is made up of 50 percent black families and 50 percent white. It is not a twentieth-century utopia. There is some interpersonal friction here as elsewhere. Still, in a community made up of equal numbers of the two races, fewer than a fifth of the whites and a third of the blacks report that this friction occurs between members of *different* races. By their own testimony, it is largely confined to disagreements *within* each racial group. Yet only one in every twenty-five whites initially *expected* relations between the races in the community to run smoothly, whereas five times as many expected serious trouble, the remainder anticipating a tolerable, if not

14. Robert K. Merton, Patricia S. West, and Marie Jahoda, *Patterns of Social Life* (New York: Columbia University Bureau of Applied Research, 1948–51); R. K. Merton, P. S. West, M. Jahoda, and H. C. Selvin, eds., *Social Policy and Social Research in Housing* (*Journal of Social Issues,* 1951).

altogether pleasant, situation. So much for expectations. Upon reviewing their actual experience, three of every four of the most apprehensive whites subsequently found that the "races get along fairly well," after all. This is not the place to report the findings of this study in detail, but substantially these demonstrate anew that under *appropriate institutional and administrative conditions,* the experience of interracial amity can supplant the fear of interracial conflict.

These changes, and others of the same kind, do not occur automatically. *The self-fulfilling prophecy, whereby fears are translated into reality, operates only in the absence of deliberate institutional controls.* And it is only with the rejection of social fatalism implied in the notion of unchangeable human nature that the tragic circle of fear, social disaster, and reinforced fear can be broken.

Ethnic prejudices do die—but slowly. They can be helped over the threshold of oblivion not by insisting that it is unreasonable and unworthy of them to survive but by cutting off the sustenance now provided them by certain social institutions.

If we find ourselves doubting the human capability to control society, if we persist in the tendency to find in the patterns of the past the chart of the future, it is perhaps time to take up anew the wisdom of Tocqueville's century-old remark: "I am tempted to believe that what we call necessary institutions are often no more than institutions to which we have grown accustomed, and that in matters of social constitution the field of possibilities is much more extensive than men living in their various societies were ready to imagine."

Nor can widespread, even typical, failures in planning human relations between ethnic groups be cited as evidence for pessimism. In the world laboratory of the sociologist, as in the more secluded laboratories of the physicist and chemist, it is the successful experiment which is decisive, not the thousand-and-one failures which preceded it. More is learned from the single success than from multiple failures. A single success proves it can be done. Thereafter, it is necessary only to learn what made it work. This, at least, is what I take to be the sociological sense of those revealing words of Thomas Love Peacock: "Whatever is, is possible."[15]

15. This concluding sentence should perhaps be amended. For during the third of a century since it was first put into print, I have been periodically asked to indicate *where* Peacock stated the epigram. The question has grown more embarrassing with every repetition. The facts of the case, so far as I am able to reconstruct them, are these: In my youth, I became strangely addicted to Peacock's satiric novels, with their graceful language, acute wit, and often alliterated titles (*Crotchet Castle, Headlong Hall, Maid Marian, and Gryll Grange*). In due course, I put away these eccentric conversational

pieces. But when I was searching for a way of summing up the reasons for believing that the self-fulfilling prophecy can be institutionally controlled, I evidently dredged up a seeming memory trace which had Peacock stating in playful and pithy style what I wanted to say. And so I emerged with the aphorism I ascribed to him: Whatever is, is possible. Later, when I found myself repeatedly asked to cite the precise source in Peacock, I paid the price of not having recorded the reference. Once again, my bedtime reading was devoted, with somewhat less pleasure than before, to Peacock's considerable writings. To no avail. The experience has begun to give me pause. Perhaps Thomas Love Peacock never delivered himself of this philosophical and sociological epigram; perhaps I am the victim of a false memory. Perhaps it was some other novelist or poet or philosopher.

Indeed, William J. Goode, a colleague of mine, learning of my failure to locate the aphorism in Peacock, became persuaded that it was originated by James Thurber. He accordingly wrote to Thurber's widow only to be told that she had never read this "law" in any of her husband's many works, that she had never heard him state it in conversations and, to close the matter once and for all, that it "really doesn't sound like him."

But if not Peacock and not Thurber, another possibility springs to mind. Perhaps I had engaged in a kind of retroactive ghostwriting. Once having hit upon the existential theorem that "whatever is, is possible," perhaps I had unaccountably (though surely not mischievously) ascribed it to that youthfully favorite novelist of mine. If so, it was a substantively inept ascription. For if Peacock had originated—or, to keep the door open, if he did in fact originate—the epigrammatic theorem, he would surely have meant it in a sense remote from my own. I had introduced it to argue that, however fanciful or repugnant a social innovation may seem to some, if it has succeeded even once, then it demonstrates that the seemingly impossible is quite possible. "A single success proves it can be done. Thereafter, it is necessary only to learn what made it work."

But Peacock would scarcely have drawn this sociological moral. His life long, he resisted innovation, social and other. All his graceful conversational novels—not least, his last, long-postponed and summational novel, *Gryll Grange*—vigorously express his hearty distaste for science and its technological offshoots.

Since the third part of a century has gone by since I assigned the theorem to Thomas Love Peacock, and since neither he nor anyone of his present-day fans has risen up to document that ascription, perhaps the statute of limitations governing intellectual property has run its course. Perhaps I should amend that concluding sentence to incorporate this short tale of its history by having it say: ". . . those revealing words which I once ascribed to Thomas Love Peacock: 'Whatever is, is possible.' "

III

SCIENCE AS A SOCIAL STRUCTURE

The Sociology of Knowledge

17

Paradigm for the Sociology of Knowledge (1945)

The Social Context

The sociology of knowledge takes on pertinence under a definite complex of social and cultural conditions.[1] With increasing social conflict, differences in the values, attitudes, and modes of thought of groups develop to the point where the orientation which these groups previously had in common is overshadowed by incompatible differences. Not only do there develop distinct universes of discourse, but the existence of any one universe challenges the validity and legitimacy of the others. The coexistence of these conflicting perspectives and interpretations within the same society leads to an active and reciprocal *distrust* between groups. Within a context of distrust, one no longer inquires into the content of beliefs and assertions to determine whether they are valid or not, one no longer confronts the assertions with relevant evidence, but introduces an entirely new question: how does it happen that these views are maintained? Thought becomes functionalized, it is interpreted in terms of its psychological or economic or social or racial sources and functions. In general, this type of functionalizing occurs when statements are doubted, when they appear so palpably implausible or absurd or biased that one need no longer examine the evidence for or against the statement but only the grounds for its being asserted at all.[2] Such alien statements are "explained by" or "imputed to" spe-

Revised from "Sociology of Knowledge," in *Twentieth-Century Sociology,* ed. George Gurvitch and Wilbert E. Moore (New York: Philosophical Library, 1945), 366–405. Reprinted by permission of the Philosophical Library.

1. See Karl Mannheim, *Ideology and Utopia* (New York: Harcourt Brace, 1946), 5–12; Pitirim A. Sorokin, *Social and Cultural Dynamics,* 4 vols. (New York: American Book Co., 1937), 2:412–13.

2. Freud had observed this tendency to seek out the "origins" rather than to test the validity of statements which seem palpably absurd to us. Thus, suppose someone maintains that the center of the earth is made of jam. "The result of our intellectual objection will be a *diversion of our interests; instead of their being directed on to the investigation itself,* as to whether the interior of the earth is really made of jam or not, *we shall wonder what kind of man it must be who can get such an idea into his head. . . .*" Sigmund Freud, *New Introductory Lectures* (New York: W. W. Norton, 1933), 49 [italics only]. On the

cial interests, unwitting motives, distorted perspectives, social position, and so on. In folk thought, this involves reciprocal attacks on the integrity of opponents; in more systematic thought, it leads to reciprocal ideological analyses. On both levels, it feeds upon and nourishes collective insecurities.

Within this social context, an array of interpretations of man and culture which share certain common presuppositions finds widespread currency. Not only ideological analysis and *Wissenssoziologie,* but also psychoanalysis, Marxism, semanticism, propaganda analysis, Paretanism, and, to some extent, functional analysis have, despite their other differences, a similar outlook on the role of ideas. On the one hand, there is the realm of verbalization and ideas (ideologies, rationalizations, emotive expressions, distortions, folklore, derivations), all of which are viewed as expressive or derivative or deceptive (of self and others) and all of which are functionally related to some substratum. On the other hand are the previously conceived substrata (relations of production, social position, basic impulses, psychological conflict, interests and sentiments, interpersonal relations, and residues). And throughout runs the basic theme of the unwitting determination of ideas by the substrata; the emphasis on the distinction between the real and the illusory, between reality and appearance in the sphere of human thought, belief, and conduct. And whatever the intention of the analysts, their analyses tend to have an acrid quality: they tend to indict, secularize, ironicize, satirize, alienate, devalue the intrinsic content of the avowed belief or point of view. Consider only the overtones of terms chosen in these contexts to refer to beliefs, ideas, and though: vital lies, myths, illusions, derivations, folklore, rationalizations, ideologies, verbal façade, pseudo-reasons, and so on.

What these schemes of analysis have in common is the practice of discounting the *face value* of statements, beliefs, and idea-systems by reexamining them within a new context which supplies the "real meaning." Statements ordinarily viewed in terms of their manifest content are debunked, whatever the intention of the analyst, by relating this content to attributes of their authors or of the society in which they live. The professional iconoclast, the trained debunker, the ideological analyst and their respective systems of thought thrive in a society where large groups of people have become alienated from common values, where separate universes of discourse are linked with reciprocal distrust. Ideological analysis systematizes the lack of faith in reigning symbols that

social level, a radical difference of outlook of various social groups leads not only to ad hominem attacks, but also to "functionalized explanations."

has become widespread; hence its pertinence and popularity. The ideological analyst does not so much create a following as he speaks for a following to whom his analyses "make sense," that is, conform to their previously unanalyzed experience. [. . .]

The "Copernican revolution" in this area of inquiry consisted in the hypothesis that not only error or illusion or unauthenticated belief but also the discovery of truth is socially (historically) conditioned. As long as attention was focused on the social determinants of ideology, illusion, myth, and moral norms, the sociology of knowledge could not emerge. It was abundantly clear that in accounting for error or uncertified opinion, some extra-theoretic factors were involved, that some special explanation was needed, since the reality of the object could not account for error. In the case of confirmed or certified knowledge, however, it was long assumed that it could be adequately accounted for in terms of a direct object-interpreter relation. The sociology of knowledge came into being with the signal hypothesis that even truths were to be held socially accountable, were to be related to the historical society in which they emerged.

To outline even the main currents of the sociology of knowledge in brief compass is to present none adequately and to do violence to all. The diversity of formulations—of a Marx or Scheler or Durkheim; the varying problems—from the social determination of categorial systems to that of class-bound political ideologies; the enormous differences in scope—from the all-encompassing categorizing of intellectual history to the social location of the thought of Negro scholars in the last decades; the various limits assigned to the discipline—from a comprehensive sociological epistemology to the empirical relations of particular social structures and ideas; the proliferation of concepts—ideas, belief-systems, positive knowledge, thought, systems of truth, superstructure, and so on; the diverse methods of validation—from plausible but undocumented imputations to meticulous historical and statistical analyses—in the light of all this, an effort to deal with both analytical apparatus and empirical studies in a few pages must sacrifice detail to scope.

To introduce a basis of comparability among the welter of studies that have appeared in this field, we must adopt some scheme of analysis. The following paradigm is intended as a step in this direction. It is, undoubtedly, a partial and, it is to be hoped, a temporary codification which will disappear as it gives way to an improved and more exacting analytical model. But it does provide a basis for taking an inventory of extant findings in the field; for indicating contradictory, contrary, and consistent results; setting forth the conceptual apparatus now in use;

determining the nature of problems which have occupied workers in this field; assessing the character of the evidence which they have brought to bear upon these problems; ferreting out the characteristic lacunae and weaknesses in current types of interpretation. Full-fledged theory in the sociology of knowledge lends itself to classification in terms of the following paradigm.[3]

Paradigm for the Sociology of Knowledge

1. Where *is the existential basis of mental productions located?*

a. *social bases:* social position, class, generation, occupational role, mode of production, group structures (university, bureaucracy, academies, sects, political parties), "historical situation," interests, society, ethnic affiliation, social mobility, power structure, social processes (competition, conflict, and so on).

b. *cultural bases:* values, ethos, climate of opinion, *Volksgeist, Zeitgeist,* type of culture, culture mentality, *Weltanschauungen,* and so on.

2. What *mental productions are being sociologically analyzed?*

a. *spheres of:* moral beliefs, ideologies, ideas, the categories of thought, philosophy, religious beliefs, social norms, positive science, technology, and so on.

b. *which aspects are analyzed:* their selection (foci of attention), level of abstraction, presuppositions (what is taken as data and what as problematical), conceptual content, models of verification, objectives of intellectual activity, and so on.

3. How *are mental productions related to the existential basis?*

a. *causal or functional relations:* determination, cause, correspondence, necessary condition, conditioning, functional interdependence, interaction, dependence, and so on.

b. *symbolic or organismic or meaningful relations:* consistency, harmony, coherence, unity, congruence, compatibility (and antonyms); expression, realization, symbolic expression, *Strukturzusammenhang,* structural identities, inner connection, stylistic analogies, logico-meaningful integration, identity of meaning, and so on.

c. *(largely) ambiguous terms to designate relations:* correspondence, reflection, bound up with, in close connection with, and so on.

4. Why *related? Manifest and latent functions imputed to these existentially conditioned mental productions.*

a. to maintain power, promote stability, orientation, exploitation, obscure actual social relationships, provide motivation, canalize behav-

3. See "Paradigms: The Codification of Sociological Theory," chap. 5 in this volume.—*Ed.*

ior, divert criticism, deflect hostility, provide reassurance, control nature, coordinate social relationships, and so on.

5. When *do the imputed relations of the existential base and knowledge obtain?*
   a. historicist theories (confined to particular societies or cultures).
   b. general analytical theories.

There are, of course, additional categories for classifying and analyzing studies in the sociology of knowledge which are not fully explored here. Thus, the perennial problem of the implications of existential influences upon knowledge for the epistemological status of that knowledge has been hotly debated from the very outset. Solutions to this problem, which assume that a sociology of knowledge is necessarily a sociological theory of knowledge, range from the claim that the "genesis of thought has no necessary relation to its validity" to the extreme relativist position that truth is "merely" a function of a social or cultural basis, that it rests solely upon social consensus and, consequently, that any culturally accepted theory of truth has a claim to validity equal to that of any other. [. . .]

The Existential Basis

A central point of agreement in all approaches to the sociology of knowledge is the thesis that thought has an existential basis insofar as it is not immanently determined and insofar as one or another of its aspects can be derived from extra-cognitive factors. But this is merely a formal consensus, which gives way to a wide variety of theories concerning the nature of the existential basis.

In this respect, as in others, Marxism is the storm center of *Wissenssoziologie*. Without entering into the exegetic problem of closely identifying Marxism—we have only to recall Marx's "*je ne suis pas Marxiste*"—we can trace out its formulations primarily in the writings of Marx and Engels. Whatever other changes may have occurred in the development of their theory during the half-century of their work, they consistently held fast to the thesis that "relations of production" constitute the "real foundation" for the superstructure of ideas. "The mode of production in material life determines the general character of the social, political and intellectual processes of life. It is not the consciousness of men that determines their existence, but on the contrary, their social existence determines their consciousness."[4] In seeking to func-

4. Karl Marx, *A Contribution to the Critique of Political Economy* (1859; reprint, Chicago: C. H. Kerr, 1904), 11–12.

tionalize ideas, that is, to relate the ideas of individuals to their sociological bases, Marx locates them within the class structure. He assumes not so much that other influences are not at all operative, but that class is a primary determinant and, as such, the single most fruitful point of departure for analysis. This he makes explicit in his first preface to *Capital:* "... here individuals are dealt with *only in so far* as they are the personifications of economic categories, embodiments of particular class-relations and class-interests."[5] In abstracting from other variables and in regarding men solely in their economic and class roles, Marx hypothesizes that these roles are primary determinants and thus leaves as an open question *the extent to which they adequately account for thought and behavior in any given case.* In point of fact, one line of development of Marxism, from the early *German Ideology* to the latter writings of Engels, consists in a progressive definition (and delimitation) of the extent to which the relations of production do actually condition knowledge and forms of thought.

However, both Marx and Engels, repeatedly and with increasing insistence, emphasized that the ideologies of a social stratum need not stem only from persons who are *objectively* located in that stratum. As early as the *Communist Manifesto,* Marx and Engels had indicated that as the ruling class approaches dissolution, "a small section . . . joins the revolutionary class. . . . Just as, therefore, at an earlier period, a section of the nobility went over to the bourgeoisie, so now a portion of the bourgeoisie goes over to the proletariat, and in particular, a portion of *the bourgeois ideologists,* who have *raised themselves* to the level of comprehending theoretically the historical movement as a whole."[6]

Ideologies are socially located by analyzing their perspectives and presuppositions and determining how problems are construed: from the standpoint of one or another class. Thought is not mechanistically located by merely establishing the class position of the thinker. It is attributed to that class for which it is "appropriate," to the class whose social situation with its class conflicts, aspirations, fears, restraints, and objective possibilities within the given sociohistorical context is being expressed. [. . .]

5. Karl Marx, *Capital* (1867; reprint, New York: International Publishers, 1937), 1: 15 [italics added]; cf. Marx and Engels, *The German Ideology* (1845; reprint, New York: International Publishers, 1939), 76; cf. Max Weber, *Gesammelte Aufsaetze zur Wissenschaftslehre,* (Tübingen: Mohr-Siebeck, 1922), 205.

6. Karl Marx and Friedrich Engels, *The Communist Manifesto,* in Karl Marx, *Selected Works,* 2 vols. (1848, reprint, Moscow: Co-operative Publishing Society, 1935), 1: 216 [italics added].

But if we cannot derive ideas from the objective class position of their exponents, this leaves a wide margin of indeterminacy. It then becomes a further problem to discover why some identify themselves with the characteristic outlook of the class stratum in which they objectively find themselves, whereas others adopt the presuppositions of a class stratum other than "their own." An empirical description of the fact is no adequate substitute for its theoretical explanation.

In dealing with existential bases, Max Scheler characteristically places his own hypothesis in opposition to other prevalent theories.[7] He draws a distinction between cultural sociology and what he calls the sociology of real factors (*Realsoziologie*). Cultural data are "ideal," in the realm of ideas and values; "real factors" are oriented toward effecting changes in the reality of nature or society. The former are defined by ideal goals or intentions; the latter derive from an "impulse structure" (*Triebstruktur,* for example, sex, hunger, power). It is a basic error, he holds, of all naturalistic theories to maintain that real factors—whether race, geopolitics, political power structure, or the relations of economic production—unequivocally determine the realm of meaningful ideas. He also rejects all ideological, spiritualistic, and personalistic conceptions which err in viewing the history of existential conditions as a unilinear unfolding of the history of mind. He ascribes complete autonomy and a determinate sequence to these real factors, though he inconsistently holds that value-laden ideas serve to guide and direct their development. Ideas as such initially have no social effectiveness. The "purer" the idea, the greater its impotence, so far as dynamic effect on society is concerned. Ideas do not become actualized, embodied in cultural developments, unless they are bound up in some fashion with interests, impulses, emotions, or collective tendencies and their incorporation in institutional structures.[8] Only then—and in this limited respect, naturalistic theories (for example, Marxism) are justified—do they exercise some definite influence. Should ideas not be grounded in the immanent development of real factors, they are doomed to become sterile Utopias.

Naturalistic theories are further in error, Scheler holds, in tacitly assuming the *independent variable* to be one and the same throughout history. There is no constant independent variable but there is, in the course of history, a definite sequence in which the primary factors prevail, a sequence that can be summed up in a "law of three phases." In

7. This account is based upon Max Scheler's most elaborate discussion, "Probleme einer Soziologie des Wissens," in his *Die Wissensformen und die Gesellschaft* (Leipzig: Der Neue-Geist Verlag, 1926), 1–229.

8. Scheler, *Die Wissensformen,* 7, 32.

the initial phase, blood-ties and associated kinship institutions constitute the independent variable; later, political power and, finally, economic factors. There is, then, no constancy in the effective primacy of existential factors but rather an ordered variability. Thus, Scheler sought to relativize the very notion of historical determinants.[9] He claims not only to have confirmed his law of the three phases inductively but to have derived it from a theory of human impulses.

Scheler's conception of *Realfaktoren*—race and kinship, the structure of power, factors of production, qualitative and quantitative aspects of population, geographical and geopolitical factors—hardly constitutes a usefully defined category. It is of small value to subsume such diverse elements under one rubric, and, indeed, his own empirical studies and those of his disciples do not profit from this array of factors. But in suggesting a variation of significant existential factors, though not in the ordered sequence which he failed to establish, he moves in the direction which subsequent research has followed.

Thus, Mannheim derives from Marx primarily by extending his conception of existential bases. Given the *fact* of multiple group affiliation, the problem becomes one of determining *which* of these affiliations are decisive in fixing perspectives, models of thought, definitions of the given, and so on. Unlike "a dogmatic Marxism," he does not assume that class position is alone ultimately determinant. He finds, for example, that an organically integrated group conceives of history as a continuous movement toward the realization of its goals, whereas socially uprooted and loosely integrated groups espouse a historical intuition which stresses the fortuitous and imponderable. It is only through exploring the variety of group formations—generations, status groups, sects, occupational groups—and their characteristic modes of thought that there can be found an existential basis corresponding to the great variety of perspectives and knowledge which actually obtain.[10]

Though representing a different tradition, this is substantially the position which had been taken by Émile Durkheim. In an early study with Mauss of primitive forms of classification, he maintained that the genesis of the categories of thought is to be found in the group structure and relations and that the categories vary with changes in the social

9. Ibid., 25–45. It should be noted that Marx had long since rejected out of hand a similar conception of shifts in independent variables which was made the basis for an attack on his *Critique of Political Economy;* see *Capital,* 1:94 n.

10. Mannheim, *Ideology and Utopia,* 247–48. In view of the recent extensive discussions of Mannheim's work, it will not be treated at length in this essay.

organization.[11] In seeking to account for the social origins of the categories, Durkheim postulates that individuals are more directly and inclusively oriented toward the groups in which they live than they are toward nature. The primarily significant experiences are mediated through social relationships, which leave their impress on the character of thought and knowledge.[12] Thus, in his study of primitive forms of thought, he deals with the periodic recurrence of social activities (ceremonies, feasts, rites), the clan structure, and the spatial configurations of group meetings as among the existential bases of thought. And, applying Durkheim's formulations to ancient Chinese thought, Granet attributes their typical conceptions of time and space to such bases as the feudal organization and the rhythmic alternation of concentrated and dispersed group life.[13]

In sharp distinction from the foregoing conceptions of existential bases is Sorokin's idealistic and emanationist theory, which seeks to derive every aspect of knowledge, not from an existential social basis, but from varying "culture mentalities." These mentalities are constructed of "major premises": thus, the ideational mentality conceives of reality as "non-material, ever-lasting Being"; its needs as primarily spiritual and their full satisfaction through "self imposed minimization or elimination of most physical needs."[14] Contrariwise, the sensate mentality limits reality to what can be perceived through the senses; it is primarily concerned with physical needs which it seeks to satisfy to a maximum, not through self-modification, but through change of the external world. The chief intermediate type of mentality is the idealistic, which represents a virtual balance of the foregoing types. It is these mentalities, that is, the major premises of each culture, from which systems of truth and knowledge are derived. And here we come to the self-contained emanationism of an idealistic position; it appears plainly tautological to say, as Sorokin does, that "in a sensate society and culture

11. Émile Durkheim and Marcel Mauss, "De quelques formes primitives de classification," *L'Année sociologique* 6 (1901–2): 1–72: ". . . even ideas as abstract as those of time and space are, at each moment of their history, in close relation with the corresponding social organization." As Marcel Granet has indicated, this paper contains some pages on Chinese thought which have been held by specialists to mark a new era in the field of sinological studies.

12. Émile Durkheim, *The Elementary Forms of the Religious Life* (1912; reprint, Glencoe, Ill.: Free Press, 1954), 443–44; see also Hans Kelsen, *Society and Nature* (Chicago: University of Chicago Press, 1943), 30.

13. Marcel Granet, *La pensée chinoise* (Paris: La Renaissance du Livre, 1934), for example, 84–104.

14. Sorokin, *Social and Cultural Dynamics*, 1:72–73.

the Sensate system of truth based upon the testimony of the organs of senses has to be dominant."[15] For sensate mentality has already been *defined* as one conceiving of "reality as only that which is presented to the sense organs."[16]

Moreover, an emanationist phrasing such as this bypasses some of the basic questions raised by other approaches to the analysis of existential conditions. Thus, Sorokin considers the failure of the sensate "system of truth" (empiricism) to monopolize a sensate culture as evidence that the culture is not "fully integrated." But this surrenders inquiry into the bases of those very differences of thought with which our contemporary world is concerned. This is true of other categories and principles of knowledge for which he seeks to apply a sociological accounting. For example, in our present sensate culture, he finds that "materialism" is less prevalent than "idealism," and that "temporalism" and "eternalism" are almost equally current; so, too, with "realism" and "nominalism," "singularism" and "universalism," and so on. Since there are these diversities within a culture, the overall characterization of the culture as sensate provides no basis for indicating which groups subscribe to one mode of thought and which to another. Sorokin does not systematically explore varying existential bases *within* a society or culture; he looks to the "dominant" tendencies and imputes these to the culture as a whole. Our contemporary society, quite apart from the *differences* of intellectual outlook of divers classes and groups, is viewed as an integral exemplification of sensate culture. On its own premises, Sorokin's approach is primarily suited for an overall characterization of cultures, not for analyzing connections between varied existential conditions and thought *within* a society. [. . .]

Relations of Knowledge to the Existential Basis

Though this problem is obviously the nucleus of every theory in the sociology of knowledge, it has often been treated by implication rather than directly. Yet each type of imputed relation between knowledge and society presupposes an entire theory of sociological method and social causation. The prevailing theories in this field have dealt with one or both of two major types of relation: causal or functional, and the symbolic or organismic or logico-meaningful.[17]

15. Ibid., 2:5.
16. Ibid., 1:73.
17. The distinction between these have long been considered in European sociological thought. The most elaborate discussion in this country is that by Sorokin in *Social and Cultural Dynamics;* see, for example, vol. 1, chaps. 1–2.

Marx and Engels, of course, dealt solely with some kind of causal relation between the economic basis and ideas, variously terming this relation as "determination, correspondence, reflection, outgrowth, dependence," and so on. In addition, there is an "interest" or "need" relation; when strata have (imputed) needs at a particular stage of historical development, there is held to be a definite pressure for appropriate ideas and knowledge to develop. The inadequacies of these divers formulations have risen up to plague those who derive from the Marxist tradition in the present day.[18]

Since Marx held that thought is not a mere "reflection" of objective class position, as we have seen, this raises anew the problem of its imputation to a determinate basis. The prevailing Marxist hypotheses for coping with this problem involve a theory of history which is the ground for determining whether the ideology is "situationally adequate" for a given stratum in the society: this requires a hypothetical construction of what men *would think and perceive* if they were able to comprehend the historical situation adequately.[19] But such insight into the situation need not *actually* be widely current within particular social strata. This then, leads to the further problem of "false consciousness," of how ideologies which are neither in conformity with the interests of a class nor situationally adequate come to prevail.

A partial empirical account of false consciousness, implied in the *Manifesto,* rests on the view that the bourgeoisie control the content of culture and thus diffuse doctrines and standards alien to the interests of the proletariat.[20] Or, in more general terms, "the ruling ideas of each age have ever been the ideas of its ruling class." But this is only a partial account; at most it deals with the false consciousness of the subordinated class. It might, for example, partly explain the fact noted by Marx that even where the peasant proprietor "does belong to the proletariat by his position he does not believe that he does." It would not, however, be pertinent in seeking to account for the false consciousness of the ruling class itself.

Another, though not clearly formulated, theme which bears upon

18. Cf. the comments of Hans Speier, "The Social Determination of Ideas," *Social Research* 5 (1938): 182–205; C. Wright Mills, "Language, Logic and Culture," *American Sociological Review* 4 (1939): 670–80.

19. Cf. the formulation of Mannheim, *Ideology and Utopia,* 175 ff.; Georg Lukács, *Geschichte und Klassenbewusstsein* (Berlin: Malik, 1923), 61 ff.; Arthur Child, "The Problem of Imputation in the Sociology of Knowledge," *Ethics* 51 (1941): 200–214.

20. Marx and Engels, *The German Ideology,* 39: "In so far as they rule as a class and determine the extent and compass of an epoch, it is self-evident that they do this in their whole range, hence among other things rule also as thinkers, as producers of ideas, and regulate the production and distribution of the ideas of their age."

the problem of false consciousness runs throughout Marxist theory. This is the conception of ideology as being an *unwitting, unconscious* expression of "real motives," these being in turn construed in terms of the objective interests of social classes. Thus, there is repeated stress on the unwitting nature of ideologies: "Ideology is a process accomplished by the so-called thinker consciously indeed but with a false consciousness. The real motives impelling him remain unknown to him, otherwise it would not be an ideological process at all. Hence he imagines false or apparent motives."[21]

The ambiguity of the term "correspondence" to refer to the connection between the material basis and the idea can only be overlooked by the polemical enthusiast. Ideologies are construed as "distortions of the social situation";[22] as merely "expressive" of the material conditions;[23] and, whether "distorted" or not, as motivational support for carrying through real changes in the society.[24] It is at this last point, when "illusory" beliefs are conceded to provide motivation for action, that Marxism ascribes a measure of independence to ideologies in the historical process. They are no longer merely epiphenomenal. They enjoy a measure of autonomy. From this develops the notion of interacting factors in which the superstructure, though interdependent with the material basis, is also assumed to have some degree of independence. Engels explicitly recognized that earlier formulations were inadequate in at least two respects: first, that both he and Marx had previously overemphasized the economic factor and understated the role of interaction;[25] and second, that they had "neglected" the formal side—the way in which these ideas develop.[26]

21. Engels' letter to Mehring, 14 July 1893, in Marx, *Selected Works,* 1:388–89; cf. Marx, *Der achtzehnte Brumaire der Louis Bonaparte* (Hamburg, 1885), 33; idem, *Critique of Political Economy,* 12.

22. Marx, *Der achtzehnte Brumaire,* 39, where the democratic Montagnards indulge in self-deception.

23. Engels, *Socialism: Utopian and Scientific,* (1880; reprint, New York: International Publishers, 1935), 26–27; Cf. Engels, *Ludwig Feuerbach and the End of Classical German Philosophy,* in vol. 2 of *Karl Marx and Freidrich Engels: Selected Works* (1886; reprint, London: Lawrence and Wishart, 1950), 122–23: "The failure to exterminate the Protestant heresy *corresponded* to the invincibility of the rising bourgeoisie. . . . Here Calvinism proved itself to be the true religious disguise of the interests of the bourgeoisie of that time" [italics mine].

24. Marx grants motivational significance to the "illusions" of the burgeoning bourgeoisie, in *Der achtzehnte Brumaire,* 8.

25. Engels, letter to Joseph Bloch, 21 September 1890, in Marx, *Selected Works,* 1:383.

26. Engels, letter to Mehring, 14 July 1893, ibid., 1:390.

The Marx-Engels views on the connectives of ideas and economic substructure hold, then, that the substructure constitutes the framework which limits the range of ideas that will prove socially effective; ideas which do not have pertinence for one or another of the conflicting classes may arise, but will be of little consequence. Economic conditions are necessary, but not sufficient, for the emergence and spread of ideas that express either the interests or outlook, or both, of distinct social strata. There is no strict determination of ideas by economic conditions, but a definite predisposition. Knowing the economic conditions, we can predict the kinds of ideas which can exercise a controlling influence in a direction that can be effective. "Men make their own history, but they do not make it just as they please; they do not make it under circumstances chosen by themselves, but under circumstances directly found, given and transmitted from the past." And in the making of history, ideas and ideologies play a definite role: consider only the view of religion as "the opiate of the masses"; consider further the importance attached by Marx and Engels to making those in the proletariat "aware" of their "own interests." Since there is no fatality in the development of the total social structure, but only a development of economic conditions which make certain lines of change *possible* and probable, idea-systems may play a decisive role in the selection of one alternative which "corresponds" to the real balance of power rather than another alternative which runs counter to the existing power situation and is therefore destined to be unstable, precarious, and temporary. There is an ultimate compulsive which derives from economic development, but this compulsive does not operate with such detailed finality that no variation of ideas can occur at all.

The Marxist theory of history assumes that, *sooner or later,* idea-systems which are inconsistent with the actually prevailing and incipient power structure will be rejected in favor of those which more nearly express the actual alignment of power. It is this view that Engels expresses in his metaphor of the "zig-zag course" of abstract ideology: ideologies may temporarily deviate from what is compatible with the current social relations of production, but they are ultimately brought back in line. For this reason, the Marxist analysis of ideology is always bound to be concerned with the "total" historical situation, in order to account both for the temporary deviations and the later accommodation of ideas to the economic compulsives. But for this same reason, Marxist analyses are apt to have an excessive degree of "flexibility," almost to the point where *any* development can be explained away as a temporary aberration or deviation; where "anachronisms" and "lags" become labels for the explaining away of existing beliefs which do not correspond to theoretical expectations; where the concept of "accident" provides a ready means of saving the theory from facts that seem to challenge its validity.[27] Once a theory includes concepts such as "lags," "thrusts," "anachronisms," "accidents," "partial independence," and "ultimate dependence," it becomes so labile and so indistinct that it can be reconciled with virtually any configuration of data. Here, as in several other theories in the sociology of knowledge, a decisive question must be raised in order to determine whether we have a genuine theory: how can the theory be invalidated? In any given historical situation, which data will contradict and invalidate the theory? Unless this can be answered directly, unless the story involves statements which can be controverted by definite types of evidence, it remains merely a pseudo-theory.

Though Mannheim has gone far toward developing actual research procedures in the substantive sociology of knowledge, he has not appreciably clarified the connectives of thought and society. As he indicates, once a thought structure has been analyzed, there arises the problem of imputing it to definite groups. This requires not only an empirical investigation of the groups or strata which prevalently think in these terms but also an interpretation of why these groups, and not others, manifest this type of thought. This latter question implies a social psychology which Mannheim has not systematically developed.

The most serious shortcoming of Durkheim's analysis lies precisely in his uncritical acceptance of a naive theory of correspondence in which the categories of thought are held to "reflect" certain features of the group organization. Thus "there are societies in Australia and North America where space is conceived in the form of an immense circle, *because* the camp has a circular form . . . the social organization has been the model for the spatial organization and a reproduction of it."[28] In similar fashion, the general notion of time is derived from the specific units of time differentiated in social activities (ceremonies, feasts, rites).[29] The category of class and the modes of classification, which involve the notion of a hierarchy, are derived from social grouping and stratification. Those social categories are then "projected into our conception of the new world."[30] In summary, then, categories "ex-

27. Cf. Weber, *Gesammelts Aufsaetze zur Wissenschaftslehre* (Tübingen: Mohr-Siebeck, 1922), 166–70.

28. Durkheim, *Elementary Forms,* 11–12.

29. Ibid., 10–11.

30. Ibid., 148.

press" the different aspects of the social order.[31] Durkheim's sociology of knowledge also suffers from his avoidance of a social psychology.

The central relation between ideas and existential factors for Scheler is interaction. Ideas interact with existential factors which serve as selective agencies, releasing or checking the extent to which potential ideas find actual expression. Existential factors do not "create" or "determine" the content of ideas; they merely account for the *difference* between potentiality and actuality; they hinder, retard, or quicken the actualization of potential ideas. In a figure reminiscent of Clerk Maxwell's hypothetical demon, Scheler states, "in a definite fashion and order, existential factors open and close the sluice-gates to the flood of ideas." This formulation, which ascribes to existential factors the function of selection from a self-contained realm of ideas is, according to Scheler, a basic point of agreement between such otherwise divergent theorists as Dilthey, Troeltsch, Max Weber, and himself.[32]

Scheler operates as well with the concept of "structural identities" which refers to common presuppositions of knowledge or belief on the one hand, and of social, economic, or political structure on the other.[33] Thus, the rise of mechanistic thought in the sixteenth century, which came to dominate prior organismic thought, is inseparable from the new individualism, the incipient dominance of the power-driven machine over the hand tool, the incipient dissolution of *Gemeinschaft* into *Gesellschaft,* production for a commodity market, rise of the principle of competition in the ethos of western society, and so on. The notion of scientific research as an endless process through which a store of knowledge can be accumulated for practical application as the occasion demands and the total divorce of this science from theology and philosophy was not possible without a rise of a new principle of infinite acquisition characteristic of modern capitalism.[34]

In discussing such structural identities, Scheler does not ascribe primacy either to the socioeconomic sphere or to the sphere of knowledge. Rather, and this Scheler regards as one of the most significant propositions in the field, both are determined by the impulse structure of the elite which is closely bound up with the prevailing ethos. Thus, modern technology is not merely the application of a pure science based on observation, logic, and mathematics. It is far more the product of an orientation toward the control of nature which defined the purposes as

31. Ibid., 440.
32. Scheler, *Die Wissensformen,* 32.
33. Ibid., 56.
34. Ibid., 25; cf. 482–84.

well as the conceptual structure of scientific thought. This orientation is largely implicit and is not to be confused with the personal motives of scientists.

With the concept of structural identity, Scheler verges on the concept of cultural integration or *Sinnzusammenhang*. It corresponds to Sorokin's conception of a "meaningful cultural system" involving "the identity of the fundamental principles and values that permeate all its parts," which is distinguished from a "causal system" involving interdependence of parts.[35] Having constructed his types of culture, Sorokin's survey of criteria of truth, ontology, metaphysics, scientific and technological output, and so on, finds a marked tendency toward the meaningful integration of these with the prevailing culture.

Sorokin has boldly confronted the problem of how to determine the *extent* to which such integration occurs, recognizing, despite his vitriolic comments on the statisticians of our sensate age, that to deal with the extent or degree of integration necessarily implies some statistical measure. Accordingly, he developed numerical indexes of the various writings and authors in each period, classified these in their appropriate category, and thus assessed the comparative frequency (and influence) of the various systems of thought. Whatever the technical evaluation of the validity and reliability of these cultural statistics, he has directly acknowledged the problem overlooked by many investigators of integrated culture or *Sinnzusammenhaengen,* namely, the approximate degree or extent of such integration. Moreover, he plainly bases his empirical conclusions very largely upon these statistics. And these conclusions again testify that his approach leads to a statement of the problem of connections between existential bases and knowledge, rather than to its solution. Thus, to take a case in point, "empiricism" is defined as the typical sensate system of truth. The last five centuries, and more particularly the last century, represent "sensate culture par excellence!"[36] Yet, even in this flood-tide of sensate culture, Sorokin's statistical indices show only some 53 percent of influential writings in the field of "empiricism." And in the earlier centuries of this sensate culture—from the late sixteenth to the mid-eighteenth—the indices of empiricism are consistently lower than those for rationalism (which is associated, presumably, with an idealistic rather than a sensate culture).[37] The object of these observations is not to raise the question whether Sorokin's conclusions coincide with his statistical data: it is not to ask

35. Sorokin, *Social and Cultural Dynamics,* vol. 4, chap. 1; vol. 1, chap. 1.

36. Sorokin, *Social and Cultural Dynamics,* 2:51.

37. Ibid., 2:30.

why the sixteenth and seventeenth centuries are said to have a dominant "sensate system of truth" in view of these data. Rather, it is to indicate that even on Sorokin's own premises, overall characterizations of historical cultures constitute merely a first step, which must be followed by analyses of deviations from the central tendencies of the culture. Once the notion of *extent* of integration is introduced, the existence of types of knowledge which are not integrated with the dominant tendencies cannot be viewed merely as a "congeries" or as "contingent." Their *social* bases must be ascertained in a fashion for which an emanationist theory does not provide.

A basic concept which serves to differentiate generalizations about the thought and knowledge of an entire society or culture is that of the "audience" or "public" or what Florian Znaniecki calls "the social circle."[38] Men of knowledge do not orient themselves exclusively toward their data nor toward the total society, but to special segments of that society with their special demands, criteria of validity, of significant knowledge, of pertinent problems, and so on. It is through anticipation of these demands and expectations of particular audiences, which can be effectively located in the social structure, that men of knowledge organize their own work, define their data, seize upon problems. Hence, the more differentiated the society, the greater the range of such effective audiences, the greater the variation in the foci of scientific attention, of conceptual formulations, and of procedures for certifying claims to knowledge. By linking each of these typologically defined audiences to their distinctive social position, it becomes possible to provide a *wissenssoziologische* account of variations and conflicts of thought within the society, a problem that is necessarily bypassed in an emanationist theory. Thus, the scientists in seventeenth-century England and France who were organized in newly established scientific societies addressed themselves to audiences very different from those of the savants who remained exclusively in the traditional universities. The direction of their efforts, toward a "plain, sober, empirical" exploration of specific technical and scientific problems, differed considerably from the speculative, unexperimental work of those in the universities. Searching out such variations in effective audiences, exploring their distinctive criteria of significant and valid knowledge,[39] relating

38. Florian Znaniecki, *The Social Role of the Man of Knowledge* (New York: Columbia University Press, 1940); cf. Robert K. Merton's essay-review of the Znaniecki book: *American Sociological Review* 6 (1941): 111–15.—*Ed.*

39. The Rickert-Weber concept of "Wertbeziehung" (relevance to value) is but a first step in this direction; there remains the further task of differentiating the various sets of values and relating these to distinctive groups or strata within the society.

these to their position within the society, and examining the socio-psychological processes through which these operate to constrain certain modes of thought constitutes a procedure which promises to take research in the sociology of knowledge from the plane of general imputation to that of testable empirical inquiry. [. . .][40]

40. This is perhaps the most distinctive variation in the sociology of knowledge now developing in American sociological circles, and may almost be viewed as an American acculturation of European approaches. This development characteristically derives from the social psychology of G. H. Mead. Its pertinence in this connection is being indicated by C. Wright Mills, Gerard de Gré, and others. See Znaniecki's conception of the "social circle," in *Social Role*. See also the beginnings of empirical findings along these lines in the more general field of public communications: Paul F. Lazarsfeld and Robert K. Merton, "Studies in Radio and Film Propaganda," *Transactions*, New York Academy of Sciences, 2d ser., 6 (1943): 58–79.

18

The Rise of Modern Science (1938)

What we call the Protestant ethic was at once a direct expression of dominant values and an independent source of new motivation. It not only led men into particular paths of activity; it exerted a constant pressure for unswerving devotion to this activity. Its ascetic imperatives established a broad base for scientific inquiry, dignifying, exalting, consecrating such inquiry. If natural philosophers had hitherto found the search for truth its own reward, they now had further grounds for disinterested zeal in this pursuit. And society, once dubious of the merits of those who devoted themselves to the "petty insignificant details of a boundless Nature," largely relinquished its doubts.

The Puritan Spur to Science

As we have seen, the capital elements of the Puritan ethic were related to the general climate of sentiment and belief. In a sense, these tenets and convictions have been accentuated through a biased selection, but this sort of bias is common to all positive inquiries. Theories which attempt to account for certain phenomena require facts, but not all facts are equally pertinent to the problem in hand. "Selection," determined by the limits of the problem, is necessary. Among the cultural variables which invariably influence the development of science are the dominant values and sentiments. At least, this is our working hypothesis. In this particular period, religion in large part made articulate much of the prevailing value-complex. For this reason, we must consider the scope and bearing of the contemporary religious convictions, since these may have been related, in one way or another, to the upsurge of science. But not all of these convictions were relevant. A certain degree

From "Motive Forces of the New Science," in *Science, Technology and Society in Seventeenth-Century England* (1938; reprint, New York: Howard Fertig, Inc., 1970), 80–102, 104–10. Reprinted by permission of Howard Fertig, Inc.

of selection is therefore necessary for the purpose of abstracting those elements which had such a perceivable relation.

Puritanism attests to the theorem that nonlogical notions with a transcendental reference may nevertheless exercise a considerable influence upon practical behavior. If the fancies of an inscrutable deity do not lend themselves to scientific investigation, human action predicated upon a particular conception of this deity does. It was precisely Puritanism which built a new bridge between the transcendental and human action, thus supplying a motive force for the new science. To be sure, Puritan doctrines rested ultimately upon an esoteric theological base but these were translated into the familiar and cogent language of the laity.

Puritan principles undoubtedly represent to some extent an accommodation to the current scientific and intellectual advance. Puritans had to find some meaningful place for these activities within their view of life. But to dismiss the relationship between Puritanism and science with this formula would be superficial. Clearly, the psychological implications of the Puritan system of values independently conduced to an espousal of science, and we would grossly simplify the facts to accord with a preestablished thesis if we failed to note the convergence of these two movements. Moreover, the changing class structure of the time reinforced the Puritan sentiments favoring science since a large proportion of Puritans came from the rising class of bourgeoisie, of merchants. They manifested their increasing power in at least three ways. First, in their positive regard for both science and technology which reflected and promised to enhance this power. Equally notable was their increasingly fervent belief in progress, a profession of faith which stemmed from their growing social and economic importance. A third manifestation was their hostility toward the existing class structure which limited and hampered their participation in political control; an antagonism which found its climax in the English Civil War.

Yet we cannot readily assume that the bourgeoisie were Puritans solely because the Puritan ethic appealed to bourgeois sentiments. The converse was perhaps even more important, as Weber has shown. Puritan sentiments and beliefs prompting rational, tireless industry were such as to aid economic success. The same considerations apply equally to the close connection between Puritanism and science: the religious movement partly "adapted" itself to the growing prestige of science but it initially involved deep-seated sentiments which inspired its followers to a profound and consistent interest in the pursuit of science.

The Puritan doctrines were nothing if not lucid. If they provided motivation for the contemporary scientists, this should be evident from their words and deeds. Not that scientists, any more than other mortals, are necessarily aware of the sentiments which invest their way of life with meaning. Nonetheless, the observer may often, though not too readily perhaps, uncover these tacit valuations and bring them to light. Such a procedure should enable us to determine whether the putative consequences of the Puritan ethic truly proved effective. Moreover, it will disclose the extent to which all this was perceived by the very persons whom it most concerned. Accordingly, we shall examine the works of the natural philosopher who "undoubtedly did more than any one of his time to make Science a part of the intellectual equipment of educated men," Robert Boyle.[1] His investigations in physics, chemistry, and physiology, to mention only the chief fields of achievement of this omnifarious experimentalist, were epochal. Add to this the fact that he was one of the individuals who attempted explicitly to establish the place of science in the scale of cultural values and his importance for our particular problem becomes manifest. But Boyle was not alone. Equally significant were John Ray, whom Haller termed, a bit effusively, the greatest botanist in the history of man; Francis Willughby, who was perhaps as eminent in zoology as was Ray in botany; John Wilkins, one of the leading spirits in the "invisible college" which developed into the Royal Society; Oughtred, Barrow, Grew, Wallis, Newton;—but a complete list would comprise a scientific register of the time. Further materials for our purpose are provided by the Royal Society which, arising about the middle of the century, provoked and stimulated scientific advance more than any other immediate factor. In this instance we are particularly fortunate in possessing a contemporary account, written under the constant supervision of the members of the society in order that it might be representative of the motives and aims of that group. This is Thomas Sprat's widely read *History of the Royal-Society of London,* published in 1667. From these works, then, and from the writings of other scientists of the period, we may glean the chief motive forces of the new science.

To the "Glory of the Great Author of Nature"

Once science has become firmly institutionalized, its attractions, quite apart from any economic benefits it may bestow, are those of all elaborated and established social activities. These attractions are essentially twofold: generally prized opportunities of engaging in socially ap-

1. J. F. Fulton, "Robert Boyle and His Influence on Thought in the Seventeenth Century," *Isis* 18 (1932): 77–102. The range of Boyle's prolific writings is shown in Professor Fulton's exemplary bibliography.

proved patterns of association with one's fellows and the consequent creation of cultural products which are esteemed by the group. Such group-sanctioned conduct usually continues unchallenged, with little questioning of its reason for being. Institutionalized values are conceived as self-evident and require no vindication.

But all this is changed in periods of sharp transition. New patterns of conduct must be justified if they are to take hold and become the foci of social sentiments. A new social order presupposes a new scheme of values. And so it was with the new science. Unaided by forces which had already gripped man's will, science could claim only a bare modicum of attention and loyalty. But in partnership with a powerful social movement which induced an intense devotion to the active exercise of designated functions, science was launched in full career.

A clear manifestation of this process is not wanting. The Protestant ethic had pervaded the realm of science and had left its indelible stamp upon the attitude of scientists toward their work. Expressing their motives, anticipating possible objections, facing actual censure, scientists found motive, sanction and authority alike in the Puritan teachings. Such a dominant force as religion in those days was not and perhaps could not be compartmentalized and delimited. Thus in Boyle's highly commended apologia of science, we read:

> [I]t will be no venture to suppose that at least in the Creating of the Sublunary World, and the more conspicuous Stars, two of God's principal ends were, the Manifestation of His own Glory, and the Good of Men.[2]

This is the motif which recurs in constant measure in the very writings which often contain considerable scientific contributions: these worldly activities and scientific achievements manifest the Glory of God and enhance the Good of Man. The juxtaposition of the spiritual and the material is characteristic and significant. This culture rested securely on a substratum of utilitarian norms which identified the useful and the true. Puritanism itself had imputed a threefold utility to science. Natural philosophy was instrumental first, in establishing practical proofs of the scientist's state of grace; second, in enlarging control of nature, and third, in glorifying God. Science was enlisted in the service of individual, society, and deity. That these were adequate grounds could not be denied. They comprised not merely a claim to legitimacy, they afforded incentives that can not be readily overestimated. One need but look through the personal correspondence of seventeenth-century scientists to realize this.

2. Robert Boyle, *Some Considerations Touching the Usefulness of Experimental Natural Philosophy*, 2d ed. (Oxford, 1664), 22.

John Wilkins proclaimed the experimental study of Nature to be a most effective means of begetting a veneration for God.[3] Francis Willughby, probably the most eminent zoologist of the time, was prevailed upon to publish his works—which his excessive modesty had led him to deem unworthy of publication—only when Ray insisted that it was a means of glorifying God.[4] And Ray's panegyric of those who honor him by studying his works was so well received that five large editions were issued in some twenty years.[5]

Many "emancipated souls" of the present day, accustomed to a radical cleavage between religion and science and largely convinced of the relative social unimportance of religion for the modern Western world, are apt to generalize this state of affairs. To them, these recurrent pious phrases signify Machiavellian tactics or calculating hypocrisy or at best merely customary usage, but nothing of deep-rooted, motivating convictions. This evidence of extreme piety leads to the charge that *qui nimium probat nihil probat.* But such an interpretation is possible only upon the basis of an unwarranted extension of twentieth-century beliefs and attitudes to seventeenth-century society. Though it always serves to inflate the ego of the iconoclast and sometimes to extol the social images of his own day, "debunking" may often supplant truth with error. As a case in point, it is difficult to believe that Boyle, who manifested his piety by expending considerable sums to have the Bible translated into foreign tongues as well as in less material ways, was simply rendering lip service to Protestant beliefs. [. . .]

In various ways, then, general religious ideas were translated into concrete policy. This was no mere intellectual exercise. Puritanism transfused ascetic vigor into activities which, in their own right, could not as yet achieve self-sufficiency. It so redefined the relations between the divine and the mundane as to move science to the front rank of social values. As it happened, this was at the immediate expense of literary, and ultimately, of religious pursuits. For if the Calvinist God is irrational in the sense that he cannot be directly grasped by the cultivated intellect, He can yet be glorified by a clear-sighted, meticulous study of His natural works. Nor was this simply a compromise with science. Puritanism differed from Catholicism, which had gradually come to tolerate science, in demanding, not merely condoning, its pursuit. An "elastic concept," the Catholic and Protestant definitions of which differed so fundamentally as to produce entirely opposed conse-

3. John Wilkins, *Principles and Duties of Natural Religion,* 6th ed. (London, 1710), 236 ff.

4. See Edwin Lankester, ed., *Memorials of John Ray* (London, 1846), 14 n.

5. John Ray, *Wisdom of God* (London, 1691), 126–29, passim.

quences, the "glorification of God" thus came to be, in Puritan hands, the "fructification of science."

"For the Comfort of Mankind"

But Protestantism had afforded further grounds for the cultivation of science. The second dominant tenet in the Puritan ethos, it will be remembered, designated social welfare, the good of the many, as a goal ever to be held in mind. Here again, the contemporary scientists adopted an objective which carried with it, in addition to its own obvious merits, a cluster of religious sentiments. Science was to be fostered and nurtured as leading to the improvement of man's lot on earth by facilitating technologic invention. The Royal Society, we are told by its worthy historian, "does not intend to stop at some particular benefit, but goes to the root of all noble inventions."[6] Further, those experiments which do not bring with them immediate gain, are not to be condemned, for as the noble Bacon had declared, experiments of light ultimately conduce to a whole troop of inventions useful to the life and state of man.[7] This power of science to better the material condition of man, he continues, is, apart from its purely mundane value, a good in the light of the Evangelical Doctrine of Salvation by Jesus Christ. [. . .]

Earlier in the century, this keynote had been sounded in the resonant eloquence of that "veritable apostle of the learned societies," Francis Bacon. Himself the initiator of no scientific discoveries; unable to appreciate the importance of his great contemporaries, Gilbert, Kepler, and Galileo; naively believing in the possibility of a scientific method that "places all wits and understandings nearly on a level"; a radical empiricist holding mathematics to be of no use in science; he was, nevertheless, highly successful in being one of the principal propagandists in favor of positive social evaluation of science and of the disclaim of sterile scholasticism. As one would expect from the son of a "learned, eloquent and religious woman, full of puritanic fervor" who was admittedly influenced by his mother's attitudes,[8] he speaks in the *Advancement of Learning* of the true end of scientific activity as the "glory of the Creator and the relief of man's estate."[9] Since, as is quite clear

6. Thomas Sprat, *The History of the Royal-Society of London,* (London, 1667), 78–79.

7. Ibid., 245, 351 ff.

8. Cf. Mary Sturt, *Francis Bacon* (London: K. Paul, Trench, Trubner and Co., 1932), 6 ff.

9. In the *Novum Organum* (London: George Routledge and Sons, n.d.), 1: aphorism LXXXIX, 114, science is characterized as the handmaid of religion since it serves to display God's power. This is not, of course, a novel contention.

from many official and private documents, the Baconian teachings constituted the basic principles on which the Royal Society was patterned, it is not strange that in the charter of the society, the same sentiment is expressed.[10] Thomas Sydenham, the zealous Puritan,[11] likewise had a profound admiration for Bacon. And, like Bacon, he was prone to exaggerate the importance of empiricism to the very point of excluding theoretical interpretation entirely. "Pure intellectual curiosity . . . seemed to him, perhaps partly owing to the Puritan strain in his character, of little importance. He valued knowledge only either for its ethical value, as showing forth the glory of the Creator or for its practical value, as promoting the welfare of man."[12] Empiricism characteristically dominated Sydenham's approach to medicine which set above all the value of clinical observation, the "repeated, constant observation of particulars." It is of some interest that the greatest clinical observers of this century, Mayerne and Sydenham, were of Puritan stock.

Throughout there was the same point-to-point correlation between the principles of Puritanism and the avowed attributes, goals, and results of scientific investigation. Such was the contention of the protagonists of science at that time. If Puritanism demands systematic, methodic labor, constant diligence in one's calling, what, asks Sprat, is more active and industrious and systematic than the Art of Experiment, which "can never be finish'd by the perpetual labours of any one man, nay, scarce by the successive force of the greatest Assembly?"[13] Here is employment enough for the most indefatigable industry since even those hidden treasures of Nature which are farthest from view may be uncovered by pains and patience.[14]

Does the Puritan eschew idleness because it conduces to sinful thoughts (or interferes with the pursuit of one's vocation)? "What room can there be for low, and little things in a mind so *usefully* and successfully employ'd [as in natural philosophy]?"[15] Are plays and playbooks

10. In the second Charter, which passed the Great Seal on 22 April 1663, and by which the Society is governed to this day, we read that the studies of its Fellows "are to be applied to further promoting by the authority of experiments the sciences of natural things and of useful arts, to the glory of God the Creator, and the advantage of the human race." *The Record of the Royal Society,* (New York: Oxford University Press, 1912), 15. Note the increased emphasis upon utilitarianism.

11. See Joseph F. Payne, *Thomas Sydenham* (New York: Longmans, Green and Co., 1900), 7–8, passim, where abundant evidence of Sydenham's sternly Puritan background is presented.

12. Ibid., p. 234.

13. Ibid., 341–42.

14. Ray, *Wisdom of God,* 125.

15. Sprat, 344–45.

pernicious and flesh-pleasing (and subversive of more serious pursuits)? Then it is the "fittest season for Experiments to arise, to teach us a Wisdom, which springs from the depths of Knowledge, to shake off the shadows, and to scatter the mists [of the spiritual distractions brought on by the Theatre]."[16] And finally, is a life of earnest activity within the world to be preferred to monastic asceticism? Then recognize the fact that the study of natural philosophy "fits us not so well for the secrecy of a Closet: It makes us serviceable to the World." In short, science embodies patterns of behavior which are congenial to Puritan tastes. Above all, it embraces two highly prized values: utilitarianism and empiricism.

In a sense this explicit coincidence between Puritan tenets and the eminently desirable qualities of science as a calling which was suggested by the historian of the Royal Society is casuistry. No doubt it is partly an express attempt to fit the scientist *qua* pious layman into the framework of the prevailing moral and social values. Since both the constitutional position and the personal authority of the clergy were much more important then than now, it probably constituted a bid for religious and social sanction. Science, no less than literature and politics, was still, to some extent, subject to approval by the clergy.

But this is not the entire explanation. Present-day discussions of "rationalization" and "derivations" have been wont to becloud certain fundamental issues. It is true that the "reasons" adduced to justify one's actions often do not account satisfactorily for this behavior. It is also an acceptable hypothesis that ideologies seldom *give rise* to action and that both the ideology and action are rather the product of common sentiments and values upon which they in turn react. But these ideas can not be ignored for two reasons. They provide clues for detecting the basic values which motivate conduct. Such signposts can not be profitably neglected. Of even greater importance is the role of ideas in directing action into *particular* channels. *It is the dominating system of ideas that determines the choice between alternative modes of action which are equally compatible with the underlying sentiments.* Without such guidance and direction, non-logical action would become, within the limits of the value-system, random.

In the seventeenth century, the frequent recourse of scientists to religious vindication suggests first of all that religion was a sufficiently powerful social force to be invoked in support of an activity which was intrinsically less acceptable at the time. It also leads the observer to the peculiarly effective religious orientation which could invest scientific

16. Sprat, 362, 365–66.

pursuits with all manner of values and could thus serve to direct the interests of believers into the channels of science.

The efforts of Sprat, Wilkins, Boyle, or Ray to justify their interest in science do not represent simply opportunistic obsequiousness, but rather an earnest attempt to justify the ways of science to God. The Reformation had transferred the burden of individual salvation from the church to the individual, and it is this "overwhelming and crushing sense of the responsibility for his own soul" which accounts in part for both the acute longing for religious justification and the intense pursuit of one's calling. If science were not demonstrably a "lawful" and desirable calling, it dare not claim the attention of those who felt themselves "ever in the Great Taskmaster's eye." It is to this intensity of feeling that such apologias were due.

Rationalism and Empiricism

The exaltation of the faculty of reason in the Puritan ethos—based partly on the conception of rationality as a curbing device of the passions—inevitably led to a sympathetic attitude toward those activities which demand the constant application of rigorous reasoning. But again, in contrast to medieval rationalism, reason is deemed subservient and auxiliary to empiricism. [. . .] It is on this point probably that Puritanism and the scientific temper are in most salient agreement, for the combination of rationalism and empiricism which is so pronounced in the Puritan ethic forms the essence of the spirit of modern science. Puritanism was suffused with the rationalism of neoplatonism, derived largely through an appropriate modification of Augustine's teachings. But it did not stop there. Associated with the designated necessity of dealing successfully with the practical affairs of life within this world—a derivation from the peculiar twist afforded largely by the Calvinist doctrine of predestination and *certitudo salutis* through successful worldly activity—was an emphasis upon empiricism. These two currents brought to converge through the ineluctable logic of an internally consistent system of theology were so associated with the other attitudes of the time as to prepare the way for the acceptance of a similar coalescence in natural science.

The Puritan insistence upon empiricism, upon the experimental approach, was intimately connected with the identification of contemplation with idleness, of the expenditure of physical energy and the handling of material objects with industry. Experiment was the scientific expression of the practical, active and methodical bents of the Puritan. This is not to say, of course, that experiment was derived in any sense from Puritanism. But it serves to account for the ardent support of the new experimental science by those who had their eyes turned toward the other world and their feet firmly planted on this. [. . .]

It was a common practice for Puritans to couple their intense scorn for a "jejeune Peripatetick Philosophy" with extravagant admiration for "mechanical knowledge," which substituted fact for fantasy. From every direction, elements of the Puritan ethic converged to reinforce this set of attitudes. Active experimentation embodied all the select virtues and precluded all the baneful vices. It represented a revolt against that Aristotelianism which was traditionally bound up with Catholicism; it supplanted passive contemplation with active manipulation; it promised practical utilities instead of sterile figments; it established in indubitable fashion the glories of His creation. Small wonder that the Puritan transvaluation of values carried with it the consistent endorsement of experimentalism.

Empiricism and rationalism were canonized, beatified, so to speak. It may very well be that the Puritan ethos did not directly influence the method of science and that this was simply a parallel development in the internal history of science, but it becomes evident that, through the psychological sanction of certain modes of thought and conduct, this complex of attitudes made an empirically founded science commendable rather than, as in the medieval period, reprehensible or at best acceptable on sufferance. In short, Puritanism altered social orientations. It led to the setting up of a new vocational hierarchy, based on criteria which inevitably bestowed prestige upon the natural philosopher. As Professor Speier has well said, "There are no activities which are honorable in themselves and are held excellent in all social structures."[17] And one of the consequences of Puritanism was the reshaping of the social structure in such fashion as to bring esteem to science. This could not but have influenced the direction of some talents into scientific fields which otherwise would have devoted to callings which were, in another social context, more highly honored.

The Shift to Science

As the full import of the Puritan ethic manifested itself—even after the political failure of the civil war which should not be erroneously identified with the collapse of Puritan influence upon social attitudes—the sciences became foci of social interest. Their new fashionableness con-

17. Hans Speier, "Honor and Social Structure," *Social Research* 2 (1935): 79.

trasts with their previous state of comparative obscurity.[18] This was not without its effects. Many, who hitherto might have turned to theology or rhetoric or philology, were directed, through the subtle, largely unperceived and newly-arisen predisposition of society, into scientific channels. Thus, Thomas Willis, whose *Cerebri Anatome* was probably the most complete and accurate account of the nervous system up to that time and whose name is immortalized in the "circle of Willis," "was originally destined to theology, but in consequence of the unfavorable conditions of that age for theological science, he turned his attention to medicine."[19]

No less indicative of a shift of interest is the lament of Isaac Barrow, when he was professor of Greek at Cambridge: "I sit lonesome as an Attic owl, who has been thrust out of the companionship of all other birds; while classes in Natural Philosophy are full."[20] Evidently, Barrow's loneliness proved too much for him, for, as is well known, in 1663, he left his chair to accept the newly established Lucasian Professorship of Mathematics, in which he was Newton's predecessor.

The science-loving amateur, so prominent a feature of the latter part of the century, is another evidence of the effect of this new attitude. Nobles and wealthy commoners turned to science, not as a means of livelihood, but as an object of devoted interest. Particularly for these individuals direct utilitarian benefits of an economic nature were a wholly negligible consideration. Science afforded them an opportunity of devoting their energies to a highly honored task; an imperative duty as the comforts of unrelieved idleness vanished from the new scale of values.

In the history of science the most famous of these amateurs is of course Robert Boyle, but perhaps the best index of their importance is to be found in their role in the formation of the Royal Society.[21] Of those who, in that "wonderful pacifick year," 1660, constituted themselves into a definite association, a considerable number—among them Lord Brouncker, Boyle, Lord Bruce, Sir Robert Moray, Dr. Wilkins, Dr. Petty, and Abraham Hill—were amateurs of this type. Hardly less assiduous were the efforts of such virtuosi as Lord Willughby, John Evelyn, Samuel Hartlib, Francis Potter, and William Molineux.

18. Cf. Sprat, *History of the Royal Society,* 403.

19. Johann H. Baas, *Outlines of the History of Medicine,* (New York, 1889), 492.

20. Quoted by Hermann Hettner, *Geschichte der englischen Literatur* (Braunschweig, 1894), 16–17.

21. Martha Ornstein, *The Rôle of the Scientific Societies in the Seventeenth Century* (Chicago: University of Chicago Press, 1928), 91 ff.

This social emphasis on science had a peculiarly fruitful effect, probably because of the general state of scientific development. The methods and objects of investigation were frequently not at many removes from daily experience, and could hence be understood not only by the especially equipped but by a large number of persons with comparatively little technical education. To be sure, dilettantish interest in science seldom enriched its fruits directly, but it did serve to establish it more firmly as a socially estimable pursuit. And this same function was performed no less ably by Puritanism. The fact that science today is largely and probably completely divorced from religious sanctions is itself of interest as an example of the process of secularization. Having grown away from its religious moorings, science has in turn become a dominant social value to which other values are subordinated. Today it is much more common to subject the most diverse beliefs to the sanctions presumably afforded by science than to those yielded by religion; the increasing reference to scientific authority in contemporary advertisements and the eulogistic connotation of the very word "scientific" are perhaps not too farfetched illustrations of the enhanced prestige of science.[22]

The Process of Secularization

The beginnings of such secularization, faintly perceptible in the latter Middle Ages, were, in one sense, emerging more fully in the Puritan ethos. But the Puritan was not simply the last of the medievalists or the first of the moderns. He was both. It was in the system of Puritan values, as we have seen, that reason and experience began to be considered as independent means of ascertaining even religious truth. Faith which is unquestioning and not "rationally weighed," proclaimed Baxter, is not faith, but a dream or fancy or opinion. In effect this grants to science a power which may ultimately limit that of religion. This unhesitant assignment of virtual hegemony to science is based on the explicit assumption of the unity of knowledge, experiential and supersensuous, so that the testimony of science must perforce corroborate religious convictions.

This conviction of the mutually confirmatory nature of reason and revelation afforded a further basis for the favorable attitude toward ex-

22. As Professor Celestin Bouglé remarks, "Science has decidedly advanced to the first rank in the table of values." *The Evolution of Values* (New York: Henry Holt and Co., 1926), 201.

perimental studies, which, it is assumed, will simply reinforce basic theological dogmas. The active pursuit of science, thus freely sanctioned by unsuspecting religionists, however, created a new tone and habit of thought. As a consequence of this change, ecclesiastics, no longer able to appeal to commonly accepted teachings of science which seem rather to contravene various theological doctrines, are likely once again to substitute authority for reason in an effort to emerge victorious from the conflict.

In one direction, then, Puritanism led inevitably to the elimination of religious restriction on scientific work. This was the distinctly modern element of Puritan beliefs. But this did *not* involve the relaxation of religious discipline over conduct; quite the converse. Compromise with the world was intolerable. It had to be conquered and controlled through direct action and this ascetic compulsion was exercised in every area of life. [. . .]

Paradoxically but inevitably, then, this religious ethic, based on rigid theological foundations, furthered the development of the very scientific disciplines which later seem to confute orthodox theology.

The articulation of these several ideas, each the focus of strong sentiments, into a system which was all the more forceful precisely because it was psychologically rather than logically coherent, led to a long chain of consequences, not least of which was the substantial destruction of this very system itself. Though the corresponding religious *ethic,* as we shall see, does not necessarily lose its effectiveness as a social force immediately upon the undermining of its theological foundations, it tends to do so in time. This sketch of the influence of science in the processes of secularization should serve to make intelligible the diverse, quite opposed roles which religion and theology may play in their relations to science.

A religion—understood here, as throughout this essay, as those ethical and moral beliefs and practices which constitute a system of faith and worship, that is, as a religious ethic—may indirectly promote the cultivation of science, although specific scientific discoveries are at the same time vehemently attacked by theologians, who suspect their possibly subversive nature. Precisely because this pattern of interlocking and contradictory forces is so often unanalyzed, it is imperative that we distinguish clearly between the intentions and aims of religious leaders and the (frequently unforeseen) consequences of their teachings. Once this pattern is clearly understood, it is not surprising or inconsistent that Luther particularly, and Melanchthon less strongly, execrated the cosmology of Copernicus. [. . .] Likewise, Calvin frowned upon the acceptance of numerous scientific discoveries of his day, whereas the religious ethic which stemmed from him inspired the pursuit of natural science.

This failure to foresee some of the most fundamental social effects of their teachings was not solely the result of the Reformers' ignorance. It was rather an outcome of that type of nonlogical thought which deals primarily with the motives rather than the probable results of behavior. Righteousness of motive is the basic concern; other considerations, including that of the probability of attaining the end, are precluded. Action enjoined by a dominant set of values *must* be performed. But, with the complex interaction which society constitutes, the effects of action ramify. They are not restricted to the specific area in which the values were originally centered, occurring in interrelated fields specifically ignored at the outset. Yet it is precisely because these fields are in fact interrelated that the further consequences in adjacent areas react upon the basic system of values. It is this usually unlooked-for reaction which constitutes a most important factor in the process of secularization, of the transformation or breakdown of value-systems. This is the essential paradox of social action—the "realization" of values may lead to their renunciation.[23] [. . .]

The Integration of Religion and Science

It is thus to the religious ethos, not the theology, that we must turn if we are to understand the integration of science and religion in seventeenth-century England.

Perhaps the most directly effective belief in this ethos for the sanction of natural science held that the study of nature enables a fuller appreciation of His works and thus leads us to admire and praise the Power, Wisdom, and Goodness of God manifested in His creation. Though this conception was not unknown to medieval thinkers, the consequences deduced from it were entirely different. In the seventeenth century, the contemporary emphasis upon empiricism led to the investigation of nature primarily through experience. This difference in interpretation of substantially the same doctrine can only be understood in the light of the different values permeating the two cultures. Cloistered contemplation was forsaken; active experimentation was introduced.

The Royal Society was of inestimable importance, both in the propagation of this new point of view and in its actual application. [. . .]

23. See chap. 15 in this volume.—*Ed.*

On such bases as these, then, religion was invoked as a sanctioning power of science. But it is necessary to place this and the similar connections previously noted in proper perspective. This is imperative if we are to correct a seeming implication of this discussion, namely, that religion was the independent and science the dependent variable during this period, although as was remarked at the outset, this is not in the least our intention.

The integration of the Puritan ethic with the accelerated development of science seems undeniable, but this is simply to maintain that they were elements of a culture which was largely centered on the values of utilitarianism and empiricism. It is perhaps not too much to say, with Lecky, that the acceptance of every great change of belief depends less upon the intrinsic force of its doctrines or the personal capabilities of its proponents than upon the previous social changes which are seen—a posteriori, it is true—to have brought the new doctrines into congruence with the dominant values of the period. The reanimation of ancient learning; the hesitant, but perceptibly defined, instauration of science; the groping, yet persistent, intensification of economic tendencies; the revolt against scholasticism;—all helped bring to a focus the social situation in which the Protestant beliefs and scientific interests found acceptance.[24] But to realize this is simply to recognize that both Puritanism and science were components of a vastly complicated system of mutually dependent factors. If some comprehensible order is to be attained, a fraction of this complex situation must be substituted for the whole; a defensible procedure only if this provisional formulation is not confused with a complete explanation. [. . .]

Religion, then, was a prime consideration and as such its teachings were endowed with a power which stands forth with striking emphasis. Moreover, there is no need of entering into the matter of the motivations of individual scientists to trace this influence, for such indications are really supererogatory for our study. Irrespective of the possibility of tracing its direct influence upon specific individuals, it is apparent that the religious ethic, considered as a social force, so consecrated science as to make it a highly respected and laudable focus of attention.

It is this *social* animus which facilitated the development of science by removing the incubus of derogatory social attitudes and instilling favorable ones instead. It is precisely this social influence which would seldom be noticed by the individual scientists upon whom it impinged. Yet since religion directly exalted science, since religion was a dominant

24. William E. H. Lecky, *A History of the Rise and Influence of Rationalism in Europe* (London, 1865), 1:6.

social force, since science was obviously held in higher social esteem during the latter part of the century, we must infer that religion played an important role in this changed attitude, particularly because of so much external corroborative evidence. This minimum of inference is inescapable.

Community of Tacit Assumption in Science and Puritanism

Up to this point we have been concerned, in the main, with the directly felt sanction of science by the Protestant ethic. Now, while this was of great importance, there was still another relationship which, subtle and difficult of apprehension though it be, was perhaps of equal significance. Puritanism was one element in the preparation of a set of largely implicit assumptions which made for the ready acceptance of the characteristic scientific temper of the seventeenth and subsequent centuries. It is not simply that Protestantism promoted free inquiry, *libre examen*, or decried monastic ascetism. These oft-mentioned characteristics touch only the bare surface of the relationship.

It has become manifest that in each age there is a system of science which rests upon a set of assumptions, usually implicit and seldom, if ever, questioned by most of the scientific workers of the time.[25] The basic assumption in modern science, that is, in the type of scientific work which becoming pronounced in the seventeenth century has since continued, "is a widespread conviction in the existence of an *Order of Things*, and, in particular, of an Order of Nature."[26] This belief, this faith, for at least since Hume it must be recognized as such, is simply "impervious to the demand for a consistent rationality."[27]

In the systems of scientific thought of Galileo, of Newton, and of their successors, the testimony of experiment is the ultimate criterion of truth, but as has been suggested, the very notion of experiment is ruled out without the prior *assumption* that Nature constitutes an intelligible order, that when appropriate questions are asked, she will answer, so to speak. Hence this assumption is final and absolute.[28] Now, as Professor Whitehead has so well indicated, this "faith in the possibility of

25. A. N. Whitehead, *Science and the Modern World* (New York: Macmillan, 1931), chap. 1; A. E. Heath in *Isaac Newton: A Memorial Volume,* ed. W. J. Greenstreet for the Mathematical Association (London: G. Bell, 1929), 133; E. A. Burtt, *The Metaphysical Foundations of Modern Physical Science* (New York: Harcourt Brace, 1927).

26. Whitehead, *Science and the Modern World,* 5.

27. Ibid., 6.

28. Cf. E. A. Burtt in *Isaac Newton: A Memorial Volume,* 139. For a classic expression of this scientific faith, see Isaac Newton's "Rules of Reasoning in Philosophy," in the *Principia,* vol. 2, trans. Andrew Motte (London, 1803), 160 ff.

science, generated antecedently to the development of modern scientific theory, is an unconscious derivative from medieval theology."[29] But this conviction, prerequisite condition of modern science though it is, was not sufficient to induce its development. What was needed was a constant interest in searching for this order of nature in an empirical and rational fashion, i.e., an *active interest* in this world and in its occurrences plus a specifically empirical approach. With Protestantism religion provided this interest—it actually imposed obligations of intense concentration on secular activity with an emphasis on experience and reason as bases for action and belief. The good works which for the sects influenced by Calvinism provided conviction of grace are not to be confused with the Catholic conception of good works. In the Puritan case it involved the notion of a transcendental God and an orientation to the "other world," it is true, but it also demanded a mastery over this world through a study of its processes; while in the Catholic instance, it demanded complete absorption, save for an unbanishable minimum, in the supersensuous, in an intuitive love of God.

It is just at this point that the Protestant emphasis upon reason and experience is of prime importance. In the Protestant system of religion, there is the unchallenged axiom, *gloria Dei,* and, as we have seen, the scheme of behavior which was nonlogically linked with this principle tends to assume a utilitarian tinge. Virtually all conceptions other than this are subject to, nay, demand, the examination of reason and experience. Even the Bible as final and complete authority was subject to the interpretation of the individual upon these bases, for though the Bible is infallible, the "meaning" of its content must be sought, as will be remembered from Baxter's discussion of this point. The similarity between the approach and intellectual attitude implicit in the religious and scientific systems is of more than passing interest. This religious point of view could not but mold an attitude of looking at the world of sensuous phenomena which was highly conducive to the willing acceptance and, indeed, preparation for, the same attitude in science. [. . .]

The conviction in immutable law is as pronounced in the doctrine of predestination as in scientific investigation: "the immutable law is there and must be acknowledged."[30] [. . .]

29. See Whitehead, *Science and the Modern World,* 19 and preceding for a discussion of this development.

30. Hermann Weber, *Die Theologie Calvins* (Berlin: Elsner, 1930), 29 "das unabänderlich Gesetz ist da und muss anerkannt werden." The significance of the doctrine of God's foreknowledge for the reinforcement of the belief in natural law is remarked by H. T. Buckle in *History of Civilization in England* (New York: Boni, 1925), on page 482. It is significant that the first writer who maintained that even lotteries are governed by purely natural laws was a Puritan minister, Thomas Gataker, in his curious little book,

The willingness of the Protestant leaders to have reason and experience "test" all religious beliefs, save the basic assumption which, just as in science, is simply accepted as a matter of faith, is in part grounded upon the previously mentioned conviction of the inherent consistency, congruence and mutually confirmatory nature of all knowledge, sensory and supersensory. It would seem, then, that there is, to some extent, a community of assumptions in Protestantism and science: in both there is the unquestioned basic assumption upon which the entire system is built by the utilization of reason and experience. Within each context there is rationality, though the bases are nonrational. The significance of this fundamental similarity is profound, though it could hardly have been consciously recognized by those whom it influenced: religion had, for whatever reasons, adopted a cast of thought which was essentially that of science so that there was a reinforcement of the typically scientific attitudes of the period. This society was permeated with attitudes toward natural phenomena which were derived from both science and religion and which unwittingly enhanced the continued prevalence of conceptions characteristic of the new science. [. . .]

On the Nature and Use of Different Kinds of Lots (London, 1619). This assumption, which ran over the barriers of religious differences, is not unrelated to the later development of "political arithmetic" by Graunt, Petty, and Halley.

19

Insiders and Outsiders (1972)

The sociology of knowledge has long been regarded as a complex and esoteric subject, remote from the urgent problems of contemporary social life. To some of us, it seems quite the other way. Especially in times of great social change, precipitated by acute social conflict and attended by much cultural disorganization, the perspectives provided by the various sociologies of knowledge bear directly upon problems agitating the society. It is then that differences in the values, commitments, and cognitive orientations of conflicting groups become deepened into basic cleavages, both social and cultural. As the society becomes polarized, so do the contending claims to truth. At the extreme, an active and reciprocal distrust between groups finds expression in intellectual perspectives that are no longer located within the same universe of discourse. The more deep-seated the mutual distrust, the more does the argument of the other appear so palpably implausible, even absurd, that one no longer inquires into substance or logical structure to assess its truth claims. Instead, one confronts the other's argument with an entirely different question: how does it happen to be advanced at all? Thought thus becomes altogether functionalized, interpreted only in terms of its presumed social or economic or psychological sources and functions. In the political arena, where the rules of the game often condone and sometimes support the practice, this involves reciprocated attacks on the integrity of the opponent; in the academic forum, where the norms are somewhat more restraining, it leads to reciprocated ideological analyses (which easily decline into innuendo). In both, the process feeds upon and nourishes collective insecurities.[1]

From "Insiders and Outsiders: A Chapter in the Sociology of Knowledge," *American Journal of Sociology* 77 (July 1972): 9–47.

1. This passage on the socio-cultural conditions making for intensified interest in the sociology of knowledge draws upon the paradigm for the sociology of knowledge which appears as chapter 17 in this volume.—*Ed.*

Social Change and Social Thought

This conception of the social sources of intensified interest in the sociology of knowledge and some of the theoretical difficulties which it fosters plainly has the character, understandably typical in the sociology of scientific knowledge, of a self-exemplifying idea. It posits reciprocal connections between thought and society, in particular the social conditions that make for or disrupt a common universe of intellectual discourse within which the most severe disagreements can take place. Michael Polanyi has noted, more perceptively than anyone else I know,[2] how the growth of knowledge depends upon complex sets of social relations based on a largely institutionalized reciprocity of trust among scholars and scientists. In one of his many passages on this theme, Polanyi observes that

> in an ideal free society each person would have perfect access to the truth: to the truth in science, in art, religion, and justice, both in public and private life. But this is not practicable; each person can know directly very little of truth and must trust others for the rest. Indeed, to assure this process of mutual reliance is one of the main functions of society. It follows that such freedom of the mind as can be possessed by men is due to the services of social institutions, which set narrow limits to man's freedom and tend to threaten it even within those limits. The relation is analogous to that between mind and body: to the way in which the performance of mental acts is restricted by limitations and distortions due to the medium which makes these performances possible.[3]

But as cleavages deepen between groups, social strata, or collectivities of whatever kind, the social network of mutual reliance is at best strained and at worst broken. In place of the vigorous but cognitively disciplined mutual checking and rechecking that operates to a significant extent, though never of course totally, within the social in-

2. See Michael Polanyi, *Personal Knowledge* (London: Routledge and Kegan Paul, 1958); *The Study of Man* (London: Routledge and Kegan Paul, 1959); *Science, Faith and Society* (Chicago: University of Chicago Press, 1964); and *The Tacit Dimension* (London: Routledge and Kegan Paul, 1967). Polanyi's detailed development of this theme over the years represents a basic contribution to the sociology of science by providing a model of the various overlapping cognitive and social structures of intellectual disciplines. John Ziman (*Public Knowledge* [Cambridge: Cambridge University Press, 1969]) has useful observations along these lines, and Donald T. Campbell ("Ethnocentrism of Disciplines and the Fish-Scale Model of Omniscience," in *Interdisciplinary Relationships in the Social Sciences,* ed. Muzafer Sherif and Carolyn W. Sherif [Chicago: Aldine Press, 1969]) has contributed some typically Campbellian (that is, imaginative and evocative) thinking on the subject, in developing his "fish-scale model" of overlapping disciplines.

3. Polanyi, *Study of Man,* 68.

stitutions of science and scholarship, there develops a strain toward separatism, in the domain of the intellect as in the domain of society. Partly grounded mutual suspicion increasingly substitutes for partly grounded mutual trust. There emerge claims to group-based truth: Insider truths that counter Outsider untruths and Outsider truths that counter Insider untruths.

In our time, vastly evident social change is being initiated and funneled through a variety of social movements. These are formally alike in their objectives of achieving an intensified collective consciousness, a deepened solidarity, and a new or renewed primary or total allegiance of their members to certain social identities, statuses, groups, or collectivities. Inspecting the familiar list of these movements centered on class, race, ethnicity, age, sex, religion, and sexual disposition, we note two other instructive similarities among them. First, the movements are principally formed on the basis of ascribed rather than acquired statuses and identities, with eligibility for inclusion being in terms of who you are rather than what you are (in the sense of status being contingent on role performance). And second, the movements largely involve the public affirmation of pride in statuses and solidarity with collectivities that have long been socially downgraded, stigmatized, or otherwise victimized in the social system. As with group affiliations generally, these newly reinforced social identities find expression in affiliative symbols of distinctive speech, bodily appearance, dress, public behavior, and, not least, assumptions and foci of thought.

The Insider Doctrine

Within this context of social change, we come upon the contemporary relevance of a long-standing problem in the sociology of knowledge: the problem of patterned differentials among social groups and strata in access to knowledge. In its strong form, the claim is put forward as a matter of epistemological principle that particular groups in each moment of history have *monopolistic access* to certain kinds of knowledge. In the weaker, more empirical form, the claim holds that some groups have *privileged access,* with other groups also being able to acquire that knowledge for themselves but at greater risk and cost.

Claims of this general sort have been periodically introduced. For one imposing and consequential example, Marx, a progenitor of the sociology of knowledge as of much else in social thought, advanced the claim that after capitalistic society had reached its ultimate phase of development, the strategic location of one social class would enable it to achieve an understanding of the society that was exempt from false consciousness.[4] For another, altogether unimposing but also consequential example involving ascribed rather than achieved status, the Nazi *Gauleiter* of science and learning, Ernst Krieck,[5] expressed an entire ideology in contrasting the access to authentic scientific knowledge by men of unimpeachable Aryan ancestry with the corrupt versions of knowledge accessible to non-Aryans. Krieck could refer without hesitation to "Protestant and Catholic science, German and Jewish science." [. . .]

For our purposes, we need not review the array of elitist doctrines which have maintained that certain groups have, on biological or social grounds, monopolistic or privileged access to new knowledge. Differing in detail, the doctrines are alike in distinguishing between Insider access to knowledge and Outsider exclusion from it.

Social Bases of Insider Doctrine

The ecumenical problem of the interaction between a rapidly changing social structure and the development of Insider and Outsider doctrines is examined here in a doubly parochial fashion. Not only are my observations largely limited to the United States in our time but they are further limited to the implications of doctrines advocated mainly by spokesmen for certain black social movements, since these movements have often come to serve as prototypical for the others (women, youth, homosexuals, and other ethnic collectivities).

Although Insider doctrines have been intermittently set forth by white elitists through the centuries, white male Insiderism in American sociology during the past generations has largely been of the tacit or de facto rather than doctrinal or principled variety. It has simply taken the form of patterned expectations about the appropriate selection of specialties and of problems for investigation. The handful of black sociologists were in large part expected, as a result of social selection and self-selection, to study problems of black life and relations between the

4. Observations on the advantaged position of the proletariat for the perception of historical and social truth are threaded throughout Marx's writings. For some of his crucial passages, see his *Poverty of Philosophy* (Moscow: Foreign Languages Press, n.d.), 125–26, for example. On Marx's thinking along these lines, Georg Lukács, in spite of his own disclaimers in the new introduction to his classic work, *History and Class Consciousness* (1923; reprint, Cambridge: M.I.T. Press, 1971), remains of fundamental importance; see especially pages 47–81 and 181–209.

5. See Ernst Krieck, *Nationalpolitische Erziehung* (Leipzig: Armanen Verlag, 1935).

races just as the handful of women sociologists were expected to study problems of women, principally as these related to marriage and the family.

In contrast to this de facto form of Insiderism, an explicitly doctrinal form has in recent years been put forward most clearly and emphatically by some black intellectuals. In its strong version, the argument holds that, as a matter of social epistemology, *only* black historians can truly understand black history, *only* black ethnologists can understand black culture, *only* black sociologists can understand the social life of blacks, and so on. In the weaker form of the doctrine, some practical concessions are made. With regard to programs of black studies, for example, it is proposed that some white professors of the relevant subjects might be brought in since there are not yet enough black scholars to staff all the proliferating programs of study. But as Nathan Hare, the founding publisher of the *Black Scholar,* stated several years ago, this is only on temporary and conditional sufferance: "Any white professors involved in the program would have to be black in spirit in order to last. The same is true for 'Negro' professors."[6] Apart from this kind of limited concession, the Insider doctrine maintains that there is a body of black history, black psychology, black ethnology, and black sociology which can be significantly advanced only by black scholars and social scientists.

In its fundamental character, this represents a major claim in the sociology of knowledge that implies the balkanization of social science, with separate baronies kept exclusively in the hands of Insiders bearing their credentials in the shape of one or another inherited status. Generalizing the specific claim, it would appear to follow that if only black scholars can understand blacks, then only white scholars can understand whites. Generalizing further from race to nation, it would then appear, for example, that only French scholars can understand French society and, of course, that only Americans, not their external critics, can truly understand American society. Once the basic principle is adopted, the list of Insider claims to a monopoly of knowledge becomes indefinitely expansible to all manner of social formations based on ascribed (and, by extension, on some achieved) statuses. It would thus seem to follow that only women can understand women—and men, men. On the same principle, youth alone is capable of understanding youth just as, presumably, only the middle-aged are able to understand

6. Nathan Hare as quoted by John H. Bunzel in "Black Studies at San Francisco State," *Public Interest* 13 (Fall 1968): 32.

their age peers. So, too, as we shift to the hybrid cases of ascribed and acquired statuses in varying mix, on the Insider principle, proletarians alone can understand proletarians, and presumably capitalists, capitalists; only Catholics, Catholics; Jews, Jews; and, to halt the inventory of socially atomized claims to knowledge with a limiting case that on its face would seem to have some merit, it would then plainly follow that only sociologists can possibly understand their fellow sociologists.[7]

In all these applications, the doctrine of extreme Insiderism represents a new credentialism. This is the credentialism of ascribed status, in which understanding becomes accessible only to the fortunate few or fortunate many who are to the manner born. It thus contrasts with the credentialism of achieved status that characterizes systems of meritocracy.[8]

Extreme Insiderism moves toward a doctrine of *social* solipsism that is isomorphic to *individual* solipsism.[9] In the first, the group or collectivity has a monopoly of knowledge about itself, just as in the second, the individual person has absolute privacy of knowledge about himself (as we recognize in that proverbial and exemplary aching tooth which he and he alone can authentically experience). And like individual solipsists who address themselves to the other minds whose existence their doctrine denies, the group solipsists (as we shall see) have an abiding commitment, not a merely empirical proclivity, to deny in continued practice what they affirm in basic principle.

The Insider doctrine can be put in the vernacular with no great loss in meaning: you have to be one in order to understand one. In less idiomatic language, the doctrine holds that one has monopolistic or privi-

7. As we shall see, this is a limiting type of case that merges into quite another type, since as a fully acquired status, rather than an ascribed one, that of the sociologist (or physician or physicist) presumably presupposes functionally relevant expertise.

8. But, as we shall see, when the extreme Insider position is transformed from a doctrine of assumptions-treated-as-established-truth into a set of questions about the distinctive roles of Insiders and Outsiders in intellectual inquiry, there develops a convergence though not coincidence between the assumptions underlying credentials based on ascribed status. In the one, early socialization in the culture or subculture is taken to provide readier access to certain kinds of understanding; in the other, the component in adult socialization represented by disciplined training in one or another field of learning is taken to provide a higher probability of access to certain other kinds of understanding.

9. As Joseph Agassi ("Privileged Access," *Inquiry* 12 [winter 1969]: 420–26) reminds us, the term "methodological solipsism" was introduced by Rudolf Carnap to designate the theory of knowledge known as sensationalism: "the doctrine that all knowledge—of the world and of one's own self—derives from sensation." The belief that all one really knows is one's subjective experience is sometimes described as the "egocentric predicament."

leged access to knowledge, or is excluded from it, by virtue of group membership or social position. For some, the notion appears in the form of a question-begging pun: Insider as Insighter, one endowed with special insight into matters necessarily obscure to others, thus possessed of penetrating discernment. Once adopted, the pun provides a specious solution, but the serious Insider doctrine has its own rationale.

In short, the Insider doctrine holds that the Outsider has a structurally imposed incapacity to comprehend alien groups, statuses, cultures, and societies. Unlike the Insider, the Outsider has neither been socialized in the group nor has engaged in the run of experience that makes up its life, and therefore cannot have the direct, intuitive sensibility that alone makes empathic understanding possible. Only through continued socialization in the life of a group can one become fully aware of its symbolisms and socially shared realities: only so can one understand the fine-grained meanings of behavior, feelings, and values; only so can one decipher the unwritten grammar of conduct and the nuances of cultural idiom. Or, to take a specific expression of this thesis by Ralph W. Conant (1968): "Whites are not and never will be as sensitive to the black community precisely because they are not part of that community." [. . .]

A somewhat less stringent version of the doctrine maintains only that Insider and Outsider scholars have significantly different foci of interest. The argument goes somewhat as follows: The Insiders, sharing the deepest concerns of the group or at the least being thoroughly aware of them, will so direct their inquiries as to have them be relevant to those concerns. So, too, the Outsiders will inquire into problems relevant to the distinctive values and interests which they share with members of *their* group. But these are bound to differ from those of the group under study if only because the Outsiders occupy different places in the social structure.

This is a hypothesis which has the not unattractive quality of being readily amenable to empirical investigation. It should be possible to compare the spectrum of research problems about, say, the black population in the country that have been investigated by black sociologists and by white ones, or say, the spectrum of problems about women that have been investigated by female sociologists and by male ones, in order to find out whether the foci of attention in fact differ and if so, to what degree and in which respects. [. . .] There is theoretical reason to suppose that the foci of research adopted by Insiders and Outsiders and perhaps their categories of analysis as well will tend to differ. At least,

Max Weber's notion of *Wertbeziehung* suggests that differing social locations, with their distinctive interests and values, will affect the selection of problems for investigation.[10]

Unlike the stringent version of the doctrine which maintains that Insiders and Outsiders must arrive at different (and presumably incompatible) findings and interpretations even when they do examine the same problems, this weaker version argues only that they will not deal with the same questions and so will simply talk past one another. With the two versions combined, the extended version of the Insider doctrine can also be put in the vernacular: one must not only be one in order to understand one; one must be one in order to understand what is most worth understanding.

Clearly, the social epistemological doctrine of the Insider links up with what Sumner long ago defined as ethnocentrism: "the technical name for [the] view of things in which one's own group is the center of everything, and all others are scaled and rated with reference to it." Sumner then goes on to include as a component of ethnocentrism, rather than as a frequent correlate of it (thus robbing his idea of some of its potential analytical power), the belief that one's group is superior to all cognate groups: "each group nourishes its own pride and vanity, boasts itself superior, exalts its own divinities, and looks with contempt on outsiders."[11] For although the practice of seeing one's own group as the center of things is empirically correlated with a belief in superiority, centrality and superiority need to be kept analytically distinct in order to deal with patterns of alienation from one's membership group and contempt for it.[12] [. . .]

There is a widespread tendency to glorify the ingroup, sometimes to that degree in which it qualifies as chauvinism: the extreme, blind, and often bellicose extolling of one's group, status, or collectivity. We need not abandon the useful concept "chauvinism" merely because it has lately become adopted as a vogue word, blunted in meaning through

10. See Weber's *Gesammelte Aufsätze zur Wissenschaftslehre* (1922; reprint, Tübingen: J. C. B. Mohr [P. Siebeck], 1951), 146–214.

11. William Graham Sumner, *Folkways* (Boston: Ginn and Co., 1907), 13.

12. By introducing their useful term "xenocentrism" to refer to both basic *and* favorable orientations to groups other than one's own, Donald P. Kent and Robert G. Burnight ("Group Centrism in Complex Societies," *American Journal of Sociology* 57 [November 1951]: 256–59) have retained Sumner's unuseful practice of prematurely combining centrality and evaluation in the one concept rather than keeping them analytically distinct. The analytical distinction can be captured terminologically by treating "xenocentrism" as the generic term, with the analytically distinct components of favorable orientation to nonmembership groups (as with the orientation of many white middle-class Americans toward blacks) being registered in the term "xenophilia" and the unfavorable orientation by Pareto's term "xenophobia."

indiscriminate use as a rhetorical weapon in intergroup conflict. Nor need we continue to confine the scope of the concept, as it was confined in its origins and later by Lasswell[13] in his short, incisive discussion of it, to the special case of the state or nation. The concept can be usefully, not tendentiously, extended to designate the extreme glorification of *any* social formation.

Chauvinism finds fullest ideological expression when groups are subject to the stress of acute conflict. Under the stress of war, for example, scientists have been known to violate the values and norms of universalism in which they were socialized, allowing their status as nationals to dominate over their status as scientists. Thus, at the outset of World War I, almost a hundred German scholars and scientists—including many of the first rank, such as Brentano, Ehrlich, Haber, Eduard Meyer, Ostwald, Planck, and Schmoller—could bring themselves to issue a manifesto that impugned the contributions of the enemy to science, charging them with nationalistic bias, logrolling, intellectual dishonesty, and, when you came right down to it, the absence of truly creative capacity. The English and French scientists were not far behind in advertising their own brand of chauvinism.[14]

Ethnocentrism, then, is not a historical constant. It becomes intensified under specifiable conditions of acute social conflict. When a nation, race, ethnic group, or other powerful collectivity has long extolled its own admirable qualities and, expressly or by implication, deprecated the qualities of others, it invites and provides the potential for counter-ethnocentrism. And when a once largely powerless collectivity acquires a socially validated sense of growing power, its members experience an intensified need for self-affirmation. Under such circumstances, collective self-glorification, found in some measure among all groups, becomes a predictable and intensified counter-response to long-standing belittlement form without.[15]

13. Harold D. Lasswell, "Chauvinism," in the *Encyclopedia of the Social Sciences* (New York: Macmillan, 1937), 3:361.

14. Current claims of Insiderism still have a distance to go, in the academic if not the political forum, to match the chauvinistic claims of those days. For collections of such documents, see Gabriel Pettit and Maurice Leudet, *Les allemands et la science* (Paris, 1916); Pierre Duhem, *La science allemande* (Paris: Hermann, 1915); Hermann Kellermann, *Der Krieg der Geister* (Weimar, 1915); and Karl Kherkhof, *Der Krieg gegen die Deutsche Wissenschaft* (Halle, 1933).

See also chaps. 20 and 21 in this volume.—*Ed.*

15. This is not a prediction after the fact. E. Franklin Frazier repeatedly made the general point, and I have examined this pattern in connection with the self-fulfilling prophecy; see Frazier, *The Negro in the United States* (New York: Macmillan, 1949), and *Black Bourgeoisie* (Glencoe, Ill.: Free Press, 1957). (Also see chap. 16 in this volume.—*Ed*)

So it is that, in the United States, the centuries-long institutionalized premise that "white (and for some, presumably only white) is true and good and beautiful" induces, under conditions of revolutionary change, the counterpremise that "black (and for some, presumably only black) is true and good and beautiful." And just as the social system has for centuries operated on the tacit or explicit premise that in cases of conflict between whites and blacks, the whites are presumptively right, so there now develops the counterpremise, finding easy confirmation in the long history of injustice visited upon American Negroes, that in such cases of conflict today, the blacks are presumptively right.

What is being proposed here is that the epistemological and ontological claims of the Insider to monopolistic or privileged access to social truth develop under particular social and historical conditions. Social groups or strata on the way up develop a revolutionary élan. The new thrust to a larger share of power and control over their social and political environment finds various expressions, among them claims to a unique access to knowledge about their history, culture, and social life.

On this interpretation, we can understand why this Insider doctrine does not argue for a black physics, black chemistry, black biology, or black technology. For the new will to control their fate deals with the social environment, not the environment of nature. There is, moreover, nothing in the segregated life experience of blacks that is said to sensitize them to the subject matters and problematics of the physical and life sciences. An Insider doctrine would have to forge genetic assumptions about racial modes of thought in order to claim, as in the case of the Nazi version it did claim, monopolistic or privileged access to knowledge in those fields of science. But the black Insider doctrine adopts an essentially social-environmental rationale, not a biologically genetic one.

The social process underlying the emergence of Insider doctrine is reasonably clear. Polarization in the social structure becomes reflected in the polarization of claims in the cognitive and ideological domain, as groups or collectivities seek to capture what Heidegger called the "public interpretation of reality."[16] With varying degrees of intent, groups in conflict want to make their interpretation the prevailing one of how things were and are and will be. The critical test occurs when the interpretation moves beyond the boundaries of the ingroup to be

16. Martin Heidegger, *Sein und Zeit* (Halle: Max Niemeyer, 1927), as cited and discussed by Karl Mannheim in *Essays on the Sociology of Knowledge* (New York: Oxford University Press, 1952), 196 ff.

accepted by Outsiders. At the extreme, it then gives rise, through identifiable processes of reference-group behavior, to the familiar case of the converted Outsider validating himself, in his own eyes and in those of others, by becoming even more zealous than the Insiders in adhering to the doctrine of the group with which he wants to identify himself, if only symbolically.[17] He then becomes more royalist than the king, more papist than the pope or, as Joyce had Buck Mulligan say of Leopold Bloom, "Greeker than the Greeks." Some white social scientists, for example, vicariously and personally guilt-ridden over centuries of white racism, are prepared to outdo the claims of the group they would symbolically join. They are ready even to surrender their hard-won expert knowledge if the Insider doctrine seems to require it. [. . .]

The black Insider doctrine links up with the historically developing social structure in still another way. The dominant social institutions in this country have long treated the racial identity of individuals as actually if not doctrinally relevant to all manner of situations in every sphere of life. For generations, neither blacks nor whites, though with notably differing consequences, were permitted to forget their race. This treatment of a social status (or identity) as relevant when intrinsically it is functionally irrelevant constitutes the very core of social discrimination. As the once firmly rooted systems of discriminatory institutions and prejudicial ideology began to lose their hold, this meant that increasingly many judged the worth of ideas on their merits, not in terms of their racial pedigree.

What the Insider doctrine of the most militant blacks proposes on the level of social structure is to adopt the salience of racial identity in every sort of role and situation, a practice long imposed upon the American Negro, and to make that identity a total commitment issuing from within the group rather than one imposed upon it from without. By thus affirming the universal saliency of race and by redefining race as an abiding source of pride rather than stigma, the Insider doctrine in effect models itself after doctrine long maintained by white racists. [. . .]

Social Structure of Insiders and Outsiders

From the discussion thus far, it should be evident that I adopt a structural conception of Insiders and Outsiders. In this conception, Insiders are the members of specified groups and collectivities or occupants of specified social statuses; Outsiders are the nonmembers. This structural

17. Robert K. Merton, *Social Theory and Social Structure,* 3d ed., rev. and enl. (New York: Free Press, 1968), 405–6.

concept comes closer to Sumner's usage in his *Folkways* than to various meanings assigned the Outsider by Nietzsche, Kierkegaard, Sartre, Camus, or, for that matter, by Colin Wilson, just as, to come nearer home, it differs from the usages adopted by Riesman, Denney, and Glazer, Price or Howard S. Becker.[18] That is to say, Insiders and Outsiders are here defined as categories in social structure, not as inside dopesters or the specially initiated possessors of esoteric information on the one hand and as social-psychological types marked by alienation, rootlessness, or rule-breaking, on the other.

In structural terms, we are all, of course, both Insiders and Outsiders, members of some groups and, sometimes derivatively, not of others; occupants of certain statuses which thereby exclude us from occupying other cognate statuses. Obvious as this basic fact of social structure is, its implications for Insider and Outsider epistemological doctrines are apparently not nearly as obvious. Else, these doctrines would not presuppose, as they typically do, that human beings in socially differentiated societies can be sufficiently located in terms of a single social status, category, or group affiliation—black or white, men or women, under thirty or older—or of several such categories, taken seriatim rather than conjointly. This neglects the crucial fact of social structure that individuals have not a single status but a status-set: a complement of variously interrelated statuses which interact to affect both their behavior and perspectives.

The structural fact of status-sets, in contrast to statuses taken one at a time, introduces severe theoretical problems for total Insider (and Outsider) doctrines of social epistemology. The array of status sets in a population means that aggregates of individuals share some statuses and not others; or, to put this in context, that they typically confront one another simultaneously as Insiders and Outsiders. Thus, if only whites can understand whites and blacks, blacks, and only men can understand men, and women, women, this gives rise to the paradox which severely limits both premises: for it then turns out, by implication, that some Insiders are excluded from understanding other Insiders with white women being condemned not to understand white men, and black men, not to understand black women, and so through the various combinations of status-subsets.

Structural analysis in terms of shared and mutually exclusive status-

18. Colin Wilson, *The Outsider* (Boston: Houghton Mifflin, 1956); David Riesman, Reuel Denney, and Nathan Glazer, *The Lonely Crowd* (New Haven: Yale University Press, 1950); Don K. Price, *The Scientific Estate* (Cambridge: Harvard University Press, 1965), 83–84; and Howard S. Becker, *Outsiders: Studies in the Sociology of Deviance* (Glencoe, Ill.: Free Press, 1963).

sets will surely not be mistaken either as advocating divisions within the ranks of collectivities defined by a single prime criterion or as predicting that such collectivities cannot unite on many issues, despite their internal divisions. Such analysis only indicates the bases of social divisions that stand in the way of enduring unity of any of the collectivities and so must be coped with, divisions that are not easily overcome as new issues activate statuses with diverse and often conflicting interests. Thus, the obstacles to a union of women in England and Northern Ireland resulting from national, political, and religious differences between them are no less formidable than the obstacles, noted by Marx, confronting the union of English and Irish proletarians. So, too, women's liberation movements seeking unity in the United States find themselves periodically contending with the divisions between blacks and whites within their ranks, just as black liberation movements seeking unity find themselves periodically contending with the divisions between men and liberated women within their ranks.[19] [. . .]

The internal differentiation of collectivities based on a single status thus provides structural bases for diverse and often conflicting intellectual and moral perspectives within such collectivities. Differences of religion or age or class or occupation work to divide what similarities of race or sex or nationality work to unite. That is why social movements of every variety that strive for unity—whether they are establishmentarian movements whipped up by chauvinistic nationals in time of war or antiestablishmentarian movements designed to undo institutionalized injustice—press for total commitments in which all other loyalties are to be subordinated, on demand, to the dominant one.

This symptomatic exercise in status-set analysis may be enough to indicate that the idiomatic expression of total Insider doctrine—one must be one in order to understand one—is deceptively simple and sociologically fallacious (just as we shall see is the case with the total Outsider doctrine). For, from the sociological perspective of the status-set, "one" is not a man *or* a black *or* an adolescent *or* a Protestant, *or* self-defined and socially defined as middle class, and so on. Sociologically, "one" is, of course, all of these and, depending on the size of the status-set, much more. Furthermore, as Simmel[20] taught us long ago, the individuality of human beings can be sociologically derived from

19. See Shirley Chisholm, "Racism and Anti-Feminism," *Black Scholar* 1 (January–February 1970): 40–45; and Linda La Rue, "The Black Movement and Women's Liberation," *Black Scholar* 1 (May 1970): 36–42.

20. Georg Simmel, *Soziologie* (Leipzig: Duncker und Humbolt, 1908), 403–54; see also Lewis A. Coser, *Georg Simmel* (Englewood Cliffs, N.J.: Prentice-Hall, 1965), 18–20.

social differentiation and not only psychologically derived from intrapsychic processes. Thus, the greater the number and variety of group affiliations and statuses distributed among individuals in a society, the smaller, on the average, the number of individuals having precisely the same social configuration.

Following out the implications of this structural observation, we note that, on its own assumptions, the total Insider doctrine should hold only for highly fragmented small aggregates sharing the same status-sets. Even a truncated status-set involving only three affiliations—WASPs, for example—would greatly reduce the number of people who, under the Insider principle, would be able to understand their fellows (WASPs). The numbers rapidly decline as we attend to more of the shared status-sets by including such social categories as sex, age, class, occupation, and so on, toward the limiting case in which the unique occupant of a highly complex status set is alone qualified to achieve an understanding of self. The tendency toward such extreme social atomization is of course damped by differences in the significance of statuses which vary in degrees of dominance, saliency, and centrality.[21] As a result, the fragmentation of the capacity for understanding that is implied in the total Insider doctrine will not empirically reach this extreme. The structural analysis in terms of status-sets, rather than in the fictional terms of individuals being identified in terms of single statuses, serves only to push the logic of Insiderism to its ultimate *methodological solipsism.*

The fact of structural and institutional differentiation has other kinds of implications for the effort to translate the Insider claim to solidarity into an Insider epistemology. Since we all occupy various statuses and have group affiliations of varying significance to us, since, in short, we individually link up with the differentiated society through our status-sets, this runs counter to the abiding and exclusive primacy of any one group affiliation. Differing situations activate different statuses which then and there dominate over the rival claims of other statuses.

This aspect of the dynamics of status-sets can also be examined from the standpoint of the differing margins of functional autonomy pos-

21. This is not the place to summarize an analysis of the dynamics of status sets that takes up variation in key statuses (dominant, central, salient) and the conditions under which various statuses tend to be activated, along lines developed in unpublished lectures by Merton (1955–71). For pertinent uses of these conceptions in the dynamics of status sets, particularly with regard to functionally irrelevant statuses, see Cynthia Epstein, *Woman's Place: Options and Limits in Professional Careers* (Berkeley: University of California Press, 1970), esp. chap. 3.

sessed by various social institutions and other social subsystems. Each significant affiliation exacts loyalty to values, standards, and norms governing the given institutional domain, whether religion, science, or economy. Sociological thinkers such as Marx and Sorokin, so wide apart in many of their other assumptions, agree in assigning a margin of autonomy to the sphere of knowledge even as they posit their respective social, economic, or cultural determinants of it. [. . .]

Along with faults of neglecting the implications of structural differentiation, status-sets, and institutional autonomy, the Insider (and comparable Outsider) doctrine has the further fault of assuming, in its claims of monopolistic or highly privileged status-based access to knowledge, that social position wholly determines cognitive perspectives. In doing so, it affords yet another example of the ease with which truths can decline into error merely by being extended well beyond the principled limits within which they have been found to hold. (There *can* be too much of a good thing.)

A longstanding conception shared by various "schools" of sociological thought holds that differences in the social location of individuals and groups tend to involve differences in their interests and value orientations (as well as the sharing of some interests and values with others). Certain traditions in the sociology of knowledge have gone on to assume that these structurally patterned differences should involve, on the *average,* patterned differences in perceptions and perspectives. And these, so the convergent traditions hold—their convergence being often obscured by diversity in vocabulary rather than in basic concept—should make for discernible differences, on the average, in the definitions of problems for inquiry and in the types of hypotheses taken as points of departure. So far, so good. The evidence is far from in, since it has also been a tradition in the sociology of scientific knowledge during the greater part of this century to prefer speculative theory to empirical inquiry. But the idea, which can be taken as a general orientation guiding such inquiry, is greatly transformed in Insider doctrine.

For one thing, that doctrine assumes total coincidence between social position and individual perspectives. It thus exaggerates into error the conception of structural analysis which maintains that there is a *tendency for, not a full determination of,* socially patterned differences in the perspectives, preferences, and behavior of people variously located in the social structure. The theoretical emphasis on tendency, as distinct from total uniformity, is basic, not casual or niggling. It provides for a range of variability in perspective and behavior among members of the same groups or occupants of the same status (differences which, as we have seen, are ascribable to social as well as psychological differentiation). At the same time, this structural conception also provides for patterned differences, *on the whole,* between the perspectives of members of different groups or occupants of different statuses. Structural analysis thus avoids what Dennis Wrong has aptly described as "the oversocialized conception of man in modern sociology."[22]

Important as such allowance for individual variability is for general structural theory, it has particular significance for a sociological perspective on the life of the mind and the advancement of science and learning. For it is precisely the individual differences among scientists and scholars that are often central to the development of the discipline. They often involve the differences between good scholarship and bad; between imaginative contributions to science and pedestrian ones; between the consequential ideas and the stillborn ones. In arguing for the monopolistic access to knowledge, Insider doctrine can make no provision for individual variability that extends beyond the boundaries of the ingroup which alone can develop sound and fruitful ideas.

Insofar as Insider doctrine treats ascribed rather than achieved statuses as central in forming perspectives, it adopts a static orientation. For with the glaring exception of age status itself, ascribed statuses are generally retained throughout the life span. Yet sociologically considered, there is nothing fixed about the boundaries separating Insiders from Outsiders. As situations involving different values arise, different statuses are activated and the lines of separation shift. Thus, for a large number of white Americans, Joe Louis was a member of an outgroup. But when Louis defeated the Nazified Max Schmeling, many of the same white Americans promptly redefined him as a member of the (national) ingroup. National self-esteem took precedence over racial separatism. That this sort of drama in which changing situations activate differing statuses in the status-set is played out in the domain of the intellect as well is the point of Einstein's ironic observation in an address at the Sorbonne: "If my theory of relativity is proven successful, Germany will claim me as a German and France will declare that I am a citizen of the world. Should my theory prove untrue, France will say that I am a German and Germany will declare that I am a Jew."[23]

Like earlier conceptions in the sociology of knowledge, recent Insider doctrines maintain that, in the end, it is a special category of In-

22. Dennis Wrong, "The Oversocialized Conception of Man in Modern Sociology," *American Sociological Review* 26 (April 1961): 183–93.

23. Philipp Frank, *Einstein: His Life and Times* (New York: Alfred A. Knopf, 1963), 144.

sider—a category that generally manages to include the proponent of the doctrine—that has sole or privileged access to knowledge. Mannheim,[24] for example, found a structural warranty for the validity of social thought in the "classless position" of the "socially unattached intellectuals" (*sozialfreischwebende Intelligenz*). In his view, these intellectuals can comprehend the conflicting tendencies of the time since, among other things, they are "recruited from constantly varying social strata and life-situations." (This is more than a little reminiscent of the argument in the *Communist Manifesto* which emphasizes that "the proletariat is recruited from all classes of the population.") Without stretching this argument to the breaking point, it can be said that Mannheim in effect claims that there is a category of socially free-floating intellectuals who are both Insiders and Outsiders. Benefiting from their collectively diverse social origins and transcending group allegiances, they can observe the social universe with special insight and a synthesizing eye. [. . .]

Outsider Doctrine and Perspectives

The strong version of the Insider doctrine, with its epistemological claim to a monopoly of certain kinds of knowledge, runs counter to a long history of thought. From the time of Francis Bacon, to reach no further back, students of the intellectual life have emphasized the corrupting influence of group loyalties upon the human understanding. Among Bacon's four Idols (or sources of false opinion), we need only recall the second, the Idol of the Cave. Drawing upon Plato's allegory of the cave in the *Republic,* Bacon undertakes to tell how the immediate social world in which we live seriously limits what we are prepared to perceive and how we perceive it. Dominated by the customs of our group, we maintain received opinions, distort our perceptions to have them accord with these opinions, and are thus held in ignorance and led into error which we parochially mistake for the truth. Only when we escape from the cave and extend our vision do we provide for access to authentic knowledge. By implication, it is through the iconoclasm that comes with changing or multiple group affiliations that we can destroy the Idol of the Cave, abandon delusory doctrines of our own group, and enlarge the prospects for reaching the truth. For Bacon, the dedicated Insider is peculiarly subject to the myopia of the cave.

24. Karl Mannheim, *Ideology and Utopia* (New York: Harcourt Brace, 1936), 10, 139, 232.

In this conception, Bacon characteristically attends only to the dysfunctions of group affiliation for knowledge. Since for him access to authentic knowledge requires that one abandon superstition and prejudice, and since these stem from groups, it would not occur to Bacon to consider the possible functions of social locations in society as providing for observability and access to particular kinds of knowledge.

In far more subtle style, the founding fathers of sociology in effect also argued against the strong form of the Insider doctrine *without turning to the equal and opposite error of advocating the strong form of the Outsider doctrine* (which would hold that knowledge about groups, unprejudiced by membership in them, is accessible only to outsiders).

The ancient epistemological problem of subject and object was taken up in the discussion of historical *Verstehen*. Thus, first Simmel and then, repeatedly, Max Weber symptomatically adopted the memorable aphorism: "one need not be Caesar in order to understand Caesar." In making this claim, they rejected, in effect, the extreme Insider thesis which asserts that one *must* be Caesar in order to understand Caesar just as they rejected the extreme Outsider thesis that one must *not* be Caesar in order to understand Caesar.

The observations of Simmel and Weber bear directly upon implications of the Insider doctrine that reach beyond its currently emphasized scope. The dedicated Insider argues that the authentic understanding of group life can be achieved only by those who are directly engaged as members in it. Taken seriously, the doctrine puts in question the validity of just about all historical writing, as Weber clearly saw.[25] If direct engagement in the life of a group is essential to understanding it, then the only authentic history is contemporary history, written in fragments by those most fully involved in making inevitably limited portions of it. Rather than constituting only the raw materials of history, the documents prepared by engaged Insiders become all there is to history. But once historians elect to write the history of a time other than their own, even the most dedicated Insider, of the national, sex, age, racial, ethnic, or religious variety, becomes the Outsider, condemned to ignorance and error. If Insiders are capable of knowing and understanding because they were actually there—in that place, in that time, and, above all else, in that active role—then all historians, black

25. See *Gesammelte Aufsätze zur Wissenschaftslehre*, 428. Having quoted the Caesar aphorism, Weber goes on to draw the implication for historiography: "Sonst wäre alle Geschichtsschreibung sinnlos."

or white, old or young, men or women, are permanently estopped from writing history of the remote time or place.

Writing some twenty years ago in another connection, Claude Lévi-Strauss noted the parallelism between history and ethnography. Both subjects, he observed,

> are concerned with societies *other* than the one in which we live. Whether this *otherness* is due to remoteness in time (however slight) or to remoteness in space, or even to cultural heterogeneity, is of secondary importance compared to the basic similarity of perspective. All that the historian or ethnographer can do, and all that we can expect of either of them, is to enlarge a specific experience to the dimensions of a more general one, which thereby becomes accessible *as experience* to men or another country of another epoch. And in order to succeed, both historian and ethnographer must have the same qualities: skill, precision, a sympathetic approach and objectivity.[26]

Our question is, of course, whether the qualities required by the historian and ethnographer as well as other social scientists are confined to or largely concentrated among Insiders or Outsiders. Simmel, and after him, Schutz, and others have pondered the roles of that incarnation of the Outsider, the stranger who moves on.[27] In a fashion oddly reminiscent of the anything-but-subtle Baconian doctrine, Simmel develops the thesis that the stranger, not caught up in commitments to the group, can more readily acquire the strategic role of the relatively objective inquirer. "He is freer, practically and theoretically," notes Simmel, "he surveys conditions with less prejudice; his criteria for them are more general and more objective ideals; he is not tied down in his action by habit, piety, and precedent."[28] Above all, and here Simmel departs from the simple Baconian conception, the objectivity of the stranger "does not simply involve passivity and detachment; it is a particular structure composed of distance and nearness, indifference and involvement." It is the stranger, too, who finds what is familiar to the

26. The essay from which this is drawn was first published in 1949 and is reprinted in Claude Lévi-Strauss, *Structural Anthropology* (New York: Basic Books, 1963); see page 16 for the quotation.

27. See Simmel, *Soziologie;* and Alfred Schutz, "The Stranger: An Essay in Social Psychology," *American Journal of Sociology* 49 (May 1944): 499–507. It is symbolically appropriate that Simmel should have been attuned to the role of the stranger as outsider. For as Lewis A. Coser has shown, Simmel's style of sociological work was significantly influenced by his role as "the Stranger in the Academy"; see Lewis Coser, *Georg Simmel* (Englewood Cliffs, N.J.: Prentice-Hall, 1965), 29–39.

28. Georg Simmel, *The Sociology of Georg Simmel,* translated, edited and with an introduction by Kurt H. Wolff (Glencoe, Ill.: Free Press, 1950), 404–5.

group significantly unfamiliar and so is prompted to raise questions for inquiry less apt to be raised at all by Insiders.

As was so often the case with Simmel's seminal mind, he thus raised a variety of significant questions about the role of the stranger in acquiring sound and new knowledge, questions that especially in recent years have begun to be seriously investigated. A great variety of inquiries into the roles of anthropological and sociological fieldworkers have explored the advantages and limitations of the Outsider as observer. Even now, it appears that the balance sheet for Outsider observers resembles that for Insider observers, both having their distinctive assets and liabilities.

Apart from the theoretical and empirical work examining the possibly distinctive role of the Outsider in social and historical inquiry, significant episodes in the development of such inquiry can be examined as "clinical cases" in point. Thus, it has been argued that in matters historical and sociological the prospects for achieving insights and understanding may actually be somewhat better for the Outsider. Soon after it appeared in 1835, Tocqueville's *Democracy in America* was acclaimed as a masterly work by "an accomplished foreigner." Tocqueville himself expressed the opinion that "there are certain truths which Americans can only learn from strangers." [. . .]

What was in the case of Tocqueville an unplanned circumstance has since often become a matter of decision. Outsiders are sought to observe social institutions and cultures on the premise that they can do so with comparative detachment. In the first decade of this century, for example, the Carnegie Foundation for the Advancement of Teaching, in its search for someone to investigate the condition of medical schools, reached out to appoint Abraham Flexner, after he had confessed to never before having been inside a school of medicine. It was a matter of policy to select a total Outsider who, as it happened, produced the uncompromising Report that did much to transform the state of American medical education at the time.

Later, casting about for a scholar who might do a thoroughgoing study of the Negro in the United States, the Carnegie Corporation searched for an Outsider, preferably one, as they put it, drawn from a country of "high intellectual and scholarly standards but with no background or traditions of imperialism." These twin conditions of course swiftly narrowed the scope of the search. Switzerland and the Scandinavian countries alone seemed to qualify, the quest ending, as we know, with the selection of Gunnar Myrdal. In the preface to *An American Dilemma,* Myrdal reflected on his status as an Outsider who, in his words, "had never been subject to the strains involved in living in a black-white society" and who "as a stranger to the problem . . . has had perhaps a greater awareness of the extent to which human valuations everywhere enter into our scientific discussion of the Negro problem.[29] [. . .]

Two observations should be made on the Myrdal episode. First, in the judgment of critical minds, the Outsider, far from being excluded from the understanding of an alien society, was able to bring needed perspectives to it. And second, that Myrdal, wanting to have both Insider and Outsider perspectives, expressly drew into his circle of associates in the study such Negro and white insiders, engaged in the study of Negro life and culture and of race relations, as E. Franklin Frazier, Arnold Rose, Ralph Bunche, Melville Herskovits, Otto Klineberg, J. G. St. Clair Drake, Guy B. Johnson, and Doxey A. Wilkerson. [. . .]

The cumulative point of this variety of intellectual and institutional cases is not—and it needs to be repeated with all possible emphasis—is *not* a proposal to replace the extreme Insider doctrine by an extreme and equally vulnerable Outsider doctrine. The intent is, rather, to transform the original question altogether. We no longer ask whether it is the Insider *or* the Outsider who has monopolistic or privileged access to social knowledge; instead, we begin to consider their distinctive and interactive roles in the process of seeking truth.

Interchange, Trade-offs, and Syntheses

The actual intellectual interchange between Insiders and Outsiders—in which each adopts perspectives from the other—is often obscured by the rhetoric that commonly attends intergroup conflict. Listening only to that rhetoric, we may be brought to believe that there really is something like "black knowledge" and "white knowledge," "man's knowledge" and "woman's knowledge" which somehow manage to be both incommensurable and antithetical. Yet the boundaries between Insiders and Outsiders are far more permeable than this allows. Just as with the process of competition generally, so with the competition of ideas. Competing or conflicting groups take over ideas and procedures from one another, thereby denying in practice the rhetoric of total incompatibility. Even in the course of social polarization, conceptions with cognitive value are utilized all apart from their source. Concepts of power structure, cooptation, the dysfunctions of established institutions, and findings associated with these concepts are utilized by social scientists,

29. Gunnar Myrdal, *An American Dilemma: The Negro Problem and Modern Democracy* (New York and London: Harper and Bros., 1944), xviii–xix.

irrespective of their social or political identities. Nathan Hare,[30] for example, who remains one of the most articulate exponents of the Insider doctrine, does not hesitate to use the notion of the self-fulfilling prophecy in trying to explain how it is, in this day and age, that organizations run by blacks find it hard to work out. As he puts it, "White people thought that we could not have any institutions which were basically black which were of good quality. This has the effect of a self-fulfilling prophecy, because if you think that black persons cannot possibly have a good bank, then you don't put your money in it. All the best professors leave black universities to go to white universities as soon as they get the chance. The blacks even do the same thing. And this makes your prediction, which wasn't true in the beginning, come out to be true."[31]

Black scholars and women scholars utilize the conception of the self-fulfilling prophecy as a matter of course whenever it seems to illuminate the condition they seek to understand. They do so without a backward glance at the functionally irrelevant circumstance that the conception was set forth and developed by scholars who happened to be neither black nor female.[32] Correlatively, white sociologists, both male and female, utilize the conception of "status without substance," not pausing to consider that it was originated by the black sociologist, Franklin Frazier.[33]

Such diffusion of ideas across the boundaries of groups and statuses has long been noted. In one of his more astute analyses, Mannheim states the general case for the emergence and spread of knowledge that transcends even profound conflicts between groups:

> Syntheses owe their existence to the same social process that brings about polarization; groups take over the modes of thought and intellectual achievements of their adversaries under the simple law of 'competition on the basis of achievement.' . . . In the socially-differentiated thought process, even the opponent is ultimately forced to adopt those categories and forms of thought which are most appropriate in a given type of world order. In the economic sphere, one of the possible results of competition is that one competitor is compelled to catch up with the other's technological advances. In just the same way, whenever groups compete for having their interpretation of reality accepted as the correct one, it may happen that one of the groups

30. See "Interview with Nathan Hare," *U.S. News and World Report* 22 (May 1967), 64–68.

31. Nathan Hare, *Black Anglo-Saxons* (London: Collier-Macmillan, 1970), 65.

32. For the paper introducing the concept of the self-fulfilling prophecy, see chap. 16 in this volume.—*Ed.*

33. See E. Franklin Frazier, *Black Bourgeoisie* (Glencoe, Ill.: Free Press, 1955).

> takes over from the adversary some fruitful hypothesis or category—anything that promises cognitive gain. . . . [In due course, it becomes possible] to find a position from which both kinds of thought can be envisaged in their partial correctness, yet at the same time also interpreted as subordinate aspects of a higher synthesis.[34]

The essential point is that, with or without intent, the process of intellectual exchange takes place precisely because the conflicting groups are in interaction. The extreme Insider doctrine, for example, affects the thinking of sociologists, black and white, who reject its extravagant claims. Intellectual conflict sensitizes them to aspects of their subject they have otherwise not taken into account. [. . .]

Perhaps enough has been said to indicate how Insider and Outsider perspectives can converge through reciprocal adoption of ideas and the developing of complementary and overlapping foci of attention in the formulation of scientific problems. But these intellectual potentials for synthesis are often curbed by the social processes dividing scientists that transform intellectual controversy into social conflict.

When a reverse transition from social conflict to intellectual controversy is achieved, when the perspectives of each group are taken seriously enough to be carefully examined rather than rejected out of hand, there can develop trade-offs between the distinctive strengths and weaknesses of Insider and Outsider perspectives that enlarge the chances for a sound and relevant understanding of social life. [. . .]

34. Karl Mannheim, *Essays on the Sociology of Knowledge* (New York: Oxford University Press, 1952), 221–23.

The Social Institution of Science

20

The Ethos of Science (1942)

[. . .] Science is a deceptively inclusive word which refers to a variety of distinct though interrelated items. It is commonly used to denote (1) a set of characteristic methods by means of which knowledge is certified; (2) a stock of accumulated knowledge stemming from the application of these methods; (3) a set of cultural values and mores governing the activities termed scientific; or (4) any combination of the foregoing. We are here concerned in a preliminary fashion with the cultural structure of science, that is, with one aspect of science as an institution. Thus, we shall consider, not the methods of science, but the mores with which they are hedged about. To be sure, methodological canons are often both technical expedients and moral compulsives, but it is solely the latter which is our concern here. This is an essay in the sociology of science, not an excursion in methodology. Similarly, we shall not deal with the substantive findings of sciences (hypotheses, uniformities, laws), except as these are pertinent to standardized social sentiments about science. This is not an adventure in polymathy.

The Ethos of Science

The ethos of science is that affectively toned complex of values and norms which is held to be binding on the man of science.[1] The norms are expressed in the form of prescriptions, proscriptions, preferences, and permissions. They are legitimatized in terms of institutional values. These imperatives, transmitted by precept and example and reinforced by sanctions are in varying degrees internalized by scientists, thus fash-

From "Science and Technology in a Democratic Order," in *Journal of Legal and Political Sociology* 1 (1942): 115–26.

1. On the concept of ethos, see William Graham Sumner, *Folkways* (Boston: Ginn, 1906), 36 ff.; Hans Speier, "The Social Determination of Ideas," *Social Research* 5 (1938): 196 ff.; Max Scheler, *Schriften aus dem Nachlass* (1933; reprint, Bern, 1957), 1: 225–62. Albert Bayet, in his book on the subject, soon abandons description and analysis for homily; see his *La morale de la science* (Paris, 1931).

ioning their scientific conscience or, if one prefers the latter-day phrase, their superego. Although the ethos of science has not been codified, it can be inferred from the moral consensus of scientists as expressed in use and wont, in countless writings on the scientific spirit and in moral indignation directed toward contraventions of the ethos.

An examination of the ethos of modern science is only a limited introduction to a larger problem: the comparative study of the institutional structure of science. Although detailed monographs assembling the needed comparative materials are few and scattered, they provide some basis for the provisional assumption that "science is afforded opportunity for development in a democratic order which is integrated with the ethos of science." This is not to say that the pursuit of science is confined to democracies. The most diverse social structures have provided some measure of support to science. We have only to remember that the Accademia del Cimento was sponsored by two Medicis; that Charles II claims historical attention for his grant of a charter to the Royal Society of London and his sponsorship of the Greenwich Observatory; that the Académie des Sciences was founded under the auspices of Louis XIV, on the advice of Colbert; that urged into acquiescence by Leibniz, Frederick I endowed the Berlin Academy, and that the St. Petersburg Academy of Sciences was instituted by Peter the Great (to refute the view that Russians are barbarians). But such historical facts do not imply a random association of science and social structure. There is the further question of the ratio of scientific achievement to scientific potentialities. Science develops in various social structures, to be sure, but which of them provide an institutional context for the fullest measure of development?

The institutional goal of science is the extension of certified knowledge. The technical methods employed toward this end provide the relevant definition of knowledge: empirically confirmed and logically consistent statements of regularities (which are, in effect, predictions). The institutional imperatives (mores) derive from the goal and the methods. The entire structure of technical and moral norms implements the final objective. The technical norm of empirical evidence, adequate and reliable, is a prerequisite for sustained true prediction; the technical norm of logical consistency, a prerequisite for systematic and valid prediction. The mores of science possess a methodologic rationale but they are also binding, not only because they are procedurally efficient, but because they are believed right and good. They are moral as well as technical prescriptions.

Four sets of institutional imperatives—universalism, communism, disinterestedness, organized skepticism—are taken to comprise the ethos of modern science.

Universalism

Universalism[2] finds immediate expression in the canon that truth-claims, whatever their source, are to be subjected to *preestablished impersonal criteria:* consonant with observation and with previously confirmed knowledge. The acceptance or rejection of claims entering the lists of science is not to depend on the personal or social attributes of their protagonists; their race, nationality, religion, class, and personal qualities are as such irrelevant. Objectivity precludes particularism. The circumstance that scientifically verified formulations refer in that specific sense to objective sequences and correlations militates against all efforts to impose particularistic criteria of validity. The Haber process cannot be invalidated by a Nuremberg decree nor can an Anglophobe repeal the law of gravitation. The chauvinist may expunge the names of alien scientists from historical textbooks but their formulations remain indispensable to science and technology. However *echtdeutsch* or hundred-percent American the final increment, some aliens are accessories before the fact of every new scientific advance. The imperative of universalism is rooted deep in the impersonal character of science.

However, the institution of science is part of a larger social structure with which it is not always integrated. When the larger culture opposes universalism, the ethos of science is subjected to serious strain. Ethnocentrism is not compatible with universalism. Particularly in times of international conflict, when the dominant definition of the situation is such as to emphasize national loyalties, the man of science is subjected to the conflicting imperatives of scientific universalism and of ethnocentric particularism. The structure of the situation in which he finds himself determines the social role that is called into play. The man of science may be converted into a man of war—and act accordingly. Thus, in 1914 the manifesto of ninety-three German scientists and scholars—among them, Baeyer, Brentano, Ehrlich, Haber, Eduard Meyer,

2. For a basic analysis of universalism in social relations, see Talcott Parsons, *The Social System* (Glencoe, Ill.: Free Press, 1951). For an expression of the belief that "science is wholly independent of national boundaries and races and creeds," see the resolution of the Council of the American Association for the Advancement of Science, *Science* 87 (1938): 10; also, "The Advancement of Science and Society: Proposed World Association," *Nature* 141 (1938): 169.

Ostwald, Planck, Schmoller, and Wassermann—unloosed a polemic in which German, French, and English men arrayed their political selves in the garb of scientists. Dispassionate scientists impugned "enemy" contributions, charging nationalistic bias, log-rolling, intellectual dishonesty, incompetence, and lack of creative capacity.[3] Yet this very deviation from the norm of universalism actually presupposed the legitimacy of the norm. For nationalistic bias is opprobrious only if judged in terms of the standard of universalism; within another institutional context, it is redefined as a virtue, patriotism. Thus in the process of condemning their violation, the mores are reaffirmed.

Even under counter-pressure, scientists of all nationalities typically adhered to the universalistic standard in more direct terms. The international, impersonal, virtually anonymous character of science was reaffirmed. (Pasteur: "Le savant a une patrie, la science n'en a pas.") Denial of the norm was conceived as a breach of faith.

Universalism finds further expression in the demand that careers be open to talents. The rationale is provided by the institutional goal. To restrict scientific careers on grounds other than lack of competence is to prejudice the furtherance of knowledge. Free access to scientific pursuits is a functional imperative. Expediency and morality coincide. Hence the anomaly of a Charles II invoking the mores of science to reprove the Royal Society for their would-be exclusion of John Graunt, the political arithmetician, and his instructions that "if they found any more such tradesmen, they should be sure to admit them without further ado."

Here again the ethos of science may not be consistent with that of the larger society. Scientists may assimilate caste standards and close their ranks to those of inferior status, irrespective of capacity or achievement. But this provokes an unstable situation. Elaborate ideologies are called forth to obscure the incompatibility of caste-mores and the institutional goal of science. Caste-inferiors must be shown to be inherently incapable of scientific work, or, at the very least, their contributions must be systematically devaluated. "It can be adduced from the history of science that the founders of research in physics, and the great discoverers from Galileo and Newton to the physical pioneers of our own time, were almost exclusively Aryans, predominantly of the Nordic race." The modifying phrase, "almost exclusively," tacitly acknowledges the insufficient basis for denying out-castes all claims to scientific achievement. Hence the ideology is rounded out by a conception of "good" and "bad" science: the realistic, pragmatic science of

3. See "Insiders and Outsiders," chap. 19 in this volume.—*Ed.*

the Aryan is opposed to the dogmatic, formal science of the non-Aryan.[4] Or, grounds for exclusion are sought in the extrascientific capacity of men of science as enemies of the state or church. Thus, the exponents of a culture which abjures universalistic standards in general feel constrained to pay lip service to this value in the realm of science. Universalism is deviously affirmed in principle and suppressed in practice.

However inadequately it may be put into practice, the ethos of democracy includes universalism as a dominant guiding principle. Democratization is tantamount to the progressive elimination of restraints upon the exercise and development of socially valued capacities. Impersonal criteria of accomplishment and not fixation of status characterize the open democratic society. Insofar as such restraints do persist, they are viewed as obstacles in the path of full democratization. Thus, insofar as laissez-faire democracy permits the accumulation of differential advantages[5] for certain segments of the population, differentials that are not bound up with demonstrated differences in capacity, the democratic process leads to increasing regulation by political authority. Under changing conditions, new technical forms of organization must be introduced to preserve and extend equality of opportunity. The political apparatus may be required to put democratic values into practice and to maintain universalistic standards.

"Communism"

"Communism" in the nontechnical and extended sense of common ownership of goods, is a second integral element of the scientific ethos. The substantive findings of science are a product of social collaboration and are assigned to the community. They constitute a common heritage in which the equity of the individual producer is severely limited. An eponymous law or theory does not enter into the exclusive possession of the discoverer and heirs, nor do the mores bestow upon them special rights of use and disposition. Property rights in science are whittled

4. *Johannes Stark, Nature* 141 (1938): 772; "Philipp Lenard als deutscher Naturforscher," *Nationalsozialistische Monatshefte* 7 (1936): 106–12. This bears comparison with Duhem's contrast between "German" and "French" science.

5. This is the first brief allusion to what became developed into the theory of the "accumulation of advantage and disadvantage." See chap. 24, "The Matthew Effect," in this volume and Harriet Zuckerman, "Accumulation of Advantage and Disadvantage: The Theory and Its Intellectual Biography," in *L'Opera di R. K. Merton e la sociologia contemporanea,* ed. Carlo Mongardini and Simonetta Tabboni (Genoa: ECIG, 1989), 133–76.—*Ed.*

down to a bare minimum by the rationale of the scientific ethic. Scientists' claims to "their" intellectual "property" are limited to those of recognition and esteem which, if the institution functions with a modicum of efficiency, are roughly commensurate with the significance of the increments brought to the common fund of knowledge. Eponymy—for example, the Copernican system, Boyle's law—is thus at once a mnemonic and a commemorative device.

Given such institutional emphasis upon recognition and esteem as the sole property right of scientists in their discoveries, the concern with scientific priority becomes a "normal" response. Those controversies over priority which punctuate the history of modern science are generated by the institutional accent on originality.[6] There issues a competitive cooperation. The products of competition are communized,[7] and esteem accrues to the producer. Nations take up claims to priority, and fresh entries into the commonwealth of science are tagged with the names of nationals: witness the controversy raging over the rival claims of Newton and Leibniz to the differential calculus. But all this does not challenge the status of scientific knowledge as common property.

The institutional conception of science as part of the public domain is linked with the imperative for communication of findings. Secrecy is the antithesis of this norm; full and open communication its enactment.[8] The pressure for diffusion of results is reenforced by the institutional goal of advancing the boundaries of knowledge and by the incentive of recognition which is, of course, contingent upon publication. A scientist who does not communicate his important discoveries to the scientific fraternity—thus, a Henry Cavendish—becomes the target for ambivalent responses. He is esteemed for his talent and, perhaps, for

6. This early observation on conflicts over priority and the following observation on peer recognition as basic in the reward system of science were developed a quarter-century later in Merton's paradigmatic essay, "Priorities in Scientific Discovery: A Chapter in the Sociology of Science," which appears, in part, as chapter 22 in this volume.—*Ed.*

7. Marked by the commercialism of the wider society though it may be, a profession such as medicine accepts scientific knowledge as common property. See R. H. Shyrock, "Freedom and Interference in Medicine," *The Annals* 200 (1938): 45. "The medical profession . . . has usually frowned upon patents taken out by medical men. . . . The regular profession has . . . maintained this stand against private monopolies ever since the advent of patent law in the seventeenth century." There arises an ambiguous situation in which the socialization of medical practice is rejected in circles where the socialization of knowledge goes unchallenged.

8. Cf. Bernal, who observes: "The growth of modern science coincided with a definite rejection of the ideal of secrecy." (J. D. Bernal, *The Social Function of Science* [New York: Macmillan, 1939], 150–51).

his modesty. But, institutionally considered, his modesty is seriously misplaced, in view of the moral compulsive for sharing the wealth of science. Layman though he is, Aldous Huxley's comment on Cavendish is illuminating in this connection: "Our admiration of his genius is tempered by a certain disapproval; we feel that such a man is selfish and anti-social." The epithets are particularly instructive for they imply the violation of a definite institutional imperative. Even though it serves no ulterior motive, the suppression of scientific discovery is condemned.

The communal character of science is further reflected in the recognition by scientists of their dependence upon a cultural heritage to which they lay no differential claims. Newton's remark—"If I have seen further it is by standing on the shoulders of giants"—expresses at once a sense of indebtedness to the common heritage and a recognition of the essentially cooperative and selectively cumulative quality of scientific achievement.[9] The humility of scientific genius is not simply culturally appropriate but results from the realization that scientific advance involves the collaboration of past and present generations. It was Carlyle, not Maxwell, who indulged in a mythopoeic conception of history.

The communism of the scientific ethos is abstractly incompatible with the definition of technology as "private property" in a capitalistic economy. Current writings on the "frustration of science" reflect this conflict. Patents proclaim exclusive rights of use and, often, nonuse. The suppression of invention denies the rationale of scientific production and diffusion, as may be seen from the court's decision in the case of *U.S. v. American Bell Telephone Co.:* "The inventor is one who has discovered something of value. It is his absolute property. He may withhold the knowledge of it from the public."[10] Responses to this conflict-situation have varied. As a defensive measure, some scientists have

9. It is of some interest that Newton's aphorism is a standardized phrase which had found repeated expression from at least the twelfth century. It would appear that the dependence of discovery and invention on the existing cultural base had been noted some time before the formulations of modern sociologists. See *Isis* 24 (1935): 107–9; 25 (1938): 451–52.

A third of a century later, Merton provided a "postscript" to this observation in the form of the book, *On the Shoulders of Giants: A Shandean Postscript,* which, first published in 1965, has since appeared in a Post-Italianate edition published by the University of Chicago Press in 1993.—*Ed.*

10. 167 U.S. 224 (1897), cited by Bernhard J. Stern, "Restraints upon the Utilization of Inventions," *The Annals* 200 (1938): 21. For an extended discussion, cf. Stern's further studies cited therein, also Walton Hamilton, *Patents and Free Enterprise,* Temporary National Economic Committee Monograph no. 31 (1941).

come to patent their work to ensure its being made available for public use. Einstein, Millikan, Compton, Langmuir have taken out patents.[11] Scientists have been urged to become promoters of new economic enterprises.[12] Others seek to resolve the conflict by advocating socialism.[13] These proposals—both those which demand economic returns for scientific discoveries and those which demand a change in the social system to let science get on with the job—reflect discrepancies in the conception of intellectual property.

Disinterestedness

Science, as is the case with the professions in general, includes disinterestedness as a basic institutional element. Disinterestedness is not to be equated with altruism nor interested action with egoism. Such equivalences confuse institutional and motivational levels of analysis.[14] A passion for knowledge, idle curiosity, altruistic concern with the benefit to humanity, and a host of other special motives have been attributed to the scientist. The quest for distinctive motives appears to have been misdirected. It is rather a distinctive pattern of institutional control of a wide range of motives which characterizes the behavior of scientists. For once the institution enjoins disinterested activity, it is to the interest of scientists to conform on pain of sanctions and, insofar as the norm has been internalized, on pain of psychological conflict.

The virtual absence of fraud in the annals of science, which appears exceptional when compared with the record of other spheres of activity, has at times been attributed to the personal qualities of scientists.[15] By implication, scientists are recruited from the ranks of those who exhibit an unusual degree of moral integrity. There is, in fact, no satisfactory evidence that such is the case; a more plausible explanation may be found in certain distinctive characteristics of science itself. Involving as it does the verifiability of results, scientific research is under the exact-

11. Hamilton, *Patents and Free Enterprise,* 154; J. Robin, *L'oeuvre scientifique: sa protection juridique* (Paris, 1928).

12. Vannevar Bush, "Trends in Engineering Research," *Sigma Xi Quarterly* 22 (1934): 49.

13. Bernal, *The Social Function of Science,* 155 ff.

14. Talcott Parsons, "The Professions and Social Structure," *Social Forces* 17 (1939): 458–59; cf. George Sarton, *The History of Science and the New Humanism* (1931; reprint, New Brunswick: Transaction Books, 1988), 130 ff. The distinction between institutional compulsives and motives is a key, though largely implicit, conception of Marxist sociology.

15. This observation is also developed in chapter 22 in this volume, in the section entitled "Fraud in Science."—*Ed.*

ing scrutiny of fellow experts. Otherwise put—and doubtless the observation can be interpreted as lese majesty—the activities of scientists are subject to rigorous policing, to a degree perhaps unparalleled in any other field of activity. The demand for disinterestedness has a firm basis in the public and testable character of science and this circumstance, it may be supposed, has contributed to the integrity of men of science. There is competition in the realm of science, competition that is intensified by the emphasis on priority as a criterion of achievement, and under competitive conditions there may well be generated incentives for eclipsing rivals by illicit means. But such impulses can find scant opportunity for expression in the field of scientific research. Cultism, informal cliques, prolific but trivial publications—these and other techniques may be used for self-aggrandizement.[16] But, in general, spurious claims appear to be negligible and ineffective. The translation of the norm of disinterestedness into practice is effectively supported by the ultimate accountability of scientists to their compeers. The dictates of socialized sentiment and of expediency largely coincide, a situation conducive to institutional stability.

In this connection, the field of science differs somewhat from that of other professions. The scientist does not stand vis-à-vis a lay clientele in the same fashion as do the physician and lawyer, for example. The possibility of exploiting the credulity, ignorance, and dependence of the layman is thus considerably reduced. Fraud, chicane, and irresponsible claims (quackery) are even less likely than among the "service" professions. To the extent that the scientist-layman relation does become paramount, there develop incentives for evading the mores of science. The abuse of expert authority and the creation of pseudosciences are called into play when the structure of control exercised by qualified compeers is rendered ineffectual.[17]

It is probable that the reputability of science and its lofty ethical status in the estimate of the layman is in no small measure due to technological achievements.[18] Every new technology bears witness to the integrity of the scientist. Science realizes its claims. However, its au-

16. See the account by Logan Wilson, *The Academic Man* (New York: Oxford University Press, 1941), 201 ff.

17. Cf. R. A. Brady, *The Spirit and Structure of German Fascism* (New York: Viking, 1937), chap. 2; Martin Gardner, *In the Name of Science* (New York: Putnam's, 1953).

18. Francis Bacon set forth one of the early and most succinct statements of this popular pragmatism: "Now these two directions—the one active, the other contemplative—are one and the same thing; and what in operation is most useful, that in knowledge is most true" (*Novum Organum* [London: George Routledge and Sons, n.d.], 2: aphorism IV).

thority can be and is appropriated for interested purposes, precisely because the laity is often in no position to distinguish spurious from genuine claims to such authority. The presumably scientific pronouncements of totalitarian spokesmen on race or economy or history are for the uninstructed laity of the same order as newspaper reports of an expanding universe or wave mechanics. In both instances, they cannot be checked by the man in the street and in both instances, they may run counter to common sense. If anything, the myths will seem more plausible and are certainly more comprehensible to the general public than accredited scientific theories, since they are closer to common-sense experience and to cultural bias. Partly as a result of scientific achievements, therefore, the population at large becomes susceptible to new mysticisms expressed in apparently scientific terms. The borrowed authority of science bestows prestige on the unscientific doctrine.

Organized Skepticism

Organized skepticism is variously interrelated with the other elements of the scientific ethos. It is both a methodological and an institutional mandate. The temporary suspension of judgment and the detached scrutiny of beliefs in terms of empirical and logical criteria have periodically involved science in conflict with other institutions. Science which asks questions of fact, including potentialities, concerning every aspect of nature and society may come into conflict with other attitudes toward these same data which have been crystallized and often ritualized by other institutions. The scientific investigator does not preserve the cleavage between the sacred and the profane, between that which requires uncritical respect and that which can be objectively analyzed.

As we have noted, this appears to be the source of revolts against the so-called intrusion of science into other spheres. Such resistance on the part of organized religion has become less significant as compared with that of economic and political groups. The opposition may exist quite apart from the introduction of specific scientific discoveries which appear to invalidate particular dogmas of church, economy, or state. It is rather a diffuse, frequently vague apprehension that skepticism threatens the current distribution of power. Conflict becomes accentuated whenever science extends its research to new areas toward which there are institutionalized attitudes or whenever other institutions extend their control over science. In modern totalitarian society, antirationalism and the centralization of institutional control both serve to limit the scope provided for scientific activity.

21

Science and the Social Order (1938)

About the turn of the century, Max Weber observed that "the belief in the value of scientific truth is not derived from nature but is a product of definite cultures."[1] We may now add: and this belief is readily transmuted into doubt or disbelief. The persistent development of science occurs only in societies of a certain order, subject to a peculiar complex of tacit presuppositions and institutional constraints. What is for us a normal phenomenon which demands no explanation and secures many self-evident cultural values has been in other times and still is in many places the abnormal and infrequent. The continuity of science requires the active participation of interested and capable persons in scientific pursuits. But this support of science is assured only by appropriate cultural conditions. It is, then, important to examine the controls that motivate scientific careers, that select and give prestige to certain scientific disciplines and reject or blur others. It will become evident that changes in institutional structure may curtail, modify, or possibly prevent the pursuit of science.[2]

Sources of Hostility toward Science

Hostility toward science may arise under at least two sets of conditions, although the concrete systems of values—humanitarian, economic, political, religious—on which it is based may vary considerably. The first involves the logical, though not necessarily empirically sound, con-

From "Science and the Social Order," *Philosophy of Science* 5 (1938): 321–37. Reprinted by permission of the Philosophy of Science Association.

1. Max Weber, *Gesammelte Aufsätze zur Wissenschaftslehre* (Tübingen: J. C. B. Mohr, 1922), 213; cf. Pitirim A. Sorokin, *Social and Cultural Dynamics,* 4 vols. (New York: American Book Company, 1937), esp. 2: chap. 2.

2. Cf. Robert K. Merton, *Science, Technology and Society in Seventeenth-Century England,* in *Osiris,* ed. G. Sarton (1938; reprint, New York: Howard Fertig, 1970), chap. 11.

clusion that the results or methods of science are inimical to the satisfaction of important values. The second consists largely of nonlogical elements. It rests upon the feeling of incompatibility between the sentiments embodied in the scientific ethos and those found in other institutions. Whenever this feeling is challenged, it is rationalized. Both sets of conditions underlie, in varying degrees, current revolts against science. It might be added that such logical and affective responses are also involved in the social approval of science. But in these instances science is thought to facilitate the achievement of approved ends and basic cultural values are felt to be congruent with those of science rather than affectively inconsistent with them. The position of science in the modern world may be analyzed, then, as a resultant of two sets of conflicting forces, approving and opposing science as a large-scale social activity.

We restrict our examination to a few conspicuous instances of hostile revaluation of the social role of science, without implying that the antiscience movement is in any sense thus localized. Much of what is said here can probably be applied to the cases of other times and places.

The situation in Nazi Germany since 1933 illustrates the ways in which logical and nonlogical processes converge to modify or curtail scientific activity. In part, the hampering of science is an unintended by-product of changes in political structure and nationalistic credo. In accordance with the dogma of race purity, practically all persons who do not meet the politically imposed criteria of "Aryan" ancestry and of avowed sympathy with Nazi aims have been eliminated from universities and scientific institutes.[3] Since these outcasts include a considerable number of eminent scientists, one indirect consequence of the racialist purge is the weakening of science in Germany.

Implicit in this racialism is a belief in race defilement through actual or symbolic contact.[4] Scientific research by those of unimpeachable "Aryan" ancestry who collaborate with non-Aryans or who even accept their scientific theories is either restricted or proscribed. A new racial-political category has been introduced to include these incorrigible scientists who were once declared to be *echt-arisch:* the category of "White Jews." A prominent member of this new race is the Nobel

3. See chap. 3 of E. Y. Hartshorne, *The German Universities and National Socialism* (Cambridge: Harvard University Press, 1937), on the purge of the universities; cf. *Volk und Werden* 5 (1937): 320–21, which refers to some of the new requirements for the doctorate.

4. This is one of the many phases of the introduction of a caste system in Germany. As R. M. MacIver has observed, "The idea of defilement is common in every caste system" (*Society* [New York: Farrar and Rinehart, 1937], 172).

Prize physicist, Werner Heisenberg, who has persisted in his declaration that Einstein's theory of relativity constitutes an "obvious basis for further research."[5]

In these instances, the sentiments of national and racial purity have prevailed over utilitarian rationality. In contrast, utilitarian considerations are foremost when it comes to official policies concerning the directions to be followed by scientific research. Scientific work which promises direct practical benefit to the Nazi party or the Third Reich is to be fostered above all, and research funds are to be reallocated in accordance with this policy. The rector of Heidelberg University announces that "the question of the scientific significance [*Wissenschaftlichkeit*] of any knowledge is of quite secondary importance when compared with the question of its utility."[6]

The general tone of anti-intellectualism, with its depreciation of the theorist and its glorification of the man of action, may have long-run rather than immediate bearing upon the place of science in Germany. For should these attitudes become fixed, the most gifted elements of the population may be expected to shun those intellectual disciplines which have become disreputable. By the late thirties, effects of this antitheoretical attitude could be detected in the allocation of academic interests in the German universities.

It would be misleading to suggest that the Nazi government has completely repudiated science and intellect. The official attitudes toward science are clearly ambivalent and unstable. (For this reason, any statements concerning science in Nazi Germany are made under correction.) On the one hand, the challenging skepticism of science interferes with the imposition of a new set of values which demand an unquestioning acquiescence. But the new dictatorships must recognize, as did Hobbes who also argued that the State must be all or nothing, that science is power. For military, economic, and political reasons, theoretical science—to say nothing of its more respectable sibling, technology—cannot be safely discarded. Experience has shown that the most esoteric researches have found important applications. Unless

5. Cf. the official organ of the SS, the *Schwarze Korps,* 15 July 1937, 2. In this issue Johannes Stark, the president of the Physikalisch-Technischen Reichsanstalt, urges elimination of such collaborations which still continue and protests the appointment of three university professors who have been "disciples" of non-Aryans. See also Hartshorne, *The German Universities,* 112–13; Alfred Rosenberg, *Wesen, Grundsätze und Ziele der Nationalsozialistischen Deutschen Arbeiterpartei* (Munich: E. Boepple, 1933), 45 ff.; Johannes Stark, "Philipp Lenard als deutscher Naturforscher," *Nationalsozialistische Monatshefte* 71 (1936): 106–11, where Heisenberg, Schrödinger, von Laue, and Planck are castigated for not having divorced themselves from the "Jewish physics" of Einstein.

6. Ernst Krieck, *Nationalpolitische Erziehung* (Leipzig: Armanen Verlag, 1935), 8.

utility and rationality are dismissed beyond recall, it cannot be forgotten that Clerk Maxwell's speculations on the ether led Hertz to the discovery that culminated in the wireless. And indeed one Nazi spokesman remarks: "As the practice of today rests on the science of yesterday, so is the research of today the practice of tomorrow."[7] Emphasis on utility requires an unbanishable minimum of interest in science which can be enlisted in the service of the State and industry. At the same time, this emphasis leads to a limitation of research in pure science.

Social Pressures on Autonomy of Science

An analysis of the role of science in the Nazi state uncovers the following elements and processes. The spread of domination by one segment of the social structure—the State—involves a demand for primary loyalty to it. Scientists, as well as all others, are called upon to relinquish adherence to all institutional norms that, in the opinion of political authorities, conflict with those of the State. The norms of the scientific ethos must be sacrificed insofar as they demand a repudiation of the politically imposed criteria of scientific validity or of scientific worth. The expansion of political control thus introduces conflicting loyalties. In this respect, the reactions of devout Catholics who resist the efforts of the political authority to redefine the social structure, to encroach upon the preserves which are traditionally those of religion, are of the same order as the resistance of scientists. From the sociological point of view, the place of science in the totalitarian world is largely the same as that of all other institutions except the newly dominant State. The basic change consists in placing science in a new social context where it appears to compete at times with loyalty to the state. Thus, cooperation with non-Aryans is redefined as a symbol of political disloyalty. In a liberal order, the limitation of science does not arise in this fashion. For in such structures, a substantial sphere of autonomy—varying in extent, to be sure—is enjoyed by nonpolitical institutions.

The conflict between the totalitarian state and the scientist derives in part, then, from an incompatibility between the ethic of science and the new political code which is imposed upon all, irrespective of occupational creed. The ethos of science[8] involves the functionally necessary demand that theories or generalizations be evaluated in terms of their logical consistency and consonance with facts. The political ethic

7. Professor Thiessen in *Wissenschaft und Vierjahresplan,* 12.

8. On the ethos of science, see chap. 20 in this volume.—*Ed.*

would introduce the hitherto irrelevant criteria of the race or political creed of the theorist. Modern science has considered the personal equation as a potential source of error and has evolved impersonal criteria for checking such error. It is now called upon to assert that certain scientists, because of their extra-scientific affiliations, are a priori incapable of anything but spurious and false theories. In some instances, scientists are required to accept the judgments of scientifically incompetent political leaders concerning *matters of science*. But such politically advisable tactics run counter to the institutionalized norms of science. These, however, are dismissed by the totalitarian state as "liberalistic" or "cosmopolitan" or "bourgeois" prejudices, inasmuch as they cannot be readily integrated with the campaign for an unquestioned political creed.[9]

From a broader perspective, the conflict is a phase of institutional dynamics. Science, which has acquired a considerable degree of autonomy and has evolved an institutional complex that engages the allegiance of scientists, now has both its traditional autonomy and its rules of the game—its ethos, in short—challenged by an external authority. The sentiments embodied in the ethos of science—characterized by such terms as intellectual honesty, integrity, organized skepticism, disinterestedness, impersonality—are outraged by the set of new sentiments that the State would impose in the sphere of scientific research. With a shift from the previous structure where limited loci of power are vested in the several fields of human activity to a structure where there is one centralized locus of authority over all phases of behavior, the representatives of each sphere act to resist such changes and to preserve the original structure of pluralistic authority. Although it is customary to think of the scientist as a dispassionate, impersonal individual, it must be remembered that the scientist, in company with all other professional workers, has a large emotional investment in his way of life, defined by the institutional norms which govern his activity. In terms of that ethos, the social stability of science can be ensured only if adequate defenses are set up against changes imposed from outside the scientific fraternity itself.

This process of preserving institutional integrity and resisting new definitions of social structure that may interfere with the autonomy of

9. Thus, says Ernst Krieck, "In the future, one will no more adopt the fiction of an enfeebled neutrality in science than in law, economy, the State or public life generally. The method of science is indeed only a reflection of the method of government" (*Nationalpolitische Erziehung*, 6). Cf. Baeumler, *Männerbund und Wissenschaft* (Berlin: Junker and Dünnhaupt, 1934), 152; Walter Frank, *Zukunft und Nation*, 10; and contrast with Max Weber's "prejudice" that "Politik gehört nicht in den Hörsaal."

science finds expression in yet another direction. It is a basic assumption of modern science that scientific propositions "are invariant with respect to the individual" and group. But in a completely politicized society—where as one Nazi theorist put it, "the universal meaning of the political is recognized"[10]—this assumption is impugned. Scientific findings are held to be merely the expression of race or class or nation. As such doctrines percolate to the laity, they invite a general distrust of science and a depreciation of the prestige of scientists, whose discoveries appear arbitrary and fickle. This variety of anti-intellectualism which threatens their social position is characteristically enough resisted by scientists. On the ideological front as well, totalitarianism entails a conflict with the traditional assumptions of modern Western science.

Functions of Norms of Pure Science

One sentiment that is assimilated by scientists from the very outset of their training pertains to the purity of science. Science must not suffer itself to become the handmaiden of theology or economy or state. The function of this sentiment is to preserve the autonomy of science. For if such extrascientific criteria of the value of science as presumable consonance with religious doctrines or economic utility or political appropriateness are adopted, science becomes acceptable only insofar as it meets these criteria. In other words, as the pure science sentiment is eliminated, science becomes subject to the direct control of other institutional agencies and its place in society becomes increasingly uncertain. The persistent repudiation by scientists of the application of utilitarian norms to their work has as its chief function the avoidance of this danger, which is particularly marked at the present time. A tacit recognition of this function may be the source of that possibly apocryphal toast at a dinner for scientists in Cambridge: To pure mathematics, and may it never be of any use to anybody!

The exaltation of pure science is thus seen to be a defense against the invasion of norms that limit directions of potential advance and threaten the stability and continuance of scientific research as a valued social activity. Of course, the technological criterion of scientific achievement also has a social function for science. The increasing comforts and conveniences deriving from technology and ultimately from science invite the social support of scientific research. They also testify

10. Alfred Baeumler, *Männerbund und Wissenschaft* (Berlin: Junker and Dünnhaupt, 1934), 152.

to the integrity of the scientist, since abstract and difficult theories which cannot be understood or evaluated by the laity are presumably proved in a fashion which can be understood by all, that is, through their technological applications. Readiness to accept the authority of science rests, to a considerable extent, upon its daily demonstration of power. Were it not for such indirect demonstrations, the continued social support of that science which is intellectually incomprehensible to the public would hardly be nourished on faith alone.

At the same time, this stress upon the purity of science has had other consequences that threaten rather than preserve the social esteem of science. It is repeatedly urged that scientists should in their research ignore all considerations other than the advance of knowledge.[11] Attention is to be focused exclusively on the scientific significance of their work with no concern for the practical uses to which it may be put or for its social repercussions generally. The customary justification of this tenet—which is partly rooted in circumstance and which, in any event, has definite social functions, as we have just seen—holds that failure to adhere to this injunction will encumber research by increasing the possibility of bias and error. But this *methodological* view overlooks the *social* results of such an attitude. The objective consequences of this attitude have furnished a further basis of revolt against science; an incipient revolt that is found in virtually every society where science has reached a high stage of development. Since scientists do not or cannot control the direction in which their discoveries are applied, they become the subject of reproach and of more violent reactions insofar as these applications are disapproved by the agents of authority or by pressure groups. The antipathy toward the technological products is projected toward science itself. Thus, when newly discovered gases or explosives are applied as military instruments, chemistry as a whole is censured by those whose humanitarian sentiments are outraged. Science is held largely responsible for endowing those engines of human destruction which, it is said, may plunge our civilization into everlasting night and confusion. Or to take another prominent instance, the

11. For example, Pareto writes, "The quest for experimental uniformities is an end in itself." See a typical statement by George A. Lundberg, "It is not the business of a chemist who invents a high explosive to be influenced in his task by considerations as to whether his product will be used to blow up cathedrals or to build tunnels through the mountains. Nor is it the business of the social scientist in arriving at laws of group behavior to permit himself to be influenced by considerations of how his conclusions will coincide with existing notions, or what the effect of his findings on the social order will be" (*Trends in American Sociology*, ed. George A. Lundberg, Read Bain, and Nels Anderson [New York: Harper, 1929], 404–5). Compare the remarks of Read Bain on the "Scientist as Citizen," *Social Forces* 11 (1933): 412–15.

rapid development of science and related technology has led to an implicitly antiscience movement by vested interests and by those whose sense of economic justice is offended. The eminent Sir Josiah Stamp and a host of less illustrious folk have proposed a moratorium on invention and discovery, in order that man may have a breathing spell in which to adjust his social and economic structure to the constantly changing environment with which he is presented by the "embarrassing fecundity of technology." These proposals have received wide publicity in the press and have been urged with unslackened insistence before scientific bodies and governmental agents. The opposition comes particularly from those representatives of labor who fear the loss of investment in skills that become obsolete before the flood of new technologies. Although these proposals probably will not be translated into action within the immediate future, they constitute one possible nucleus about which a revolt against science in general may materialize. It is largely immaterial whether these opinions which make science ultimately responsible for undesirable situations are valid or not. W. I. Thomas's sociological theorem—"If men define situations as real, they are real in their consequences"—is much in point here.

In short, this basis for the revaluation of science derives from what I have called elsewhere the "imperious immediacy of interest."[12] Concern with the primary goal, the furtherance of knowledge, is coupled with a disregard of the consequences that lie outside the area of immediate interest, but these social results react so as to interfere with the original pursuits. Such behavior may be rational in the sense that it may be expected to lead to the satisfaction of the immediate interest. But it is irrational in the sense that it defeats other values that are not, at the moment, paramount but are nonetheless an integral part of the social scale of values. Precisely because scientific research is not conducted in a social vacuum, its effects ramify into other spheres of value and interest. Insofar as these effects are deemed socially undesirable, science is charged with responsibility. The goods of science are no longer considered an unqualified blessing. Examined from this perspective, the tenet of pure science and disinterestedness has helped to prepare its own epitaph.

Battle lines are drawn in terms of the question: can a good tree bring forth evil fruit? Those who would cut down or stunt the tree of knowledge because of its accursed fruit are met with the claim that the evil fruit has been grafted on the good tree by the agents of state and economy. It may salve the conscience of the individual man of science

12. See chap. 15 in this volume.—*Ed.*

to hold that an inadequate social structure has led to the perversion of his discoveries. But this will hardly satisfy an embittered opposition. Just as the *motives* of scientists may range from a passionate desire in the furtherance of knowledge to a profound interest in achieving personal fame and just as the *functions* of scientific research may vary from providing prestige-laden rationalizations of the existing order to enlarging our control of nature, so may other social *effects* of science be considered pernicious to society or result in the modification of the scientific ethos itself. There is a tendency for scientists to assume that the social effects of science *must* be beneficial in the long run. This article of faith performs the function of providing a rationale for scientific research, but it is manifestly not a statement of fact. It involves the confusion of truth and social utility which is characteristically found in the nonlogical penumbra of science. [. . .]

22

The Reward System of Science (1957)

Like other institutions, the institution of science has developed an elaborate system for allocating rewards to those who variously live up to its norms. Of course, this was not always so. The evolution of this system has been the work of centuries, and it will of course never be finished. [. . .] With the growth and professionalization of science, the system of honorific rewards has become diversely elaborated, and apparently at an accelerated rate.

Heading the list of the immensely varied forms of recognition long in use is eponymy,[1] the practice of affixing the name of the discovering scientist to all or part of what has been found, as with the Copernican system, Hooke's law, Planck's constant, or Halley's comet. In this way, scientists leave their signatures indelibly in history; their names enter into all the scientific languages of the world.

At the rugged and thinly populated peak of this system of eponymy are the men who have put their stamp upon the science and thought of their age. Such men are naturally in very short supply, and these few sometimes have an entire epoch named after them, as when we speak of the Newtonian epoch, the Darwinian era, or the Freudian age.

The gradations of eponymy have the character of a Guttman scale in which those men assigned highest rank are also assigned lesser degrees of honorific recognition. Accordingly, these peerless scientists are typically included also in the next highest ranks of eponymy, in which they

From "Priorities in Scientific Discovery: A Chapter in the Sociology of Science," *American Sociological Review* 22, no. 6 (December 1957): 635–59. Reprinted by permission of the American Sociological Association.

1. In his dedication to "The Starry Messenger," announcing his discovery of the satellites of Jupiter, Galileo begins with a paean to the practice of eponymy which opens with these words: "Surely a distinguished public service has been rendered by those who have protected from envy the noble achievements of men who have excelled in virtue, and have thus preserved from oblivion and neglect those names which deserve immortality" (Stillman Drake, *Discoveries and Opinions of Galileo,* [New York: Doubleday, 1957], 23). He then proceeds to call the satellites "the Medicean Stars" in honor of the Grand Duke of Tuscany, who soon becomes his patron.

are credited with having fathered a new science or a new branch of science (at times, according to the heroic theory, through a kind of parthenogenesis for which they apparently needed no collaborators). Of the illustrious Fathers of this or that science (or of this or that specialty), there is an end, but an end not easily reached. Consider only these few, culled from a list many times this length:

Morgagni, the Father of Pathology;
Cuvier, the Father of Paleontology;
Faraday, the Father of Electrotechnics;
Daniel Bernoulli, the Father of Mathematical Physics;
Bichat, the Father of Histology;
van Leeuwenhoek, the Father of Protozoology and Bacteriology;
Jenner, the Father of Preventive Medicine;
Chladni, the Father of Modern Acoustics;
Herbart, the Father of Scientific Pedagogy;
Wundt, the Father of Experimental Psychology;
Pearson, the Father of Biometry;
and, of course,
Comte, the Father of Sociology.

In a science as far-flung and differentiated as chemistry, there is room for several paternities. If Robert Boyle is the undisputed Father of Chemistry (and, as his Irish epitaph has it, also the Uncle of the Earl of Cork), then Priestley is the Father of Pneumatic Chemistry, Lavoisier the Father of Modern Chemistry, and the nonpareil Willard Gibbs, the Father of Physical Chemistry.

On occasion, the presumed father of a science is called upon, in the persons of his immediate disciples or later adherents, to prove his paternity, as with Johannes Müller and Albrecht von Haller, who are severally regarded as the Father of Experimental Physiology.

Once established, this eponymous pattern is stepped up to extremes. Each new specialty has its own parent, whose identity is often known only to those at work within the specialty. Thus, Manuel Garcia emerges as the Father of Laryngoscopy, Adolphe Brongiart as the Father of Modern Paleobotany, Timothy Bright as the Father of Modern Shorthand, and Father Johann Dzierson (whose important work may have influenced Mendel) as the Father of Modern Rational Beekeeping.

Sometimes, a particular form of a discipline bears eponymous witness to the man who first gave it shape, as with Hippocratic medicine, Aristotelian logic, Euclidean geometry, Boolean algebra, and Keynesian economics. Most rarely, the same individual acquires a double immortality, both for what he achieved and for what he failed to achieve, as in the cases of Euclidean and non-Euclidean geometries, and Aristotelian and non-Aristotelian logics.

In rough hierarchic order, the next echelon is comprised by thousands of eponymous laws, theories, hypotheses, instruments, constants, and distributions. No short list can hope to be representative of the wide range of these scientific contributions that have immortalized the scientists who made them. But a few examples in haphazard array might include the Brownian movement, the Zeeman effect, Rydberg's constant, Moseley's atomic number, and the Lorenz curve or, to come closer home, where we refer only to assured contemporary recognition rather than to possibly permanent fame, the Spearman rank-correlation coefficient, the Rorschach ink-blot, the Thurstone scale, the Bogardus social-distance scale, the Gini coefficient, the Bales categories of interaction, the Guttman scalogram, and the Lazarsfeld latent-structure analysis. [. . .]

Eponymy is only the most enduring and perhaps most prestigious kind of recognition institutionalized in science. Were the reward system confined to this, it would not provide for the many other distinguished scientists without whose work the revolutionary discoveries could not have been made. Graded rewards in the coin of the scientific realm—honorific recognition by fellow-scientists—are distributed among the stratified layers of scientific accomplishment. Merely to list some of these other but still considerable forms of recognition will perhaps be enough to remind us of the complex structure of the reward system in science.

In recent generations, the Nobel Prize, with nominations for it made by scientists of distinction throughout the world, is perhaps the preeminent token of recognized achievement in science.[2] There is also an iconography of fame in science, with medals honoring famous scientists and the recipients of the award alike (as with the Rumford medal and the Arago medal). Beyond these are memberships in honorary academies of sciences (for example, the National Academy of Sciences, the Royal Society, and the French Academy of Sciences) and fellowships in national and local societies. In those nations that still preserve a titled aristocracy, scientists have been ennobled, as in England since the time when Queen Anne added laurels to her crown by knighting Newton, not, as might be supposed, because of his superb administra-

2. On the machinery and results of the Nobel Prize awards, see Bernard Barber, *Science and the Social Order* (Glencoe, Ill.: Free Press, 1952), 108 ff.; and Leo Moulin, "The Nobel Prizes for the Sciences, 1901–50," *British Journal of Sociology* 6 (September 1955): 246–63; Harriet Zuckerman, *Scientific Elite: Nobel Laureates in the United States* (New York: Free Press, 1977).

tive work as Master of the Mint, but for his scientific discoveries. These things move slowly; it required almost two centuries before another Queen of England would, in 1892, confer a peerage of the realm upon a man of science for his work in science, and thus transform the preeminent Sir William Thomson into the no less eminent Lord Kelvin.[3] Scientists themselves have distinguished the stars from the supporting cast by issuing directories of "starred men of science," and universities have been known to accord honorary degrees to scientists along with the larger company of philanthropists, industrialists, businessmen, statesmen, and politicians.

Recognition is finally allocated by those guardians of posthumous fame, the historians of science. From the most disciplined scholarly works to the vulgarized and sentimentalized accounts designed for the millions, great attention is paid to priority of discovery, to the iteration and reiteration of "firsts." In this way, many historians of science help maintain the prevailing institutional emphasis on the importance of priority. [. . .]

Although scientific knowledge is impersonal in the sense that its claim to truth must be assessed entirely apart from its source, the historian of science is called upon to prevent scientific knowledge from sinking (or rising) into anonymity, to preserve the collective memory of its origins. Anonymous givers have no place in this scheme of things. Eponymity, not anonymity, is the standard. And, as we have seen, outstanding scientists, in turn, labor hard to have their names inscribed in the golden book of firsts.

Seen in composite, from the eponyms enduringly recording the names of scientists in the international language of science to the immense array of parochial and ephemeral prizes, the reward system of science reinforces and perpetuates the institutional emphasis upon originality. It is in this specific sense that originality can be said to be a major institutional goal of modern science, at times the paramount one, and recognition for originality a derived, but often as heavily emphasized, goal. In the organized competitions to contribute to scientific knowledge, the race *is* to the swift, to they who get there first with their contributions in hand.

Institutional Norm of Humility

If the institution of science placed great value *only* on originality, scientists would perhaps attach even more importance to recognition of pri-

3. For caustic comment on the lag in according such recognition to scientists, see excerpts from newspapers of the day in Silvanus P. Thompson, *The Life of William Thomson: Baron Kelvin of Largs* (London: Macmillan, 1910), 2:906–7.

ority than they do. But, of course, this value does not stand alone. It is only one of a complex set making up the ethos of science—disinterestedness, universalism, organized skepticism, communism of intellectual property, and humility being some of the others.[4] Among these, the socially enforced value of humility is in most immediate point, serving, as it does, to reduce the misbehavior of scientists below the rate that would occur if importance were assigned only to originality and the establishing of priority.

The value of humility takes diverse expression. One form is the practice of acknowledging the heavy indebtedness to the legacy of knowledge bequeathed by predecessors. This kind of humility is perhaps best expressed in the epigram Newton made his own: "If I have seen further, it is by standing on the shoulders of giants" (this, incidentally, in a letter to Hooke who was then challenging Newton's priority in the theory of colors).[5] Exploring the literature of a field of science becomes not only an instrumental practice, designed to learn from the past, but a commemorative practice, designed to pay homage to those who have prepared the way for one's work.

Humility is expected also in the form of the scientist's insisting upon his personal limitations and the limitations of scientific knowledge altogether. Galileo taught himself and his pupils to say, "I do not know." Perhaps another often-quoted image by Newton most fully expresses this kind of humility in the face of what is yet to be known:

> I do not know what I may appear to the world, but to myself I seem to have been only like a boy playing on the seashore, and diverting myself in now and then finding a smoother pebble or a prettier shell than ordinary, whilst the great ocean of truth lay all undiscovered before me.[6]

If this contrast between public image ("what I may appear to the world") and self-image ("but to myself I seem") is fitting for the greatest among scientists, it is presumably not entirely out of place for the

4. See chap. 20 on the ethos of science in this volume.—*Ed.*

5. Alexandre Koyré, "An Unpublished Letter of Robert Hooke to Isaac Newton," *Isis* 43 (December 1952): 312–37, at 315. See Robert K. Merton, *On the Shoulders of Giants* (1965, 1985; reprint, Chicago: University of Chicago Press, 1993).—*Ed.*

6. David Brewster, *Memoirs of the Life, Writings, and Discoveries of Sir Isaac Newton* (Edinburgh: Thomas Constable, 1855), 2:407. For our purposes, unlike those of the historian, it is a matter of indifference whether Newton actually felt modest or was merely conforming to expectation. In either case, he expresses the norm of personal humility, which is widely held to be appropriate. I. B. Cohen (*Franklin and Newton* [Philadelphia: American Philosophical Society, 1956] 47–58, passim) repeatedly and incisively makes the point that both admirers and critics of Newton have failed to make the indispensable distinction between what he said and what he did.

rest. The same theme continues unabated. Laplace, the Newton of France, in spite of what has been described as "his desire to shine in the constantly changing spotlight of public esteem," reportedly utters an epigrammatic paraphrase of Newton in his last words, "What we know is not much; what we do not know is immense."[7] Lagrange summarizes his lifetime of discovery in the one phrase, "I do not know."[8] And Lord Kelvin, at the Jubilee celebrating his fifty years as a distinguished scientist in the course of which he was honored by scores of scientific societies and academies, characterizes his lifelong effort to develop a grand and comprehensive theory of the properties of matter by the one word, "Failure."[9]

Like all human values, the value of modesty can be vulgarized and run into the ground by excessive and thoughtless repetition. It can become merely conventional, emptied of substance and genuine feeling. There really *can* be too much of a good thing. [. . .]

It would appear, then, that the institution of science, like other institutions, incorporates potentially incompatible values: among them, the value of originality, which leads scientists to want their priority to be recognized, and the value of humility, which leads them to insist on how little they have been able to accomplish. These values are not real contradictories, of course—"'tis a poor thing, but mine own"—but they do call for opposed kinds of behavior. To blend these potential incompatibles into a single orientation, to reconcile them in practice, is no easy matter. Rather, as we shall now see, the tension between these kindred values—kindred as Cain and Abel were kin—creates an inner conflict among men of science who have internalized both of them and generates a distinct ambivalence toward the claiming of priorities.

Ambivalence about Priority

The components of this ambivalence are fairly clear.[10] After all, to insist on one's originality by claiming priority is not exactly humble and to

7. Eric T. Bell, *Men of Mathematics* (New York: Simon and Schuster, 1937), 172. Bell refers to "a common and engaging trait of the truly eminent scientist in his frequent confession of how little he knows." What he describes as a trait of the scientist can also be seen as an expectation on the part of the community of scientists. It is not that many scientists *happen* to be humble men; they are *expected* to be humble. See Eric T. Bell, "Mathematics and Speculation," *The Scientific Monthly* 32 (March 1931): 193–209, at 204.

8. See "Specified Ignorance," chap. 4 in this volume.—*Ed.*

9. G. F. Fitzgerald, *Lord Kelvin, 1846–99,* Jubilee commemoration volume, with an essay on his works, 1899; S. P. Thompson, *Life of William Thomson,* 2:984.

10. See "Sociological Ambivalence," chap. 11 in this volume.—*Ed.*

dismiss one's priority by ignoring it is not exactly to affirm the value of originality. As a result of this conflict, scientists come to despise themselves for wanting that which the institutional values of science have led them to want.

With the rare candor that distinguishes him, Darwin so clearly exhibits this agitated ambivalence in its every detail that this one case can be taken as paradigmatic for many others (which are matters of less-detailed and less candid record). In his *Autobiography,* he writes that, even before his historic voyage on the Beagle in 1831, he was "ambitious to take a fair place among scientific men—whether more ambitious or less so than most of my fellow-workers, I can form no opinion."[11] A quarter of a century after this voyage, he is still wrestling with his ambition, exclaiming in a letter that "I wish I could set less value on the bauble fame, either present or posthumous than I do, but not, I think, to any extreme degree."[12]

Two years before the traumatizing news from Wallace, reporting his formulation of the theory of evolution, Darwin writes his now-famous letter to Lyell, explaining that he is not quite ready to publish his views, as Lyell had suggested he do in order not to be forestalled, and again expressing his uncontrollable ambivalence in these words: "I rather hate the idea of writing for priority, yet I certainly should be vexed if any one were to publish my doctrines before me."[13]

And then, in June 1858, the blow falls. What Lyell warned would happen and what Darwin could not bring himself to believe could happen, as all the world knows, did happen. Here is Darwin writing Lyell of the crushing event:

> [Wallace] has today sent me the enclosed, and asked me to forward it to you. It seems to me well worth reading. Your words have come true with a vengeance—that I should be forestalled. . . . I never saw a more striking coincidence; if Wallace had my MS. sketch written out in 1842, he could not have made a better short abstract! Even his terms now stand as heads of my chapters. . . . So all my originality, whatever it may amount to, will be smashed.[14]

Humility and disinterestedness urge Darwin to give up his claim to priority; the wish for originality and recognition urges him that all need not be lost. At first, with typical magnanimity, but without pretense of

11. Francis Darwin, ed., *Life and Letters of Charles Darwin* (New York: Appleton, 1925), 54.
12. Ibid., 452.
13. Ibid., 426–27.
14. Ibid., 473.

equanimity, he makes the desperate decision to step aside altogether. A week later, he is writing Lyell again; perhaps he might publish a short version of his long-standing text, "a dozen pages or so." And yet, he says in his anguished letter, "I cannot persuade myself that I can do so honourably." Torn by his mixed feelings, he concludes his letter, "My good dear friend, forgive me. This is a trumpery letter, influenced by trumpery feelings." And in an effort finally to purge himself of his feelings, he appends a postscript, "I will never trouble you or Hooker on the subject again."[15]

The next day he writes Lyell once more, this time to repudiate the postscript. Again, he registers his ambivalence: "It seems hard on me that I should lose my priority of many years' standing, but I cannot feel at all sure that this alters the justice of the case. First impressions are generally right, and I at first thought it would be dishonourable in me now to publish."[16]

As fate would have it, Darwin is just then prostrated by the death of his infant daughter. He manages to respond to the request of his friend Hooker and sends the Wallace manuscript and his own original sketch of 1844, "solely," he writes, "that you may see by your own handwriting that you did read it. . . . Do not waste much time. It is miserable in me to care at all about priority."[17]

Other members of the scientific community do what the tormented Darwin will not do for himself. Lyell and Hooker take matters in hand and arrange for that momentous session in which both papers are read at the Linnean Society. And as they put it in their letter prefacing the publication of the joint paper of "Messrs. C. Darwin and A. Wallace," "in adopting our present course . . . we have explained to him [Darwin] that we are not solely considering the relative claims to priority of himself and his friend, but the interests of science generally."[18] Despite this disclaimer of interest in priority, be it noted that scientific *knowledge* is not the richer or the poorer for having credit given where credit is due; it is the social *institution* of science and individual men of science that would suffer from repeated failures to allocate credit justly.

This historic and not merely historical episode so plainly exhibits the ambivalence occasioned by the double concern with priority and mod-

15. Ibid., 474–75.

16. Ibid., 475.

17. Ibid., 476.

18. C. Darwin and A. R. Wallace, "On the Tendency of Species to Form Varieties and on the Perpetuation of Varieties and Species by Natural Means of Selection," communicated by Sir C. Lyell and J. D. Hooker (read 1 July 1858), *Journal of the Linnean Society* 3 (1859): 45.

esty that it need not be examined further. Had the institutionalized emphasis on originality been alone in point, the claim to priority would have invited neither self-blame nor self-contempt; publication of the long-antecedent work would have proclaimed its own originality. But the value of originality was joined with the value of humility and modesty. To insist on priority would be to trumpet one's own excellence, but scientific peers and friends of the discoverers, acting as a third party in accord with the institutional norms, could with full propriety announce the joint claims to originality that the discoverers could not bring themselves to do. Underneath it all lies a deep and agitated ambivalence about priority. [. . .]

The institutional values of modesty and humility are apparently not always enough to counteract both the institutional emphasis upon originality and the actual workings of the system of allocating rewards. Originality, as exemplified by the new idea or the new finding, is more readily observable by others in science and is more fully rewarded than the often unobservable kind of humility that keeps independent discoverers from reporting that they too had had the same idea or the same finding. Moreover, after publication by another, it is often difficult, if not impossible, to demonstrate that one had independently arrived at the same result. For these and other reasons, it is generally an unequaled contest between the values of recognized originality and of modesty. Great modesty may elicit respect, but great originality promises everlasting fame.

In short, the social organization of science allocates honor in a way that tends to vitiate the institutional emphasis upon modesty. It is this, I believe, which goes far toward explaining why so many scientists, even those who are ordinarily of the most scrupulous integrity, will go to great lengths to press their claims to priority of discovery. As I have often suggested, perhaps too often, any *extreme* institutional

> emphasis upon achievement—whether this be scientific productivity, accumulation of wealth or, by a small stretch of the imagination, the conquests of a Don Juan—will attenuate conformity to the institutional norms governing behavior designed to achieve the particular form of 'success,' especially among those who are socially disadvantaged in the competitive race.[19]

Or more specifically and more completely, great concern with the goal of recognition for originality can generate a tendency toward sharp practices just inside the rules of the game or sharper practices far out-

19. Robert K. Merton, *Social Theory and Social Structure* (Glencoe, Ill.: Free Press, 1957), 166.

side. That this has been the case with the behavior of scientists who were all-out to have their originality recognized, the rest of this chapter will try to show.

Types of Response to Cultural Emphasis on Originality

Fraud in Science

The extreme form of deviant behavior in science would of course be the use of fraud to obtain credit for an original discovery. [. . .]

At the extreme are hoaxes and forgery: the concocting of false data in science and learning—or, more accurately, in pseudoscience and anti-scholarship. Literary documents have been forged in abundance, at times by men of previously unblemished reputation, in order to gain money or fame. [. . .]

The pressure to demonstrate the truth of a theory or to produce a sensational discovery has occasionally led to the faking of scientific evidence. The biologist Paul Kammerer produced specimens of spotted salamanders designed to prove the Lamarckian thesis experimentally; was thereupon offered a chair at the University of Moscow where in 1925 the Lamarckian views of Michurin held reign; and upon proof that the specimens were fakes, attributed the fraud to a research assistant. Most recently, the Piltdown man—that is, the skull and jaw from which his existence was inferred—has been shown, after forty years of uneasy acceptance, to be a carefully contrived hoax.[20]

Excessive concern with "success" in scientific work has on occasion led to the types of fraud which the nineteenth-century mathematician and inventor of calculating machines, Charles Babbage, picturesquely described as "trimming" and "cooking." The trimmer clips off "little bits here and there from observations which differ most in excess from the mean, and [sticks] . . . them on to those which are too small . . . [for the unallowable purpose of] 'equitable adjustment.'" The cook makes "multitudes of observations" and selects only those which agree with a hypothesis, and, as Babbage says, "the cook must be very unlucky if he cannot pick out fifteen or twenty which will do for serving up."[21] This eagerness to demonstrate a thesis can, on occasion, lead even truth to be fed with cooked data, as it did for the neurotic scientist, described

20. William L. Straus, Jr., "The Great Piltdown Hoax," *Science* 119 (26 February 1954): 265–69.

21. Charles Babbage, *Reflections on the Decline of Science in England* (London: B. Fellowes, 1830), 174–83.

by Lawrence Kubie, "who had proved his case, but was so driven by his anxieties that he had to bolster an already proven theorem by falsifying some quite unnecessary additional statistical data."[22]

The great cultural emphasis upon recognition for original discovery can lead by gradations from these rare practices of outright fraud to more frequent practices just beyond the edge of acceptability, sometimes without scientists being aware that they have exceeded allowable limits. Scientists may find themselves reporting only "successful experiments or results, so-called, and neglecting to report 'failures.'" Alan Gregg, that informed observer of the world of medical research, practice, and education, reports the case of

> the medical scientist of the greatest distinction who told me that during his graduate fellowship at one of the great English universities he encountered for the first time the idea that in scientific work one should be really honest in reporting the results of his experiments. Before that time he had always been told and had quite naturally assumed that the point was to get his observations and theories accepted by others, and published.[23]

Yet these deviant practices should be seen in perspective. What evidence there is suggests that they are extremely infrequent, and this temporary focus upon them will surely not be distorted into regarding the exceptional case as the typical. Apart from the moral integrity of scientists themselves—and this is, of course, the major basis for honesty in science—there is much in the social organization of science that provides a further compelling basis for honest work. Scientific research is typically, if not always, under the exacting scrutiny of fellow experts, involving, as it usually though not always does, the verifiability of results by others. Scientific inquiry is in effect subject to rigorous policing, to a degree perhaps unparalleled in any other field of human activity. Personal honesty is supported by the public and testable character of science. As Babbage remarked, "the cook would [at best] procure a temporary reputation . . . at the expense of his permanent fame."

Competition in the realm of science, intensified by the great emphasis on original and significant discoveries, may occasionally generate incentives for eclipsing rivals by illicit or dubious means. But this seldom occurs in the form of preparing fraudulent data; instead, it appears in quite other forms of deviant behavior involving spurious claims to discovery. More concretely, it is an occasional theft rather

22. Lawrence S. Kubie, "Some Unsolved Problems of the Scientific Career," parts 1 and 2, *American Scientist* 41 (1953): 596–613; 42 (1954): 104–12.

23. Alan Gregg, *Challenges to Contemporary Medicine* (New York: Columbia University Press, 1956), 115.

than forgery, and more often, libel and slander rather than theft that are found on the small seamy side of science.

Plagiary: Fact and Slander

Deviant behavior most often takes the form of occasional plagiaries and many slanderous charges or insinuations of plagiary. The historical record shows relatively few cases (and of course the record may be defective) in which one scientist actually pilfered another. We are assured that in the *Mécanique céleste* (until then, outranked only by Newton's *Principia*) "theorems and formulae are appropriated wholesale without acknowledgement" by Laplace.[24] [. . .] It is true also that Robert Boyle, not impressed by the thought that theft of his ideas might be a high tribute to his talent, was in 1688 driven to the desperate expedient of printing "An advertisement about the loss of many of his writings," later describing the theft of his work and reporting that he would from then on write only on loose sheets, in the hope that these would tempt thieves less than "bulky packets" and going on to say that he was resolved to send his writings to press without extensive revision in order to avoid prolonged delays.[25] But even with such cases of larceny on the grand scale, the aggregate of demonstrable theft in modern science is not large.

What does loom large is the repeated practice of charging others with pilfering scientific ideas. Falsely accused of plagiarizing Harvey in physiology, Snell in optics, and Harriot and Fermat in geometry, Descartes in turn accuses Hobbes and the teenage Pascal of plagiarizing him.[26] To maintain his property, Descartes implores his friend Mersenne, "I also beg you to tell him [Hobbes] as little as possible about what you know of my unpublished opinions, for if I'm not greatly mis-

24. As stated by the historian of astronomy, Agnes Mae Clerke, in her article on Laplace in the eleventh edition of the *Encyclopaedia Britannica.*

25. The account by A. M. Clerke in the article on Boyle in the *Dictionary of National Biography* is somewhat mistaken in attributing charges of plagiary to the published advertisement. This speaks only of losses of manuscript through "unwelcome accidents" (e.g., the upsetting of corrosive liquors over a file of manuscripts) and at most hints at less impersonal sources of loss. But a later unpublished paper by Boyle, dug up by his biographer Birch, is leveled against the numerous plagiarists of his works. This document, running to three folio pages of print, is a compendium of the ingenious devices for thievery developed by the grand larcenists of seventeenth-century science. See *The Works of the Honourable Robert Boyle,* 6 vols. With the Life of the author, by J. Birch (London, 1772), 1:cxxv–ccxxviii, ccxxii–ccxxiv.

26. For the case of Harvey, see A. R. Hall, *The Scientific Revolution, 1500–1800* (London: Longmans, Green, 1954), 148; for Hobbes, see Descartes, *Oeuvres,* ed. Charles Adam and Paul Tannery, vol. 3, *Correspondance* (Paris, 1899), 283 ff.; for Pascal, see ibid., vol. 5 (1903), 366.

taken, he is a man who is seeking to acquire a reputation at my expense and through shady practices."[27] All unknowing that the serene and unambitious Gauss had long since discovered the method of least squares, Legendre, himself "a man of the highest character and scrupulously fair," practically accuses Gauss of having filched the idea from him and complains that Gauss, already so well-stocked with momentous discoveries, might at least have had the decency not to adopt his brainchild.[28] [. . .]

Or to turn to our own province, Comte, tormented by the suggestion that his law of three stages had really been originated by St. Simon, denounces his one-time master and describes him as a "superficial and depraved charlatan.[29] Again, to take Freud's own paraphrase, Janet claims that "everything good in psychoanalysis repeats, with slight modifications, the views of Janet—everything else in psychoanalysis being bad."[30] Freud refuses to lock horns with Janet in what he describes as "gladiator fights in front of the noble mob," but some years later, his disciple, Ernest Jones, reports that at a London congress he has "put an end to" Janet's pretensions, and Freud applauds him in a letter that urges him to "strike while the iron is hot," in the interests of "fair play."[31]

So the almost changeless pattern repeats itself. Two or more scientists quietly announce a discovery. Since it is often the case that these are truly independent discoveries, with each scientist having separately exhibited originality of mind, the process is sometimes stabilized at that point, with due credit to both, as in the instance of Darwin and Wal-

27. Descartes, *Oeuvres,* 3:320.

28. Bell, *Men of Mathematics,* 259–60. Legendre seems to have been particularly sensitive to these matters, perhaps because he was often victimized; note Clerke's remark that between Laplace and Legendre "there was a feeling of 'more than coldness,' owing to his appropriation, with scant acknowledgment, of the other's labors." *Encyclopaedia Britannica,* vol. 16, 202.

29. Frank E. Manuel, *The New World of Henri Saint-Simon* (Cambridge: Harvard University Press, 1956), 340–42; also Richard L. Hawkins, *Auguste Comte and the United States* (Cambridge: Harvard University Press, 1936), 81–82, as cited by Manuel.

30. Sigmund Freud, *History of the Psychoanalytic Movement* (London: Hogarth Press, 1949); also, Freud, *An Autobiographical Study* (London: Hogarth Press, 1948), 54–55, where he seeks "to put an end to the glib repetition of the view that whatever is of value on psycho-analysis is merely borrowed from the ideas of Janet. . . . Historically, psycho-analysis is completely independent of Janet's discoveries, just as in its content it diverges from them and goes far beyond them." For Janet's not always delicate insinuations, see his *Psychological Healing* (New York: Macmillan, 1925), 1:601–40.

31. Ernest Jones, *Sigmund Freud: Life and Work* (London: Hogarth Press, 1955), 2:112.

lace. But since the situation is often ambiguous with the role of each not easy to demonstrate, and since each *knows* that he had himself arrived at the discovery, and since the institutionalized stakes of reputation are high and the joy of discovery immense, this is often not a stable solution. One or another of the discoverers—or frequently, his colleagues or fellow-nationals—suggests that he rather than his rival was really first, and that the independence of the rival is at least unproved. Then begins the familiar deterioration of standards governing conflictful interaction: the other side, grouping its forces, counters with the opinion that plagiary had indeed occurred, that let him whom the shoe fits wear it and furthermore, to make matters quite clear, the shoe is on the other foot. Reinforced by group loyalties and often by chauvinism, the controversy gains force, mutual recriminations of plagiary abound, and there develops an atmosphere of thoroughgoing hostility and mutual distrust.

On some occasions, this can lead to outright deceit in order to buttress valid claims, as with Newton in his controversy with Leibniz over the invention of the calculus. When the Royal Society finally established a committee to adjudicate the rival claims, Newton, who was then president of the Royal Society, packed the committee, helped direct its activities, anonymously wrote the preface for the second published report—the draft is in his handwriting—and included in that preface a disarming reference to the old legal maxim that "no one is a proper witness for himself [and that] he would be an iniquitous Judge, and would crush underfoot the laws of all the people, who would admit anyone as a lawful witness in his own cause."[32] We can gauge the immense pressures for self-vindication that must have operated for such a man as Newton to have adopted these means for defense of his valid claims. It was not because Newton was so weak but because the institutionalized values were so strong that he was driven to such lengths.

This interplay of offensive and defensive maneuvers—no doubt students of the theory of games can recast it more rigorously—thus gives further emphasis to priority. Scientists try to exonerate themselves in advance from possible charges of filching by going to great lengths to establish their priority of discovery. Often this kind of anticipatory de-

32. There is a sizable library discussing the Newton-Leibniz controversy. I have drawn chiefly upon Louis T. More (*Isaac Newton* [New York: Scribner's, 1934]), who devotes the whole of chapter 15 to the subject; Augustus de Morgan, *Essays on the Life and Works of Newton* (Chicago: Open Court, 1914), esp. Appendix 2; and Brewster, *Memoirs of Newton,* chapter 22; cf. Cohen, *Franklin and Newton,* who is properly critical of More's biography at various points (e.g., 84–85).

fense produces the very result it was designed to avoid by inviting others to show that prior announcement of publication need not mean there was no plagiary.

The effort to safeguard priority and to have proof of one's integrity has led to a variety of institutional arrangements designed to cope with this strain on the system of rewards. In the seventeenth century, for example, and even as late as the nineteenth, discoveries were sometimes reported in the form of anagrams—as with Galileo's "triple star" of Saturn and Hooke's law of tension—for the double purpose of establishing priority of conception and of yet not putting rivals to one's original ideas, until they had been further worked out. Then, as now, complex ideas were quickly published in abstracts, as when Halley urged Newton to do so in order to secure "his invention to himself till such time as he would be at leisure to publish it."[33] There is also the longstanding practice of depositing sealed and dated manuscripts with scientific academies in order to protect both priority and idea.[34] Scientific journals often print the date on which the manuscript of a published article was received, thus serving, even apart from such intent, to register the time it first came to notice. Numerous personal expedients have been developed: for example, letters detailing one's own ideas are sent off to a potential rival, thus disarming him; preliminary and confidential reports are circulated among a chosen few; personal records of research are meticulously dated (as by Kelvin). Finally, it has often been suggested that the functional equivalent of a patent office be established in science to adjudicate rival claims to priority.[35]

In prolonged and yet overly quick summary, these are some of the forms of deviance invited by the institutional emphasis on priority and some of the institutional expedients devised to reduce the frequency of these deviations. But as we would expect from the theory of alternative responses to excessively emphasized goals, other forms of behavior, verging toward deviance though still well within the tacit rules and not

33. Thomas Birch, *The History of the Royal Society of London* (London: 1756–57), 4:437.

34. For a recent instance, see the episode described by Wiener in which the race between Bouligand and Wiener to contribute new concepts "in potential theory" ended in a "dead heat," since Bouligand has submitted his "results to the [French] Academy in a sealed envelope, after a custom sanctioned by centuries of academy tradition" (Norbert Wiener, *I Am a Mathematician* [New York: Doubleday, 1956], 92).

35. J. Hettinger, "Problems of Scientific Property and Its Solution," *Science Progress* 26 (January 1932): 449–61; also the paper by Dr. A. I. Sotesi, of the New York Academy of Medicine, cited by Bernhard J. Stern in *Social Factors in Medical Progress* (New York: Columbia University Press, 1927), 108.

as subject to moral disapproval as the foregoing, have also made their appearance.

Alternative Responses to Emphasis on Originality

The large majority of scientists, like the large majority of artists, writers, doctors, bankers and bookkeepers, have little prospect of great and decisive originality. For most of us artisans of research, getting things into print becomes a symbolic equivalent to making a significant discovery. Nor could science advance without the great unending flow of papers reporting careful investigations, even if these are routine rather than distinctly original. The indispensable reporting of research can, however, become converted into an itch to publish that, in turn, becomes aggravated by the tendency, in many academic institutions, to transform the sheer number of publications into a ritualized measure of scientific or scholarly accomplishment.

The urge to publish is given a further push by the moral imperative of science to make one's work known to others; it is the obverse to the culturally repudiated practice of jealously hoarding scientific knowledge for oneself. As Joseph Priestley liked to say, "whenever he discovered a new fact in sience, he instantly proclaimed it to the world in order that other minds might be employed upon it besides his own."[36] Indeed, John Aubrey, that seventeenth-century master of the thumbnail biography and member of the Royal Society, could extend the moral imperative for communication of knowledge to justify even plagiary if the original author will not put his ideas into print. In his view it was better to have scientific goods stolen and circulated than to have them lost entirely.

To this point (and I provide comfort by reporting that the end of the chapter is in sight), we have examined types of deviant responses to the institutional emphasis on priority that are *active* responses: the fabrication of "data," aggressive self-assertion, the denouncing of rivals, plagiary, and charges of plagiary. Other scientists have responded to the same pressures *passively* or at least by internalizing their aggressions and directing them against themselves. Since these passive responses,

36. Priestley's remark as paraphrased by his longtime friend, T. L. Hawkes, and reported by George Wilson, *Life of the Honourable Henry Cavendish* (London, 1851), 111. The seventeenth-century Dutch genius of microscopy, Anton van Leeuwenhoek, also adopted a policy, as he described it, that "whenever I found out anything remarkable, I have thought it my duty to put down my discovery on paper, so that all ingenious people might be informed thereof" (Quoted by Ralph H. Major, *A History of Medicine* [Oxford: Blackwell Scientific Publications, 1954], 1:531). The same sentiment was expressed by Saint-Simon, among many others. Cf. Manuel, *Saint-Simon*, 63–64.

unlike the active ones, are private and often not publicly observable, they seldom enter the historical record. This need not mean, of course, that passive withdrawal from the competition for originality in science is infrequent; it might simply mean that the scientists responding in this fashion do not come to public notice, unless they do so after their accomplishments have qualified them for the pages of history.

Chief among these passive deviant responses is what I have described, on occasion, as *retreatism,* the abandoning of the once-esteemed cultural goal of originality and of practices directed toward reaching that goal. In such instances, scientists withdraw from the field of inquiry, either by giving up science altogether or by confining themselves to some alternative role in it, such as teaching or administration. (This does not say, of course, that teaching and administration do not have their own attractions, or that they are less significant than inquiry; I refer here only to the scientists who reluctantly abandon their research because it does not measure up to their own standards of excellence.) [. . .]

Perhaps the most telling instance of retreatism in mathematics is that of Janos Bolyai, inventor of one of the non-Euclidean geometries. The young Bolyai tries to obey his mathematician father who, out of the bitter fruits of his own experience, warns his son to give up any effort to prove the postulate on parallels—or, as his father more picturesquely put it, to "detest it just as much as lewd intercourse; it can deprive you of all your leisure, your health, your rest, and the whole happiness of your life." He dutifully becomes an army officer instead, but his daemon does not permit the twenty-one-year-old Bolyai to leave the postulate alone. After years of work, he develops his geometry, sends the manuscript to his father who in turn transmits it to Gauss, the prince of mathematicians, for a magisterial opinion. Gauss sees in the work proof of authentic genius, writes the elder Bolyai so, and adds, in all truth, that he cannot express his enthusiasm as fully as he would like, for "to praise it, would be to praise myself. Indeed, the whole contents of the work, the path taken by your son, the results to which he is led, coincide almost entirely with my meditations, which have occupied my mind partly for the last thirty or thirty-five years. . . . I am very glad that it is just the son of my old friend, who takes the precedence of me in such a remarkable manner." Delighted by this accolade, the elder Bolyai sends the letter to his son, innocently saying that it is "very satisfactory and redounds to the honor of our country and our nation." Young Bolyai reads the letter, but has no eye for the statements which say that his ideas are sound, that in the judgment of the incomparable Gauss he is blessed with genius. He sees only that Gauss has anticipated him. For a time, he believes that his father must have previously confided his ideas to Gauss who had thereupon made them his own.[37] His priority lost, and, with the further blow, years later, of coming upon Lobachevsky's non-Euclidean geometry, he never again publishes any work in mathematics.[38] [. . .]

Conclusion

The interpretation I have tried to develop here is not, I am happy to say, a new one. Nor do I consider it fully established and beyond debate. [. . .]

In short review, the interpretation is this: Like other social institutions, the institution of science has its characteristic values, norms, and organization. Among these, the emphasis on the value of originality has a self-evident rationale, for it is originality that does much to advance science. Like other institutions also, science has its system of allocating rewards for performance of roles. These rewards are largely honorific, since even today, when science is largely professionalized, the pursuit of science is culturally defined as being primarily a disinterested search

37. The principal source on the Bolyais, including the germane correspondence, is Paul Stäckel, *Wolfgang und Johann Bolyai, Geometrische Untersuchungen,* 2 vols. (Leipzig, 1913), which was not available to me at this writing. An excellent short account is provided by Roberto Bonola, *Non-Euclidean Geometry,* trans. H. S. Carslaw, 2d ed. (La Salle, Ill.: Open Court Publishing Company, 1938), 96–113; see also Dirk J. Struik, *A Concise History of Mathematics* (New York: Dover Publications, 1948), 2:251–54; Franz Schmidt, "Lebensgeschichte des Ungarischen Mathematikers Johann Bolyai de Bolya," *Abhandlungen zur Geschichte der Mathematik* 8 (1898): 135–46.

38. Two letters provide context for Bolyai's great fall from the high peak of exhilaration into the slough of despond. In 1823 he writes his father: ". . . the goal is not yet reached, but I have made such wonderful discoveries that I have been almost overwhelmed by them, and it would be the cause of continual regret if they were lost. When you will see them, you too will recognize it. In the meantime I can say only this: *I have created a new universe from nothing.* All that I have sent you till now is but a house of cards compared to the tower. I am as fully persuaded that it will bring me honor, as if I had already completed the discovery." And just as, a generation later, Lyell was prophetically to warn Darwin of being forestalled, so does the elder Bolyai warn the younger: "If you have really succeeded in the question, it is right that no time be lost in making it public, for two reasons: first, because ideas pass easily from one to another, who can anticipate its publication; and secondly, there is some truth in this, that many things have an epoch, in which they are found at the same time in several places, just as the violets appear on every side in spring. Also every scientific struggle is just a serious war, in which I cannot say when peace will arrive. Thus we ought to conquer when we are able, since the advantage is always to the first comer." (Quoted by Bonola, *Non-Euclidean Geometry,* 98, 99.) Small wonder that though young Bolyai continued to work sporadically in mathematics, he never again published the results of his work.

See chapter 23 on scientists' recognition that discoveries are often made independently and on this phenomenon activating a race for priority.—*Ed.*

for truth and only secondarily a means of earning a livelihood. In line with the value-emphasis, rewards are to be meted out in accord with the measure of accomplishment. When the institution operates effectively, the augmenting of knowledge and the augmenting of personal fame go hand in hand; the institutional goal and the personal reward are tied together. But these institutional values have the defects of their qualities. The institution can get partly out of control, as the emphasis upon originality and its recognition is stepped up. The more thoroughly scientists ascribe an unlimited value to originality, the more they are in this sense dedicated to the advancement of knowledge, the greater is their involvement in the successful outcome of inquiry and their emotional vulnerability to failure.

Against this cultural and social background, one can begin to glimpse the sources, other than idiosyncratic ones, of the misbehavior of individual scientists. The culture of science is, in this measure, pathogenic. It can lead scientists to develop an extreme concern with recognition which is in turn the validation by peers of the worth of their work. Contentiousness, self-assertive claims, secretiveness, lest one be forestalled, reporting only the data that support an hypothesis, false charges of plagiarism, even the occasional theft of ideas and, in rare cases, the fabrication of data,—all these have appeared in the history of science and can be thought of as deviant behavior in response to a discrepancy between the enormous emphasis in the culture of science upon original discovery and the actual difficulty many scientists experience in making an original discovery. In this situation of stress, all manner of adaptive behaviors are called into play, some of these being far beyond the mores of science.

All this can be put more generally. We have heard much in recent years about the dangers brought about by emphasis on the relativity of values, about the precarious condition of a society in which men do not believe in values deeply enough and do not feel strongly enough about what they do believe. If there is a lesson to be learned from this review of some consequences of a belief in the absolute importance of originality, perhaps it is the old lesson that unrestricted belief in absolutes has its dangers too. It can produce the kind of fanatic zeal in which anything goes. In its way, the absolutizing of values can be just as damaging as the decay of values to the life of man in society.

23

Multiple Discoveries in Science (1961)

The Self-Exemplifying Hypothesis of Multiples

At the root of a sociological theory of the development of science is the strategic fact of the multiple and independent appearance of the same scientific discovery—what I shall, for convenience, hereafter describe as a multiple. Ever since 1922 American sociologists have properly associated the theory with William F. Ogburn and Dorothy S. Thomas, who did so much to establish it in sociological thought.[1] On the basis of their compilation of some 150 cases of independent discovery and invention, they concluded that the innovations became virtually inevitable as certain kinds of knowledge accumulated in the cultural heritage and as social developments directed the attention of investigators to particular problems.

Appropriately enough, this is a hypothesis confirmed by its own history. (Almost, as we shall see, it is a Shakespearean play within a play.) For this idea of the sociological significance of multiple independent discoveries and inventions has been periodically rediscovered over a span of centuries. Today I shall not reach back of the nineteenth century for cases. Let us begin, then, with 1828, when Macaulay, in his essay on Dryden, observes that the independent invention of the calculus by Newton and Leibniz belongs to a larger class of instances in which the same invention or discovery had been made by scientists working apart from one another. For example, Macaulay tells us in truly Macaulayan prose and with the unmistakable Macaulayan flair,

> We are inclined to think that, with respect to every great addition which has been made to the stock of human knowledge, the case has been similar: that without Copernicus we should have been Copernicans—that without

From "Singletons and Multiples in Scientific Discovery," American Philosophical Society *Proceedings* 105, no. 5 (October 1961): 470–86.

1. William F. Ogburn and Dorothy S. Thomas, "Are Inventions Inevitable?" *Political Science Quarterly* 37 (March 1922) 83–98; William F. Ogburn, *Social Change* (New York: Heubsch, 1922), 90–122.

> Columbus America would have been discovered—that without Locke we should have possessed a just theory of the origin of human ideas.[2]

This is not the time to examine in detail the many occasions on which the fact of multiples with its implications for a theory of scientific development has been noted; on the evidence, often independently noted and set down in print. Working scientists, historians and sociologists of science, biographers, inventors, lawyers, engineers, anthropologists, Marxists and anti-Marxists, Comteans and anti-Comteans have time and again, though with varying degrees of perceptiveness, called attention both to the fact of multiples and to some of its implications. [. . .]

Nevertheless, the fact of multiple discoveries in science continues to be regarded by some, including minds of a high order, as something surpassing strange and almost unexplainable. Here is the great pathologist and historian of medicine, William Henry Welch, on the subject:

> The circumstances that a long-awaited discovery or invention has been made by more than one investigator, independently and almost simultaneously, and with varying approach to completeness, is a curious and not always explicable phenomenon familiar in the history of discovery.[3]

Other scholars tacitly assume that the pattern of multiples is both curious and distinctive of their own field of inquiry, if not entirely confined to it. As one example, consider the observation by the notable historian of geometry, Julian Lowell Coolidge:

> It is a curious fact in the history of mathematics that discoveries of the greatest importance were made simultaneously by different men of genius.[4]

And recently, the sociologist Talcott Parsons is recorded as having described the threefold, or possibly fivefold, discovery of "the internalization of values and culture as part of the personality" as "a very remarkable phenomenon because all of these people were independent of each other and their discovery is . . . fundamental."[5]

In part, of course, observations of this kind are merely casual remarks, not to be taken literally. But I should like now to develop the

2. *Miscellaneous Works of Lord Macaulay,* ed. Lady Trevelyan (New York: Harper, 1880), 1:110–11.

3. William Henry Welch, *Papers and Addresses* (Baltimore: Johns Hopkins Press, 1920), 3:229.

4. Julian Lowell Coolidge, *A History of Geometrical Methods* (Oxford: Clarendon Press, 1940), 122.

5. Talcott Parsons, in *Alpha Kappa Deltan: A Sociological Journal* 29 (winter 1959): 3–12, at 9–10.

hypothesis that, far from being odd or curious or remarkable, the pattern of independent multiple discoveries in science is in principle the dominant pattern rather than a subsidiary one. It is the singletons—discoveries made only once in the history of science—that are the residual cases, requiring special explanation. Put even more sharply, the hypothesis states that all scientific discoveries are in principle multiples, including those that on the surface appear to be singletons.

Evidence on the Hypothesis of Multiples

Stated in this extreme form, the hypothesis must at first sound extravagant, not to say incorrigible, removed from any possible test of competent evidence. For if even historically established singletons are declared to be multiples-in-principle—potential multiples that happened to emerge as singletons—it would seem that this is a self-sealing hypothesis, immune to investigation. And yet, it may be that things are not really as bad as all that.

An incorrigible hypothesis is, of course, not an hypothesis at all, but only a dogma or perhaps an incantation. I suggest, however, that, far from being incorrigible and therefore outrageous, this hypothesis of multiples is actually held much of the time by working scientists. The evidence for this is ready to hand and once its pertinence is seen, it can be gathered in abundance. Here, then, are ten kinds of related evidence that bears upon the hypothesis that discoveries in science are in principle multiples, with the singletons being the exceptional type requiring special explanation.

First is the class of discoveries long regarded as singletons that turn out to be rediscoveries of previously unpublished work. Cases of this kind abound. But here, I allude only to two notable instances: Cavendish and Gauss. Much of Cavendish's vast store of unpublished experiments and theories became progressively known only after his death in 1810, as Harcourt published some of his work in chemistry in 1839; Clerk Maxwell, his work in electricity in 1879; and Thorpe, his complete chemical and dynamical researches in 1921. But in the meanwhile, many of Cavendish's unpublished discoveries were made independently by contemporary and later investigators, among them, Black, Priestley, John Robison, Charles, Dalton, Gay-Lussac, Faraday, Boscovich, Larmor, Pickering, to cite only a few. And in most cases, the rediscoveries were regarded as singletons until Cavendish's records were belatedly published. The case of Gauss, as we know, is much the same. Loath to rush into print, Gauss crowded his notebooks with mathematical inventions and other discoveries that turned up independently in work by

Abel, Bolyai, Jacobi, Laplace, Galois, Dedekind, Franz Neumann, Grassmann, Hamilton, and others. Again, presumed singletons turned out to be multiples, as once unpublished work became known. Far from being exceptions, Cavendish and Gauss are instances of a larger class.

What holds for unpublished work often holds also for work which, though published, proves relatively neglected or inaccessible, owing either to its being at odds with prevailing conceptions, or its difficulty of apprehension, or its having been printed in little-known journals, and so on. Here, again, singletons become redefined as multiples when the earlier work is belatedly identified. In this class of cases, to choose among the most familiar, we need only recall Mendel and Gibbs. The case of Mendel[6] is too well known to need review; that of Gibbs almost as familiar, since Ostwald, in his preface to the German edition of the *Studies in Thermodynamics,* remarked, in effect, that "it is easier to rediscover Gibbs than to read him."[7]

These are all cases of seeming singletons which then turn out to have been multiples or rediscoveries. Other, more compelling classes of evidence bear upon the apparently incorrigible hypothesis that singletons, rather than multiples, are the exception requiring distinctive explanation and that discoveries in science are, in principle, potential multiples. These next classes of evidence are all types of forestalled multiples, discoveries that are historically identified as singletons only because the public report of the discovery forestalled others from making it independently. These are the cases of which it can be said: There, but for the grace of swift diffusion, goes a multiple.[8]

Second, then, and in every one of the sciences, including the social sciences, there are reports in print stating that a scientist has discontinued an inquiry, well along toward completion, because a new publication has anticipated both the hypothesis and the design of inquiry into the hypothesis. The frequency of such instances cannot be firmly estimated, of course, but I can report having located many.

Third, and closely akin to the foregoing type, are the cases in which the scientist, though forestalled, goes ahead to report an original, albeit anticipated, work. We can all call to mind those countless footnotes in the literature of science that announce with chagrin, "Since completing

6. See Hugo Iltis, *Life of Mendel* (New York: W. W. Norton, 1932); Conway Zirkle, "Gregor Mendel and His Precursors," *Isis* 42 (June 1951): 97–104.

7. This is the entirely apt paraphrase by Muriel Rukeyser in *Willard Gibbs* (New York: Doubleday Doran, 1942), 314.

8. It is only appropriate that the original saying—"There, but for the grace of God, . . ."—should itself be, with minor variations, a repeatedly reinvented expression.

this experiment, I find that Woodworth (or Bell or Minot, as the case may be) had arrived at this conclusion last year and that Jones did so fully sixty years ago." No doubt many of us here today have experienced one or more of these episodes in which we find that our best and, strictly speaking, our most original inquiries have been anticipated. [...]

Fourth, these publicly recorded instances of forestalled multiples do not, of course, begin to exhaust the presumably great, perhaps vast, number of unrecorded instances. Many scientists cannot bring themselves to report in print that they were forestalled. These cases are ordinarily known only to a limited circle, closely familiar with the work of the forestalled scientists. Interview studies of communication among scientists have begun to identify the frequency of such ordinarily unknown forestalling of multiples. Systematic field studies of this kind have turned up large proportions of what is often described as "unnecessary duplication" in research resulting from imperfections in the channels of communication between contemporary scientists. One such study[9] of American and Canadian mathematicians, for example, found 31 percent of the more productive mathematicians reporting that delayed publication of others' work had resulted in such "needless duplication," that is, in multiples.

Fifth, we find seeming singletons repeatedly turning out to be multiples, as friends, enemies, coworkers, teachers, students, and casual scientific acquaintances have reluctantly or avidly performed the service of a candid friend by acquainting an elated scientist with the fact that his original finding or idea is not the singleton he had every reason to suppose it to have been, but rather a doubleton or larger multiple, with the result that this latest independent version of the discovery never found its way into print. So, the young W. R. Hamilton hits upon and develops an idea in optics and as he plaintively describes the episode:

> A fortnight ago I believed that no writer had ever treated of Optics on a similar plan. But within that period, my tutor, the Reverend Mr. Boyton, has shown me in the College Library a beautiful memoir of Malus on the subject. . . . With respect to those results which are common to both, it is proper to state that I have arrived at them in my own researches before I was aware of his.[10]

What his tutor did for Hamilton, others have done for innumerable scientists through the years. The diaries, letters, and memoirs of scien-

9. See Herbert Menzel, *Review of Studies in the Flow of Information Among Scientists,* Columbia University Bureau of Applied Social Research, a report prepared for the National Science Foundation (January 1960), 1:21, 2:48.

10. Robert Perceval Graves, *Life of Sir William Rowan Hamilton,* 3 vols. (Dublin: Hodges, Figgis, 1882), 1:177.

tists are crowded with cases of this pattern (and with accounts of how they variously responded to these carriers of bad news).

Sixth, the pattern of forestalled multiples emerges as part of the oral tradition rather than the written one in still another form: as part of lectures. Here again, one instance must stand for many. Consider only the famous lectures of Kelvin at the Johns Hopkins where, it is recorded, he enjoyed "the surprise of finding [from members of his audience] that some of the things he was newly discovering for himself had already been discovered and published by others."[11]

A seventh type of pattern, tending to convert potential multiples into singletons, so far as the formal historical record goes, occurs when scientists have been diverted from a clearly developed program of investigation which, from all indications, was pointed in the direction successfully taken up by others. It is, of course, conjecture that the discoveries actually made by others would in fact have been made by the first but diverted investigator. But consider how such a scientist as Sir Ronald Ross, persuaded that his discoveries of the malarial parasite and the host mosquito were only the beginning, reports his conviction that, but for the interference with his plan by the authorities who employed him, he would have gone on to the discoveries made by others:

> The great treasure-house had been opened, but I was dragged away before I could handle the treasures. Scores of beautiful researches now lay open to me. I should have followed the "vermicule" in the mosquito's stomach—that was left to Robert Koch. I intended to mix the "germinal threads" with birds' blood—that was left to Schaudinn. I wished to complete the cycle of the human parasites—that was left to the Italians and others.[12]

Conjectural, to be sure, but with some indications that extraneous circumstances terminated a program of research that would have resulted in some of these discoveries becoming multiples rather than remaining adventitious singletons.

These several patterns of forestalled multiples, however, provide us with only sketchy evidence bearing on the apparently incorrigible hypothesis that multiples, both potential and actual, are the rule in scientific discovery and singletons the exception requiring special explanation. I turn now to evidence of quite another sort, the behavior of scientists themselves and the assumptions underlying that behavior.

11. Silvanus P. Thompson, *The Life of William Thomson, Baron Kelvin of Largs* (London: Macmillan, 1910), 2:815–16.

12. Ronald Ross, *Memoirs, with a Full Account of the Great Malaria Problem and Its Solution* (London: John Murray, 1923), 313.

And here I suggest that, far from being outrageous, the hypothesis is in fact commonly adopted as a working assumption by scientists themselves. I suggest that in actual practice, scientists, and perhaps especially the greatest among them, themselves assume that singleton discoveries are imminent multiples. Granted that it is a difficult and unsure task to infer beliefs from behavior; almost as difficult and unsure as to infer behavior from beliefs. But in this case, we shall see that the behavior of scientists clearly testifies to their underlying belief that discoveries in science are potential multiples.

After all, scientists have cause to know that many discoveries are made independently. They not only know it, but act on it. Since the culture of science puts a premium not only on originality but on chronological firsts in discovery, this awareness of multiples understandably activates a rush to ensure priority. Numerous expedients have been developed to ensure not being forestalled: for example, letters detailing one's new ideas or findings are dispatched to a potential rival, thus disarming him; preliminary reports are circulated; personal records of research are meticulously dated (as by Abel or Kelvin).

The race to be first in reporting a discovery testifies to the assumption that if the one scientist does not soon make the discovery, another will. This, then, provides an eighth kind of evidence bearing on our hypothesis [evidence set out in chapter 22 of this volume. The many instances detailed there are quite typical.] Norbert Wiener is no more circumstantial and outspoken about his experience than were Wallis, Wren, Huygens, Newton, the Bernoullis, and an indefinitely large number of other scientists through the centuries whose diaries, autobiographies, letters, and notes testify to the same effect.

In all this, I exclude those cases in which scientists move to establish their priority only to ensure that their discoveries not be diffused in the community of scientists before their own creative role in them is made eminently visible or to ensure that they not be later accused of having derived their own ideas from fellow-scientists who have borrowed them or cases in which, like that of Priestley, scientists publish quickly in order to advance science rapidly by making their work available to others at once. In this class of cases pertinent to the hypothesis, I refer only to those in which the rush to establish priority is avowedly motivated by the concern not to be forestalled, for this alone is competent evidence that scientists in fact assume that their initial singletons are destined not to remain singletons for long; that, in short, a multiple is definitely in the making.

But ninth, not all scientists who see themselves involved in a potential multiple are prepared to be outspoken about the matter. In many cases of this sort, their scientific colleagues, or kin, are. As we have seen [in the preceding chapter], the elder Bolyai, himself a mathematician of some consequence, prophetically warned his son that "no time be lost in making it [his non-Euclidean geometry] public, for two reasons:

> first, because ideas pass easily from one to another, who can anticipate its publication, and secondly, there is some truth in this, that many things have an epoch, in which they are found at the same time in several places, just as the violets appear on every side in spring. . . . Thus we ought to conquer when we are able, for the advantage is always to the first comer.[13] [. . .]

Gauss supplies us with another striking instance of the scientist's or mathematician's firm belief that a discovery or invention is not reserved to oneself alone. In 1795, at the ripe age of eighteen, he works out the method of least squares. To him the method seems to flow so directly from antecedent work that he is persuaded others must already have hit upon it; he is willing to bet, for example, that Tobias Mayer must have known it.[14] In this he was, of course, mistaken, as he learned later; his invention of least squares had not been anticipated. Nevertheless, he was abundantly right in principle: the invention was bound to be a multiple. As things turned out, it proved to be a quadruplet, with Legendre inventing it independently in 1805 before Gauss had got around to publishing it, and with Daniel Huber in Basel and Robert Adrain in the United States coming up with it a little later.[15]

There is a final and perhaps most decisive kind of evidence that the community of scientists does in fact assume that discoveries are potential multiples. This evidence is provided by the institutional expedients designed to protect the scientist's priority of conception. Since the seventeenth century, scientific academies and societies have established the practice of having sealed and dated manuscripts deposited with them in order to protect both priority and idea. As this was described in the early minutes of the Royal Society,

> When any fellow should have a philosophical notion or invention, not yet made out, and desire that the same sealed up in a box might be deposited

13. The letter is quoted in Roberto Bonola, *Non-Euclidean Geometry* 2d ed. (La Salle, Ill.: Open Court Publishing Co., 1938), 98–99.

14. C. A. F. Peters, ed., *Briefwechsel zwischen C. F. Gauss and H. C. Schumacher* (Altona: Gustav Esch, 1860), 3:387.

15. Waldo G. Dunnington, *Carl Friedrich Gauss* (New York: Exposition Press, 1955), 19. Adrian, the outstanding American mathematician of his day, was involved in several multiples. See J. L. Coolidge, "Robert Adrain and the Beginning of American Mathematics," *American Mathematical Monthly* 33 (February 1926): 61–76.

> with one of the secretaries, till it could be perfected, and so brought to light, this might be allowed for the better securing inventions to their authors.[16]

From at least the sixteenth century and as late as the nineteenth, it will also be remembered, discoveries were often reported in the form of anagrams—as with Galileo's "triple star" of Saturn and Hooke's law of tension—for the double purpose of establishing priority of conception and yet of not putting rivals on to one's original ideas, until they had been worked out further. From the time of Newton, scientists have printed short abstracts for the same purpose.[17] These and comparable expedients all testify that scientists, even those who manifestly subscribe to the contrary opinion, in practice assume that discoveries are potential multiples and will remain singletons only if prompt action forestalls the later independent discovery. It would appear, then, that what might first have seemed to be an incorrigible, perhaps outrageous, hypothesis about multiples in science is in fact widely assumed by scientists themselves.

A great variety of evidence—I have here set out only ten related kinds—testifies, then, to the hypothesis that, once science has become institutionalized, and significant numbers are at work on scientific investigation, the same discoveries will be made independently more than once and that singletons can be conceived of as forestalled multiples.

Sociological Theory of Genius in Science

I now return to the last part of the sociological theory of scientific development, dealing with the role of genius in that development. As I have intimated, the hypothesis of multiples has long been tied to the companion hypothesis that the great scientists, the undeniable geniuses, are altogether dispensable, for had they not lived, things would have turned out pretty much as they actually did. For generations the debate has waxed hot and heavy on this point. Scientists, philosophers, historians, sociologists, and psychologists have all at one time or another taken a polemical position in the debate. Emerson and Carlyle, Spencer and William James, Ostwald and de Candolle, Galton and Cooley—these are only a few among the many who have placed

16. Thomas Birch, *The History of the Royal Society of London* (London: A. Millar, 1756), 2:30. The French Academy of Sciences made extensive use of this arrangement; among the many documents deposited under seal was Lavoisier's on combustion; see Lavoisier, *Oeuvres de Lavoisier,* fasc. 2, *Correspondance,* ed. René Fric (Paris: Michel, 1957), 388–89.

17. See Birch, *History of the Royal Society* 4:437.

the social theory in opposition to the theory that provides ample space for the individual of scientific genius. That so many acute minds should have for so long regarded this as an authentic debate must not keep us from noticing how the issues have been falsely drawn; and that once the two theories are clearly stated, there is no necessary opposition between them. Instead, it is proposed that once scientific genius is conceived of sociologically, rather than, as the practice has commonly been, psychologically, the two ideas of the environmental determination of discovery can be consolidated into a single theory. Far from being incompatible, the two complement one another.

In this enlarged sociological conception, scientists of genius are precisely those whose work in the end would be eventually rediscovered. These rediscoveries would be made not by a single scientist but by an entire corps of scientists. On this view, the individual of scientific genius is the functional equivalent of a considerable array of other scientists of varying degrees of talent. On this hypothesis, the undeniably large stature of great scientists remains acknowledged. It is not cut down to size in order to fit a Procrustean theory of the environmental determination of scientific discovery. At the same time, this enlarged conception does not abandon the sociological theory of discovery in order to provide for the indisputable, great differences between scientists of large talent and of small; it does *not,* in the phrase of Bacon, "place all wits and understandings nearly on a level."

This enlarged sociological conception holds that great scientists will be repeatedly involved in multiples. First, because the genius will make many scientific discoveries altogether; and since each of these is, on the first part of the theory, a potential multiple, some will have become actual multiples. Second, this means that each scientist of genius will contribute the functional equivalent to the advancement of science of what a considerable number of other scientists will have contributed in the aggregate, some of these having been caught up in the repeated multiples in which the genius was actually involved.

In a word, the greatest men of science have been involved in a multiplicity of multiples. This is true for Galileo and Newton, for Faraday and Clerk Maxwell, for Hooker, Cavendish, and Stensen, for Gauss and Laplace, for Lavoisier, Priestley, and Scheele—in short, for all those whose place in the pantheon of science is beyond dispute, however much they may differ in the measure of their genius.

Once again, I can only allude to the pertinent evidence rather than report it in full. But consider the case of Kelvin, by way of illustration. After examining some 400 of his 661 scientific communications and addresses, Dr. Elinor Barber and I find him testifying to at least 32 multiple discoveries in which he eventually found that his independent discoveries had also been made by others. These 32 multiples involved an aggregate of 30 other scientists, some, like Stokes, Green, Helmholtz, Cavendish, Clausius, Poincaré, Rayleigh, themselves men of undeniable genius, others, like Hankel, Pfaff, Homer Lane, Varley and Lamé being men of talent, no doubt, but still not of the highest order. The great majority of these multiples of Kelvin were doublets, but some were triplets and a few, quadruplets. For the hypothesis that each of these discoveries was destined to find expression, even if the genius of Kelvin had not obtained, there is the best of traditional proof: each was in fact made by others. Yet Kelvin's stature as a scientist remains undiminished. For it required a considerable number of others to duplicate these thirty-two discoveries which Kelvin himself made.

Following out the logic of this kind of act, we can set up a matrix of multiple discoveries, with the entries in the matrix indicating the particular scientists involved in each of the multiples. Some of these others are themselves geniuses, in turn often involved in still other multiples. Others in the matrix are scientists of somewhat less talent who, on the average, are involved in fewer multiples. And toward the lower end of the scale of demonstrated scientific talent are the far more numerous scientists who in the aggregate are indispensable to the advancement of science and whose one moment of prime achievement came when they found for themselves one of the many discoveries that the scientific genius had made independently of them.

To continue for a moment with the specimen case of Kelvin, these thirty-two multiples are of course only a portion of the multiples in which he was eventually involved. For, as I have said, they are only the ones which Kelvin himself found to have been made by others. Beyond these are the discoveries by Kelvin that were only later made independently by others. Of these we do not yet have a firm estimate. And beyond these still are what I have described as the forestalled multiples: the discoveries of Kelvin which were not, so far as the record shows, made independently by others but which, on our hypothesis, would have been made had it not been for the widespread circulation of Kelvin's prior findings. Yet, even on this incomplete showing, it would seem that this one man of scientific genius was, in a reasonably exact sense, functionally equivalent to a sizable number of other scientists. And still, by the same token, his individual accomplishments in science remain undiminished when we note that he was not individually indispensable for these discoveries (since they were in fact made by others). This is the sense in which an enlarged sociological theory can take account both of the environmental determination of discovery while still providing for great variability in the intellectual stature of individual scientists. [. . .]

What has been found to hold for Kelvin is being found to hold for other scientists of the first rank who are now being examined in the light of the theory. They are all scientists of multiple multiples; their undeniable stature rests in doing individually what must otherwise be done and, as we have reason to infer, at a much slower pace, by a substantial number of other scientists, themselves of varying degrees of demonstrated talent. The sociological theory of scientific discovery has no need, therefore, to retain the false disjunction between the cumulative development of science and the distinctive role of the scientific genius.

There is perhaps time for a few needed and self-imposed caveats. For I cannot escape the uneasy sense that this short though, you will grant me, not entirely succinct, summary of masses of data on scientific discovery must lend itself to misunderstanding. This is so, if only because so much has unavoidably been left unsaid. As a preventative to such misunderstanding, I conclude by listing some *seeming* implications which are anything but implicit in what I have managed to report.

First, in presenting this modified version of a three-century-old conception of the course of scientific discovery, I do not imply that all discoveries are inevitable in the sense that, come what may, they will be made, at the time and the place, if not by the individual(s) who in fact made them. Quite the contrary: there are, of course, cases of scientific discoveries that could have been made generations, even centuries, before they were actually made, in the sense that the principal ingredients of these discoveries were long present in the culture. This recurrent fact of long-delayed discovery raises distinctive problems for the theory advanced here, but these are not unsolvable problems.[18]

Second, and perhaps contrary to the impression I have given, the theory rejects the pointless practice of what I have called "adumbrationism," that is, the practice of claiming to find dim anticipations of current scientific discoveries in older, and preferably ancient, work by the expedient of excessively liberal interpretations of what is being said

18. This passage only touches upon what has been conceptualized and investigated as a "postmature discovery": "For a discovery to qualify as postmature, for it to evoke surprise from the pertinent scientific community, it must have three attributes. In retrospect, it must be judged to have been technically achievable with methods then available; . . . capable of being expressed in terms comprehensible to working scientists at the time; and its implications must have been capable of being appreciated." Harriet Zuckerman and Joshua Lederberg, "Postmature Scientific Discovery?" *Science* 324 (1986): 629–63.—*Ed.*

now and of what was said then. The theory is not a twentieth-century version of the seventeenth- and eighteenth-century quarrel between the ancients and the moderns.

Third, the theory is not another version of Ecclesiastes, holding that "there is no new thing under the sun." The theory provides for the growth, differentiation, and development of science just as it allows for the fact that new increments in science are in principle or in fact repeated increments. It allows also for occasional mutations in scientific theory which are significantly new even though they are introduced by more than one scientist.

Fourth, the theory does not hold that to be truly independent, multiples must be chronologically simultaneous. This is only the limiting case. Even discoveries far removed from one another in calendrical time may be instructively construed as "simultaneous" or nearly so in social and cultural time, depending upon the accumulated state of knowledge in the several cultures and the structures of the several societies in which they appear.

Fifth, the theory allows for differences in the probability of actual rather than potential multiples according to the character of the particular discovery. Discoveries in science are of course not all of a piece. Some flow directly from antecedent knowledge in the sense that they are widely visible implications of what has gone just before. Other discoveries involve more of a leap from antecedent knowledge, and these are perhaps less apt to be actual multiples. But it is suggested that, in the end, these too manifest the same process of scientific development as the others.

Sixth, and above all, the theory rejects the false disjunction between the social determination of scientific discovery and the role of genius or "great man" in science. By conceiving scientific genius sociologically, as one who in his own person represents the functional equivalent of a number and variety of often lesser talents, the theory maintains that the genius plays a distinctive role in advancing science, often accelerating its rate of development and sometimes, by the excess of authority attributed to him, slowing further development.

Seventh and finally, the diverse implications of the theory are subject to methodical investigation. The basic materials for such study can be drawn from both historical evidence and from field inquiry into the experience of contemporary scientists. What Bacon obliquely noticed and many others recurrently examined can become a major focus in the contemporary sociology of science.

24

The Matthew Effect, II (1988)

[. . .] We begin by noting a theme that runs through Harriet Zuckerman's hours-long interviews with Nobel laureates in the early 1960s. It is repeatedly suggested in these interviews that eminent scientists get disproportionately great credit for their contributions to science while relatively unknown ones tend to get disproportionately little for their occasionally comparable contributions. As a laureate in physics put it: "The world is peculiar in this matter of how it gives credit. It tends to give the credit to already famous people."[1] Nor are the laureates alone in stating that the more prominent scientists tend to get the lion's share of recognition; less notable scientists in a cross-sectional sample studied by Warren O. Hagstrom have reported similar experiences.[2] But it is the eminent scientists, not least those who have received the ultimate contemporary accolade, the Nobel Prize, who provide presumptive evi-

From "The Matthew Effect in Science, II: Cumulative Advantage and the Symbolism of Intellectual Property," *Isis* 79 (1988): 607–23. This paper is an extension of "The Matthew Effect in Science: The Reward and Communications Systems of Science are Considered," *Science* 159 (5 January 1968): 56–63. As is Merton's practice, he has returned to the paradigmatic idea several times. The fragments reproduced here are from the most recent, extended formulation. —*Ed.*

1. Harriet Zuckerman, "Nobel Laureates: Sociological Studies of Scientific Collaboration" (Ph.D. diss., Columbia University, 1965). The later fruits of Zuckerman's research appear in Zuckerman, *Scientific Elite: Nobel Laureates in the United States,* 2d ed. (New York: Transaction Publishers, 1996); an account of the procedures adopted in these tape-recorded interviews appears in Zuckerman, "Interviewing an Ultra-Elite," *Public Opinion Quarterly* 36 (1972): 159–75. This is occasion for repeating what I have noted in reprinting the original "Matthew Effect in Science": "It is now [1973] belatedly evident to me that I drew upon the interview and other materials of the Zuckerman study to such an extent that, clearly, the paper should have appeared under joint authorship." A sufficient sense of distributive and commutative justice requires one to recognize, however belatedly, that to write a scientific or scholarly paper is not necessarily sufficient grounds for designating oneself as its sole author.

2. Warren O. Hagstrom, *The Scientific Community* (New York: Basic Books, 1965), 24–25.

dence of this pattern. For they testify to its occurrence, not as aggrieved victims, which might make their testimony suspect, but as "beneficiaries," albeit sometimes embarrassed and unintentional ones.

The claim that prime recognition for scientific work, by informed peers and not merely by the inevitably uninformed lay public, is skewed in favor of established scientists requires, of course, that the nature and quality of these diversely appraised contributions be identical or at least much the same. That condition is approximated in cases of full collaboration and in cases of independent multiple discoveries. The distinctive contributions of collaborators are often difficult to disentangle; independent multiple discoveries, if not identical, are at least enough alike to be defined as functional equivalents by the principals involved or by their informed peers.

In papers jointly published by scientists of markedly unequal rank and reputation, another laureate in physics reports, "the man who's best known gets more credit, an inordinate amount of credit." Or as a laureate in chemistry put it, "If my name was on a paper, people would remember *it* and not remember who else was involved."[3] The biological scientists R. C. Lenontin and J. L. Hubby have lately reported a similar pattern of experience with a pair of their collaborative papers, which have been cited often enough to qualify as "citation classics" (as designated by the institute for Scientific Information). One paper was cited some 310 times; the other, some 525 times. The first paper described a method; the second

> gave the detailed result of the application of the method to natural populations. The two papers were a genuinely collaborative effort in conception, execution, and writing and clearly form an indivisible pair, . . . published back-to-back in the same issue of the journal. The order of authors was alternated, with the biochemist, Hubby, being the senior author in the method paper and the population geneticist, Lewontin, as senior author in the application paper. Yet paper II has been cited over 50 percent more frequently than paper I. Citations to paper I virtually never stand alone but are nearly always paired with a citation to II, but the reverse is not true. Why? We seem to have a clear-cut case of Merton's "Matthew Effect"—that the already better known investigator in a field gets the credit for joint work, irrespective of the order of authors on the paper, and so gets even better known by an autocatalytic process. In 1966 Lewontin had been a professional for a dozen years and was well known among population geneticists, to whom the paper was addressed, while Hubby's career had been much shorter and was known chiefly to biochemical geneticists. As a result, population geneticists have consistently regarded Lewontin as the senior member

3. Zuckerman, *Scientific Elite,* 140, 228.

> of the team and given him undue credit for what was a completely collaborative work that would have been impossible for either one of us alone.[4]

At the extreme, such misallocation of credit can occur even when a published paper bears only the name of a hitherto unknown and uncredentialed scientist. Consider this observation by the invincible geneticist and biochemist, J. B. S. Haldane (whose *not* having received a Nobel Prize can be cited as prime evidence of the fallibility of the judges sitting in Stockholm). Speaking with Ronald Clark of S. K. Roy, his talented Indian student who had conducted important experiments designed to improve strains of rice, Haldane observed that

> Roy himself deserved about 95 per cent of the credit. . . . "The other 5 per cent may be divided between the Indian Statistical Institute and myself," he added. "I deserve credit for letting him try what I thought was a rather ill-planned experiment, on the general principle that I am not omniscient." But [Haldane] had little hope that credit would be given that way. "Every effort will be made here to crab his work," he wrote. "He has not got a Ph.D. or even a first-class M.Sc. So either the research is no good, or I did it."[5]

It is such patterns of the misallocation of recognition for scientific work that I described as "the Matthew effect." The not quite foreordained term derives, of course, from the first book of the New Testament, the Gospel according to Matthew (13:12 and 25:29). In the stately prose of the King James Version, created by what must be one of the most scrupulous and consequential teams of scholars in Western history, the well-remembered passage reads: "For unto everyone that hath shall be given, and he shall have abundance, but from him that hath not shall be taken away even that which he hath."

Put in less stately language, the Matthew effect is the accruing of large increments of peer recognition to scientists of great repute for particular contributions in contrast to the minimizing or withholding of such recognition for scientists who have not yet made their mark. The biblical parable generates a corresponding sociological parable. For this is the form, it seems, that the distribution of psychic income and cognitive wealth in science also takes. How this comes to be and with what consequences for the fate of individual scientists and the advancement of scientific knowledge are the questions at hand.

4. R. C. Lewontin and J. L. Hubby, "Citation Classic," *Current Contents/Life Sciences* 43 (28 October 1985): 16.

5. Ronald W. Clark, *J. B. S.: The Life and Work of J. B. S. Haldane* (New York: Coward-McCann, 1969), 247.

The Accumulation of Advantage and Disadvantage for Scientists

Taken out of its spiritual context and placed in a wholly secular context, the Matthew doctrine would seem to hold that the posited process must result in a boundlessly growing inequality of wealth, however wealth is construed in any sphere of human activity.[6] Conceived of as a locally ongoing process and not as a single event, the practice of giving unto everyone that hath much while taking from everyone that hath little will lead to the rich getting forever richer while the poor become poorer. Increasingly absolute and not only relative deprivation would be the continuing order of the day. But as we know, things are not as simple as that. After all, the extrapolation of local exponentials is notoriously misleading. In noting this, I do not intend nor am I competent to assess the current economic theory of the distribution of wealth and income. Instead, I shall report what a focus upon the skewed distribution of peer recognition and research productivity in science has led some of us to identify as the processes and consequences of the accumulation of advantage and disadvantage in science. [. . .]

I shall only remind you of a few of the marked inequalities and strongly skewed distributions of productivity and resources in science, and then focus on the consequences of the bias in favor of precocity that is built into our institutions for detecting and rewarding talent, an institutionalized bias that may help bring about severe inequalities during the life course of scholars and scientists.

First, then, a quick sampling of the abundance of conspicuously skewed distributions and inequalities identifiable at a given time:

> The total number of scientific papers published by scientists differs enormously, ranging from the large proportion of Ph.D.s who publish one paper or none at all to the rare likes of William Thomson, Lord Kelvin, with his six hundred plus papers, or the mathematician Arthur Cayley, publishing a paper every few weeks throughout his work life for a total of almost a thousand.[7]
>
> The skewed distribution in the sheer number of published papers is best approximated by variants of Alfred J. Lotka's "inverse square law" of scientific productivity, which states that the number of scientists with *n* publi-

6. Cf. this section with chap. 22 in this volume, "The Reward System of Science." —*Ed.*

7. Silvanus P. Thomson, *The Life of William Thompson, Baron Kelvin of Largs* (London: Macmillan, 1910), 2:1225–74; J. D. North, "Arthur Cayley," *Dictionary of Scientific Biography,* ed. Charles C. Gillispie (New York: Scribners, 1970–1980), 3:163.

cations is inversely proportional to n^2. In a variety of disciplines, this works out to some 5 or 6 percent of the scientists who *publish at all* producing about half of all papers in their discipline.[8]

The distributions are even more skewed in the use of scientists' work by their peers, as that use is crudely indexed by the number of citations to it. Much the same distribution has been found in various data sets: typical is Garfield's finding that, for an aggregate of some nineteen million articles published in the physical and biological sciences between 1961 and 1980, 0.3 percent were cited more than one hundred times; another 2.7 percent between twenty-five and one hundred times; and, at the other extreme, some 58 percent of those that were cited at all were cited only once in that twenty-year period.[9] This inequality, you will recognize, is steeper than Pareto-like distributions of income.

When it comes to *changes* in the extent of inequalities of research productivity and recognition during the course of an individual's work life as a scientist, the needed longitudinal data are much more scarce. Again, a few suggestive findings must serve.

In their simulation of longitudinal data (through disaggregation of a cross section of some two thousand American biologists, mathematicians, chemists, and physicists into several strata by career age), Paul D. Allison and John A. Stewart found "a clear and substantial rise in inequality for both [the number of research publications in the preceding five years and the number of citations to previously published work] from the younger to the older strata, strongly supporting the accumulative advantage hypothesis.[10]

Allison and Stewart also confirmed the Zuckerman-Merton hypothesis that decreasing research productivity with increasing age results largely from differing rates of attrition in research roles and that this approximates an all-or-none phenomenon. The hypothesis held that "the more productive scientists, recognized as such by the reward system of science, tend to persist in their research roles," while those with declining research productivity tend to shift to other indispensable roles in science, not excluding the conventionally maligned role of research administrator.[11]

8. Alfred J. Lotka, "The Frequency Distribution of Scientific Productivity," *Journal of the Washington Academy of Sciences* 16 (1926): 317–23; and Derek J. de Solla Price, *Little Science, Big Science . . . and Beyond* (New York: Columbia University Press, 1986), 38–42.

9. Eugene Garfield, *The Awards of Science and Other Essays* (Philadelphia: ISI Press, 1985), 176.

10. Paul D. Allison and John A. Stewart, "Productivity Differences among Scientists: Evidence for Accumulative Advantage," *American Sociological Review* 39 (1974): 596–606.

11. Ibid.; see also Harriet Zuckerman and Robert K. Merton, "Age, Aging and Age Structure in Science" (1972) reprinted in Robert K. Merton, *The Sociology of Science: Theoretical and Empirical Investigations* (Chicago: University of Chicago Press, 1973), 497–559, at 519–37.

> Derek Price pointedly reformulated and developed that hypothesis, "because there is a very large but decreasing chance that any given researcher will discontinue publication, the group of workers that reaches the [research] front during a particular year will decline steadily in total output as time goes on. Gradually, one after another, they will drop away from the research front. Thus the yearly output of the group as a whole will decline [and now comes the essential point Zuckerman and I tried to emphasize], even though any given individual within it may produce at a steady rate throughout his [or her] entire professional lifetime. We need, therefore, to distinguish this effect [of mortality at the research front] from any differences in the actual rates of productivity at different ages among those that remain at the front.[12]

With regard to the Matthew effect and associated cumulation of advantage,

> Stephen Cole found, in an ingeniously designed study of a sample of American physicists, that the greater their authors' scientific reputation, the more likely that papers of roughly equal quality (as assessed by the later number of citations to them) will receive rapid peer recognition (by citation within a year after publication). Prior repute of authors somewhat advances the speed of diffusion of their contributions.[13]
>
> Cole also found that it is a distinct advantage for physicists of still small reputation to be located in the departments most highly rated by peers: their new work diffuses more rapidly through the science networks than comparable work by their counterparts in peripheral university departments.[14]

Accumulation of Advantage and Disadvantage among the Young

I now focus on the special problems in the accumulation of advantage and disadvantage that derive from an institutionalized bias in favor of precocity. The advantages that come with early accomplishment taken as a sign of things to come stand in Matthew-like contrast to the situation confronted by young scientists whose work is judged as ordinary.[15] Such early prognostic judgments, I suggest, lead in some un-

12. Derek J. de Solla Price, "The Productivity of Research Scientists," *1975 Yearbook of Science and the Future* (Chicago: Encyclopaedia Britannica, 1975), 409–21, at 414. Stephen Cole's studies of age cohorts in various sciences confirm this pattern of a steady rate of publication by a significant fraction of scientists; see Cole, "Age and Scientific Performance," *American Journal of Sociology* 84 (1979): 958–77.

13. Stephen Cole, "Professional Standing and the Reception of Scientific Discoveries," *American Journal of Sociology* 76 (1970): 286–306, at 291–92.

14. Ibid., 292.

15. Jonathan R. Cole and Stephen Cole, *Social Stratification in Science* (Chicago: University of Chicago Press, 1973), 112–22, passim.

known fraction of cases to inadvertent suppression of talent through the process of the self-fulfilling prophecy. Moreover, this is more likely to be the case in a society, such as American society, where educational institutions are so organized as to put a premium on relatively *early* manifestations of ability—in a word, on precocity. Since it was that wise medical scientist Alan Gregg who led me to become aware of this bias institutionalized in our educational system, and since I cannot improve on his formulation, I transmit it here in the thought that you, too, may find it revealing:

> By being generous with time, yes, lavish with it, Nature allows man an extraordinary chance to learn. What gain can there be, then, in throwing away this natural advantage by rewarding precocity, as we certainly do when we gear the grades in school to chronological age by starting the first grade at the age of six and college entrance for the vast majority at seventeen and a half to nineteen? *For, once you have most of your students the same age, the academic rewards*—from scholarships to internships and residencies—go to those who are uncommonly bright *for their age*. In other words, you have rewarded precocity, which may or may not be the precursor of later ability. So, in effect, you have unwittingly belittled man's cardinal educational capital—time to mature.[16]

The social fact noted by Gregg is of no small consequence for the collective advancement of knowledge as well as for distributive justice. As he goes on to argue, "precocity thus may succeed in the immediate competitive struggle, but, in the long run, at the expense of mutants having a slower rate of development but greater potentialities."[17] By suggesting that there are such slow-starting mutants who have *greater* potentialities than some of the precocious, Gregg is plainly assuming part of what he then concludes. But, as I noted almost thirty years ago, Gregg's

> argument cuts deeply, nevertheless. For we know only of the "late bloomers" who have eventually come to bloom at all; we don't know the potential late bloomers who, cut off from support and response in their youth, never manage to come into their own at all. Judged ordinary by comparison with their precocious "age-peers," they are treated as youth of small capacity. They slip through the net of our institutional sieves for the location of ability, since this is a net that makes chronological age the basis for assessing relative ability. Treated by the institutional system as mediocrities with little

16. Alan Gregg, *For Future Doctors* (Chicago: University of Chicago Press, 1957), 125–26 [italics added].

17. Ibid., 125.

> promise of improvement, many of these potential late bloomers presumably come to believe it of themselves and act accordingly. At least what little we know of the formation of self-images suggests that this is so. For most of us most of the time, and not only the so-called "other-directed men" among us, tend to form our self-image—our image of potentiality and achievement—as a reflection of the images others make plain they have of us. *And it is the images that institutional authorities have of us that in particular tend to become self-fulfilling images: if the teachers, inspecting our Iowa scores and our aptitude-test figures and comparing our record with [those] of our "age-peers," conclude that we're run-of-the-mine and treat us accordingly, then they lead us to become what they think we are.*[18]

Of even more direct import for our immediate subject is the further observation back then that the institutionalized bias toward precocity, noted by Gregg, may have notably different consequences for comparable youngsters in differing social classes and ethnic groups.

> The potential late bloomers in the less privileged social strata are more likely to lose out altogether than their counterparts in the middle and upper strata. If poor [youths] are not precocious, if they don't exhibit great ability early in their lives and so are not rewarded by scholarships and other sustaining grants, they drop out of school and in many instances never get to realize their potentialities. The potential late bloomers among the well-to-do have a better prospect of belated recognition. Even if they do poorly in their school work at first, they are apt to go on to college in any case. The values of their social class dictate this as the thing to do, and their families can see them through. By remaining in the system, they can eventually come to view. But many of their [presumably] more numerous counterparts in the lower strata are probably lost for good. The bias toward precocity in our institutions thus works profound [and ordinarily hidden] damage on the [potential] late bloomers with few economic or social advantages.[19]

Such differential outcomes need not be intended by the people engaged in running our educational institutions and thereby affecting patterns of social selection. And it is such unanticipated and unintended consequences of purposive social action—in this case, rewarding primarily early signs of ability—that tend to persist. For they are *latent,* not manifest, social problems, that is, social conditions and processes that are at odds with certain interests and values of the society but

18. This sociological extension of Gregg's biopsychosocial observation remains as formulated in 1960: Robert K. Merton, "'Recognition' and 'Excellence': Instructive Ambiguities," in *Recognition of Excellence: Working Papers,* ed. Adam Yarmolinsky (Glencoe, Ill.: Free Press, 1962); reprinted in Merton, *Sociology of Science,* 419–38, at 428 [italics added].

19. Ibid., 428–29.

are not generally recognized as being so.[20] In identifying the wastage that results from marked inequalities in the training and exercise of socially prized talent, social scientists bring into focus what has been experienced by many as only a personal problem rather than a social problem requiring new institutional arrangements for its reduction or elimination.

Mutatis mutandis, what holds for the accumulation of advantage and disadvantage in the earliest years of education would hold also at a later stage for those youngsters who have made their way into fields of science and scholarship but who, not having yet exhibited prime performance, are shunted off into the less stimulating milieus for scientific work, with their limited resources. Absent or in short supply are the resources of access to needed equipment, an abundance of able assistance, time institutionally set aside for research, and, above all else perhaps, a cognitive microenvironment composed of colleagues at the research front who are themselves evokers of excellence, bringing out the best in the people around them. Not least is the special resource of being located at strategic nodes in the networks of scientific communication that provide ready access to information at the frontiers of research. By hypothesis, some unknown fraction of the unprecocious workers in the vineyards of science are caught up in a process of cumulative disadvantage that removes them early on from the system of scientific work and scholarship.

Other social and cognitive contexts may make for such patterned differentials of cumulative advantage and disadvantage. Harriet Zuckerman suggests, as an example, that just as class origins may differentially affect the rates at which potential late bloomers remain in the educational system long enough to bloom, so academic disciplines may differ in an unplanned tolerance for late blooming. Disciplines in which scholars often develop comparatively late—say, the humanities—presumably provide greater opportunities for late bloomers than those in which early maturation is more common—say, mathematics and the physical and biological sciences. Generalized, these conjectures hold that *contextual differences* such as social class or fields of intellectual activity as well as *individual differences* in the pattern of intellectual

20. On the first concept see Robert K. Merton, "The Unanticipated Consequences of Purposive Social Action," *American Sociological Review* 1 (1936): 894–904; on the concept of manifest and latent social problems see Robert K. Merton, *Social Research and the Practicing Professions,* ed. Aaron Rosenblatt and Thomas F. Gieryn (Cambridge, Mass.: Abt Books, 1982), 43–99, esp. 55 ff. See also chaps. 15 and 16 in this volume.—*Ed.*

growth affect the likelihood of success and failure for potential late bloomers.[21]

Differences in individual capabilities aside, then, processes of accumulative advantage and disadvantage accentuate inequalities in science and learning: inequalities of peer recognition, inequalities of access to resources, and inequalities of scientific productivity. Individual self-selection and institutional social selection interact to affect successive probabilities of being variously located in the opportunity structure of science. When the scientific role performance of individuals measures up to or conspicuously exceeds the standards of a particular institution or discipline—whether this be a matter of ability or chance—there begins a process of cumulative advantage in which those individuals tend to acquire successively enlarged opportunities for advancing their work (and the rewards that go with it) even further.[22] Since elite institutions have comparatively large resources for advancing research in certain domains, talent that finds its way into these institutions early has the enlarged potential of acquiring differentially accumulating advantages. The systems of reward, allocation of resources, and other elements of social selection thus operate to create and to maintain a class structure in science by providing a stratified distribution of chances among scientists for significant scientific work. [. . .]

Accumulation of Advantage and Disadvantage among Scientific Institutions

Skewed distributions of resources and productivity that resemble those we have noted among individual scientists are found among scientific institutions. These inequalities also appear to result from self-augmenting processes. Clearly, the centers of historically demonstrated accomplish-

21. Harriet Zuckerman, "Accumulation of Advantage and Disadvantage: The Theory and Its Intellectual Biography" in *L'Opera di R. K. Merton e la sociologia contemporanea,* ed. Carlo Mongardini and Simonetta Tabboni (Genoa: ECIG, 1989), 153–76.

22. In terms of a clinical rather than statistical sociology, I have tried to trace the process of accumulation of advantage in the academic life course of the historian of science and my longtime friend, Thomas S. Kuhn, as I have done more recently in tracking my own experience as apprentice to the then world dean of the history of science who has been honored by the establishment of the George Sarton Chair in the History of Science at the University of Ghent. For the case of Kuhn see Robert K. Merton, *The Sociology of Science: An Episodic Memoir* (Carbondale: Southern Illinois University Press, 1979), 71–109; for my own case see Merton, "George Sarton: Episodic Recollections by an Unruly Apprentice," *Isis* 76 (1985): 470–86.

ments in science attract far larger resources of every kind, human and material, than research organizations that have not yet made their mark. These skewed distributions are well known and need only bare mention here.

> In 1981, some 28 percent of the $4.4 billion of federal support for academic research and development went to just ten universities.[23]
>
> Universities with great resources and prestige in turn attract disproportionate shares of the presumably most promising students (subject to the precocity restriction we have noted): in 1983, two-thirds of the five hundred National Science Foundation graduate fellows elected to study at just fifteen universities.[24]
>
> Those concentrations have been even more conspicuous in the case of outstanding scientists. Zuckerman found, for example, that at the time they did the research that ultimately brought them the Nobel Prize, 49 percent of the future American laureates working in universities were in just five of them: Harvard, Columbia, Rockefeller, Berkeley, and Chicago. By way of comparison, these five universities comprised less than 3 percent of all faculty members in American universities.[25]
>
> Zuckerman also found that these resource-full and prestige-full universities seem able to spot and to retain these prime movers in contemporary science. For example, they kept 70 percent of the future laureates they had trained, in comparison with 28 percent of the other Ph.D.s they had trained. Much the same pattern, though less markedly, held for a larger set of sixteen elite institutions.[26]

But enough about these details of great organizational inequalities in science. This only raises the question anew: If the processes of accumulating advantage and disadvantage are truly at work, why are there not even greater inequalities than have been found to obtain?

Countervailing Processes

Or to put the question more concretely and parochially, why have not Harvard, rich in years—350 of them—and in much else, and Columbia, with its 230 years, and, to remain parochial, the Rockefeller, with

23. National Science Foundation, *Federal Support to Universities, Colleges, and Selected Nonprofit Institutions, Fiscal Year 1981* (Washington, D.C.: U.S. Government Printing Office, 1983), 79–80.

24. National Science Foundation, *Grants and Awards for Fiscal Year 1983* (Washington, D.C.: U.S. Government Printing Office, 1984), 215–17.

25. Zuckerman, *Scientific Elite*, 171.

26. Ibid., chap. 5.

its 75 years of prime reputation both as research institute and graduate university, jointly garnered just about *all* the American Nobel laureates rather than a "mere" third of them within five years after their receiving the prize?[27] Put more generally, why do the posited processes of accumulating advantage and disadvantage not continue without assignable limit?

Even Thomas Macaulay's ubiquitous schoolboy would nowadays know that exponential processes do not continue endlessly. Yet some of us make sensible representations of growth processes within a local range and then mindlessly extrapolate them far outside that range. As Derek Price was fond of saying in this connection, if the exponential rate of growth in the number of scientists during the past half century were simply extrapolated, then every man, woman, and child–to say nothing of their cats and dogs—would have to end up as scientists. Yet we have an intuitive sense that somehow they will not.

In much the same way, every schoolgirl knows that when two systems grow at differing exponential rates, the gap between them swiftly and greatly widens. Yet we sometimes forget that as such a gap approaches a limit, other forces come into play to constrain still further concentrations and inequalities of whatever matters are in question. Such countervailing processes that close off the endless accumulation of advantage have not yet been systematically investigated for the case of science—more particularly, for the distribution of human and material resources in universities and of scientific productivity within them. Still, I would like to speculate briefly about the forms countervailing processes might take.

Consider for example the notion of an excessive density of talent. It is not a frivolous question to ask: How much concentrated talent can a single academic department or research unit actually stand? How many prime movers in a particular research area can work effectively in a single place? Perhaps there really can be too much of an abstractly good thing.

Think a bit about the patterned motivations of oncoming talents as they confront a high density of talented masters in the same department or research unit. The more autonomous among them might not entirely enjoy the prospect of remaining in the vicinity and, with the Matthew effect at work, in the shadow of their masters, especially if they felt, as youth understandably often comes to feel—sometimes with ample grounds—that those masters have seen their best days. Correlatively, some of the firmly established masters, in a pattern of master-

27. Ibid., 241.

apprentice ambivalence, might not relish the thought of having exceedingly talented younger associates in their own or competing research terrains, who they perceive might subject them to premature replacement, at least in local peer esteem, when, as anyone can see, they, the masters, are still in their undoubted prime.[28] Not every one of us elders has the same powers of critical self-appraisal, and the same largeness of spirit, as Isaac Barrow, the first occupant of the Lucasian Chair of Mathematics at Cambridge, who, it has been said, stepped down from that special chair at the advanced age of thirty-nine in favor of his twenty-seven-year-old student—a chap named Isaac Newton. In our time, of course, at least during the years of seemingly limitless academic affluence and expansion, Barrow would have stayed on and Newton would have been given a new chair. But again, as we have ample cause to know, continued expansion of that kind in any one institution also has its limits.

Apart from such forces generated *within* universities that make for dispersion of human capital in science and learning, there is also the system process of social and cognitive competition *among* universities. Again, a brief observation must stand for a detailed analysis. Entering into that external competition is the fact that the total resources available to a university or research institute must be allocated somehow amongst its constituent units. Some departments wax poor even in rich universities. This provides opportunities to institutions of considerably smaller resources and reputation. These may elect to concentrate their limited resources in particular fields and departments and so to provide competitively attractive microenvironments to talents of the first class in those fields.

As another countervailing process, populist and democratic values may be called into play in the wider society, external to academic institutions and to science, and lead governmental largesse to be more widely spread in a calculated effort to counteract cumulating advantage in the great centers of learning and research.

But enough of such speculations. I must not further defer examination of the symbolism of intellectual property in science by continuing with observations on countervailing forces that emerge to curb the accumulation of advantage that might otherwise lead to a permanent institutional monopoly or sustained oligopoly in fields of science and the

28. Robert K. Merton and Elinor Barber, "Sociological Ambivalence" (1963), reprinted in Robert K. Merton, *Sociological Ambivalence* (New York: Free Press, 1976), 3–31, esp. 4–6; Vanessa Merton, Robert K. Merton, and Elinor Barber, "Client Ambivalence in Professional Relationships," in *New Directions in Helping*, ed. B. M. DePaulo et al. (New York: Academic Press, 1983), 2:13–44, at 26–27.

sustained domination of a few individuals in those fields. Just as there is reason to know that the preeminence of individual scientists will inexorably come to an end, so there is reason to expect that various preeminent departments of science will decline while others rise in the fullness of time.

The Symbolism of Intellectual Property in Science

To explore the forms of inequality in science registered by such concepts as the Matthew effect and the accumulation of advantage, we must have some way of thinking about the distinctive equivalents in the domain of science of income, wealth, and property found in the economic domain. How do scientists manage to perceive one another simultaneously as peers and as unequals, in the sense of some being first among equals—*primus inter pares,* as the ancients liked to say? What is the distinctive nature of the coin of the realm and of intellectual property in science?

The tentative answer to the coinage question I proposed back in 1957 seems to have gained force in light of subsequent work in the sociology of science.[29] The system of coinage is taken to be based on the public recognition of one's scientific contributions by qualified peers. That coinage comes in various denominations: largest in scale and shortest in supply is the towering recognition symbolized by eponyms for an entire epoch in science, as when we speak of the Newtonian, Darwinian, Freudian, Einsteinian, or Keynesian eras. A considerable plane below, though still close to the summit of recognition in our time, is the Nobel Prize. Other forms and echelons of eponymy, the practice of affixing the names of scientists to all or part of what they have contributed, comprise thousands of eponymous laws, theories, theorems, hypotheses, and constants, as when we speak of Gauss's theorems, Planck's constant, the Heisenberg uncertainty principle, a Pareto distribution, a Gini coefficient, or a Lazarsfeld latent structure. Other forms of peer recognition distributed to far larger numbers take further graded forms: election to honorific scientific societies, medals and awards of various kinds, named chairs in institutions of learning and research, and, moving to what is surely the most widespread and altogether basic form of scholarly recognition, that which comes with having one's work *used and explicitly acknowledged* by one's peers.

I shall argue that cognitive wealth in science is the changing stock of knowledge, while the socially based psychic income of scientists takes

29. See chap. 22 in this volume.—*Ed.*

the form of pellets of peer recognition that aggregate into reputational wealth. This conception directs us to the question of the distinctive character of intellectual property in science.

As I suggested at the outset, it is only a seeming paradox that, in science, one's private property is established by giving its substance away. For in a long-standing social reality, only when scientists have published their work and made it generally accessible, preferably in the public print of articles, monographs, and books that enter the archives, does it become legitimately established as more or less securely theirs. That is, after all, what we mean by the expression "scientific contribution": an offering that is accepted, however provisionally, into the common fund of knowledge.

That crucial element of free and open communication is what I have described as the norm of "communism" in the social institution of science[30]—with Bernard Barber going on to propose the less connotational term "communality."[31] Indeed, long before the nineteenth-century Karl Marx adopted the watchword of a fully realized communist society—"from each according to his abilities, to each according to his needs"—this was institutionalized practice in the communication system of science. This is not a matter of human nature, of nature-given altruism. Institutionalized arrangements have evolved to motivate scientists to contribute freely to the common wealth of knowledge according to their trained capacities, just as they can freely take from that common wealth what they need. Moreover, since a fund of knowledge is not diminished through exceedingly intensive use by members of the scientific collectivity—indeed, it is presumably augmented—that virtually free and common good is not subject to what Garrett Hardin has aptly analyzed as "the tragedy of the commons": first the erosion and then the destruction of a common resource by the individually rational and collectively irrational exploitation of it.[32] In the commons of science it is structurally the case that the give and the take both work to enlarge the common resource of accessible knowledge.

The structure and dynamics of this system are reasonably clear. Since positive recognition by peers is the basic form of *extrinsic* reward in science, all other extrinsic rewards, such as monetary income from science-connected activities, advancement in the hierarchy of scientists, and enlarged access to human and material scientific capital, derive

30. See chaps. 20 and 21 in this volume.—*Ed.*

31. Bernard Barber, *Science and the Social Order* (Glencoe, Ill.: Free Press, 1952), 130–32.

32. Garrett Hardin, "The Tragedy of the Commons," *Science* 162 (1968): 1243–47.

from it. But, obviously, peer recognition can be widely accorded only when the correctly attributed work is widely known in the pertinent scientific community. Along with the motivating *intrinsic* reward of working on a scientific problem and solving it, this kind of extrinsic reward system provides great incentive for engaging in the often arduous and tedious labors required to produce results that enlist the attention of qualified peers and are put to use by some of them.

This system of open publication that makes for the advancement of scientific knowledge requires normatively guided reciprocities. It can operate effectively only if the practice of making one's work communally accessible is supported by the correlative practice in which scientists who make use of that work acknowledge having done so. In effect, they thus reaffirm the property rights of the scientist to whom they are then and there indebted. This amounts to a pattern of legitimate *ap*propriation as opposed to the pattern of illegitimate *ex*propriation (plagiary).

We thus begin to see that the institutionalized practice of citations and references in the sphere of learning is not a trivial matter. While many a general reader—that is, the lay reader located outside the domain of science and scholarship—may regard the lowly footnote or the remote endnote or the bibliographic parenthesis as a dispensable nuisance, it can be argued that these are in truth central to the incentive system and an underlying sense of distributive justice that do much to energize the advancement of knowledge.

As part of the intellectual property system of science and scholarship, references and citations serve two types of functions: instrumental cognitive functions and symbolic institutional functions. The first of these involves directing readers to the sources of knowledge that have been drawn upon in one's work. This enables research-oriented readers, if they are so minded, to assess for themselves the knowledge claims (the ideas and findings) in the cited source; to draw upon other pertinent materials in that source that may not have been utilized by the citing intermediary publication; and to be directed in turn by the cited work to other, prior sources that may have been obliterated by their incorporation in the intermediary publication.

But citations and references are not only essential aids to scientists and scholars concerned to verify statements or data in the citing text or to retrieve further information. They also have not-so-latent symbolic functions. They maintain intellectual traditions and provide the peer recognition required for the effective working of science as a social activity. All this, one might say, is tucked away in the aphorism that Newton made his own in that famous letter to Hooke where he wrote: "If I have seen further, it is by standing on ye sholders of Giants."[33] The very form of the scientific article as it has evolved over the last three centuries normatively requires authors to acknowledge on whose shoulders they stand, whether these be the shoulders of giants or, as is often the case, those of men and women of science of approximately average dimensions for the species *scientificus*. Thus, in our brief study of the evolution of the scientific journal as a sociocognitive invention, Harriet Zuckerman and I have taken note of how Henry Oldenburg, the editor of the newly invented *Transactions of the Royal Society* in seventeenth-century England, induced the emerging new breed of scientist to abandon a frequent long-standing practice of sustained secrecy and to adhere instead to "the new form of free communication through a motivating exchange: open disclosure in exchange for institutionally guaranteed honorific property rights in the new knowledge given to others."[34]

That historically evolving set of complementary role obligations has taken deep institutional root. A composite cognitive and moral framework calls for the systematic use of references and citations. As with all normative constraints in society, the depth and consequential force of the moral obligation to acknowledge one's sources become most evident when the norm is violated (and the violation is publicly visible). The failure to cite the original text that one has quoted at length or drawn upon becomes socially defined as theft, as intellectual larceny or, as it is better known since at least the seventeenth century, as plagiary. Plagiary involves expropriating the one kind of private property that even the dedicated abolitionist of private productive property, Karl Marx, passionately regarded as inalienable (as witness his preface to the first edition of *Capital* and his further thunderings on the subject throughout that revolutionary work).

To recapitulate: the bibliographic note, the reference to a source, is not merely a grace note, affixed by way of erudite ornamentation. (That it can be so used, or abused, does not of course negate its core uses.) The reference serves both instrumental and symbolic functions in the

33. George Sarton was long interested in the history of the aphorism. Since it says much in little about one of the ways in which scientific knowledge grows, I indulged in a Shandean account of its historical adventures: Robert K. Merton, *On the Shoulders of Giants* (1965; reprint, New York: Harcourt Brace Jovanovich, 1985; reprint, Chicago: University of Chicago Press, 1993).

34. Harriet Zuckerman and Robert K. Merton, "Patterns of Evaluation in Science: Institutionalization, Structure and Functions of the Referee System," *Minerva* 9 (1971): 66–100.

transmission and enlargement of knowledge. Instrumentally, it tells us of work we may not have known before, some of which may hold further interest for us; symbolically, it registers in the enduring archives the intellectual property of the acknowledged source by providing a pellet of peer recognition of the knowledge claim, accepted or expressly rejected, that was made in that source.

Intellectual property in the scientific domain that takes the form of recognition by peers is sustained, then, by a code of common law. This provides socially patterned incentives, apart from the intrinsic interest in inquiry, for attempting to do good scientific work and for giving it over to the common wealth of science in the form of an open contribution available to all who would make use of it, just as the common law exacts the correlative obligation on the part of the users to provide the reward of peer recognition by reference to that contribution. Did space allow—which, happily for you, it does not—I would examine the special case of tacit citation and of "obliteration by incorporation" (or, even more briefly, OBI): the obliteration of the sources of ideas, methods, or findings by their being anonymously incorporated in current canonical knowledge.[35] Many of these cases of seemingly unacknowledged intellectual debt, it can be shown, are literally exceptions that prove the rule, that is to say, they are no exceptions at all since the references, however tacit, are evident to knowing peers.

Once we understand that the sole property right of scientists in their discoveries has long resided in peer recognition of it and in derivative collegial esteem, we begin to understand better the concern of scientists to get there first and to establish their priority.[36] That concern then becomes identifiable as a "normal" response to institutionalized values. The complex of validating the worth of one's work through appraisal by competent others and the seeming anomaly, even in a capitalistic society, of publishing one's work without being directly recompensed for each publication have made for the growth of public knowledge and the eclipse of private tendencies toward hoarding private knowledge (secrecy), still much in evidence as late as the seventeenth century. Current renewed tendencies toward secrecy, and not alone in what Henry

35. I easily resist the temptation to begin a discourse on this pattern in the transmission of knowledge. Short proleptic discussions of "obliteration by incorporation" are found in Robert K. Merton, *Social Theory and Social Structure,* 3d ed., rev. and enl. (New York: Free Press, 1968), 25–38; Robert K. Merton, foreword to Eugene Garfield, *Citation Indexing: Its Theory and Application in Science, Technology, and Humanities* (New York: Wiley, 1979); Eugene Garfield, *Essays of an Information Scientist* (Philadelphia: ISI Press, 1977), 396–99.

36. See chap. 22 in this volume.—*Ed.*

Etzkowitz has described as "entrepreneurial science,"[37] will, if extended and prolonged, introduce major change in the institutional and cognitive workings of science.

Since I have imported, not altogether metaphorically, such categories as intellectual property, psychic income, and human capital into this account of the institutional domain of science, it is perhaps fitting to draw once again upon a chief of the tribe of economists for a last word on our subject. Himself an inveterate observer of human behavior rather than only of economic numbers, and also himself a practitioner of science who keeps green the memory of those involved in the genealogy of an idea, Paul Samuelson cleanly distinguishes the gold of scientific fame from the brass of popular celebrity. This is how he concluded his presidential address, a quarter century ago, to an audience of fellow economists: "Not for us is the limelight and the applause [of the world outside ourselves]. But that doesn't mean the game is not worth the candle or that we do not in the end win the game. In the long run, the economic scholar works for the only coin worth having—our own applause."[38]

37. Henry Etzkowitz, "Entrepreneurial Scientists and Entrepreneurial Universities in American Academic Science," *Minerva* 21 (1983): 198–233.

38. Paul Samuelson, "Economics and the History of Ideas" (delivered in 1961), reprinted in *The Collected Scientific Papers of Paul A. Samuelson,* ed. Joseph E. Stiglitz (Cambridge: MIT Press, 1966), 2:1499–516.

CODA

A Life of Learning (1994)

I doubt that any of my learned predecessors experienced as much harmless pleasure as mine when *they* were asked to give the Haskins Lecture. After all, none of them were sociologists, happy to learn that their work was thought humanistic enough to warrant this great honor. And surely, none of them had their lecture mark the seventy-fifth anniversary of ACLS and also take place in their hometown.

Other coincidences of time and place deepen my pleasure in this meeting. For one, this new Benjamin Franklin Hall of the American Philosophical Society happens to be within walking distance of the house in which I was born almost eighty-four years ago. For quite another, the daunting invitation to give the Haskins Lecture reached me just as I was preparing a new edition of my prodigal brainchild, *On the Shoulders of Giants*. And naturally, *OTSOG*, as I have come to call it in a breath-saving acronym, draws often upon Haskins's magisterial work, *The Renaissance of the Twelfth Century*.

But enough. Now that I have subjected you to this brief recital of coincidences, some of you no doubt ache to remind me that the humanist Plutarch anticipated this sort of thing when he observed: "Fortune is ever changing her course and time is infinite; so it is no great wonder that many coincidences should occur. . . ." And no doubt others of you would prefer to draw upon the mathematical statisticians, Persi Diaconis and Frederick Mosteller, who conclude that "we are swimming in an ocean of coincidences. Our explanation is that *nature* and we ourselves are creating these, sometimes causally, and also partly through perception and partly through objective accidental relationships." As will soon become plain, I am inclined to agree with both the humanist and the scientists.

After much ego-centered meditation about the Haskins Lecture, I

From the Charles Homer Haskins Lecture presented at the annual meeting of the American Council of Learned Societies, Philadelphia, April 1994. Published as Occasional Paper 25 (New York: ACLS, 1994), Reprinted by permission of the American Council of Learned Societies.

have come to two conclusions: one, that my life of learning has been largely shaped by a long series of chance encounters and consequential choices, and not by anything like a carefully designed plan. The other that, in my case at least, "the Child is [truly] father of the Man," a conclusion that invokes Wordsworth and Laurence Sterne rather more than Sigmund and Anna Freud. Those conclusions will lead me to focus this evening, far more than I had at first intended, on my early years. And since few, if any, of you gracing this ACLS celebration will have known the vanished world of my distant youth and since my word portraits of that world are bound to be imperfect, I shall resort from time to time to the use of more lifelike visuals, pictures from a family album.

I

My very first chance encounter occurred, of course, with my birth. For who or what dictated that I, and not another, should be born to my loving mother and father? Not the genetic me but the entire me as I have come to be. As it happens, my first appearance also involved a coincidence of time and place, for I was a Yankee-Doodle-baby, born on Independence Day eight blocks from Independence Square. This I report on the firm testimony of my mother, who was presumably close at hand. As she vividly described it more than once, the event took place in the family house well before midnight of July 4th—while local patriots were still noisily celebrating the holiday. It did *not* take place on July 5th, as mistakenly recorded on the birth certificate after a forgetful lapse of a month by the family doctor who helped bring me into the world; said doctor plainly being a latter-day version of Tristram Shandy's accoucheur, Dr. Slop. My parents did not discover the error until they needed evidence that I was old enough to enter public school; by that time, the bureaucratic damage had been done. Ever since, I've had two birthdays a year: July 4th for the family and July 5th on public documents (until, in a much-delayed show of independence, I recently began to set the record straight).

(Incidentally, the same sort of thing also happened to Saul Bellow. His birth certificate has him born on July 10th although he generally lists it as *June* 10th, since his mother insisted that it *was* June. And yet, his impending biographer James Atlas tells us, Bellow entered that misconceived July birth-date on his application for a Guggenheim fellowship just as I did, in turn, on my own Guggenheim application. A continuing reign of bureaucratic error.)

At any rate, here at least is visual evidence of my having appeared at all (figure 1). Followed by apparent evidence of my being oriented

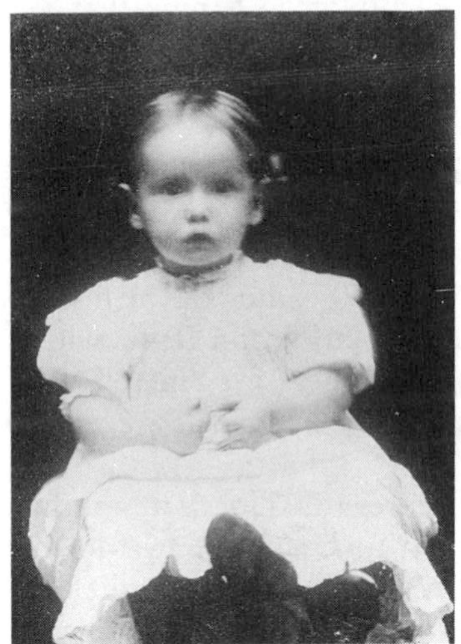

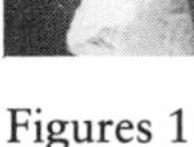

Figures 1

Figure 2

to the glories of the book years before I began my formal schooling (figure 2). I suppose that my mother was making a statement by placing her only son in that Little Lord Fauntleroy garb.

The document wrongly attesting the time of my birth sensitized me early on to an elementary rule of historical method: when reconstructing the past, draw gratefully on archival documents but beware of taking them at face value. So it was that decades later, when I became apprenticed at Harvard to the pioneering historian of science, George Sarton, I found myself resonating to his cautionary remark that even "the dates printed on the covers of periodicals are often inaccurate." Just as I resonated later to the infectious seventeenth-century John Aubrey who, while doing strenuous field work in English cemeteries to discover when little lives were *actually* rounded with a sleep, concluded that even epitaphs etched on tombstones might deceive, as, for example, the epitaph which asked passers-by to "Pray for the soul of Constantine Darrel Esq. who died Anno Domini 1400 and his wife, who died Anno Domini 1495." But no more about rules of historical evidence and back, for another Shandean moment, to my birth.

That event received *no* public notice. Not, I believe, because it was obscured by another historic event that same day: the battle for the heavyweight championship of the world between the "black giant" Jack Johnson and the "white giant" Jim Jeffries (if I may adopt Jack

London's description of that pugilistic pair). Nor do I think that the *Philadelphia Inquirer* failed to record my arrival simply because it was busy reporting that "not since October of 1907 has the financial district been thrown into such a state of demoralization . . . by the panicky markets in stocks." Nor again, do I believe for even a moment that word of my birth went unnoticed simply because "mid-summer clearance sales" had the ladies hurrying to Philadelphia's Lit Brothers for their pick of "$6 dresses marked down to $3.50" while the men were off to Blum Brothers, just two blocks away at Market and Tenth, where they could find "white serge suits with black stripes" for a mere $10—both of these being obvious good buys in a consumer society even for that distant time.

Not at all. I suspect that my birth went unregarded for quite another reason. It was probably because, as a *New Yorker* profile by Morton Hunt put it some thirty-five years ago, I was born "almost at the bottom of the social structure" in the slums of South Philadelphia to working-class Jewish immigrants from Eastern Europe. But since a proper slum involves wretched over-crowding in dismal housing, perhaps our family situation did not truly qualify as slum-like. After all, upon being delivered by our own Dr. Slop, I found myself at ease in the ample six-room quarters above my father's newly acquired milk-butter-and-egg shop located at 828 South Third Street. When the uninsured shop was destroyed by fire a few years later and the family's fortunes declined, my father became a carpenter's assistant in the Philadelphia Navy Yard and we moved into a smaller red-brick row house. There too, I had no cause to feel deprived—or, as the sociologists now say, I did not experience "relative deprivation." Our house had an occasionally used parlor and a diversely used dining room—where, for example, I developed a slender interest in technology by building a crystal radio set, followed by a peanut-tube set and ultimately by a grand heterodyne set. The coal-burning stove in the kitchen provided heat for the entire house. The gas-lighting served admirably for years and, having nothing better, we made do with the privy in the backyard. In short, we were living the lives of those who would come to be known as "the deserving poor," fueled with the unquestioned premise that things would somehow get better, surely so for the children.

(As you see from figure 3, I still have a picture of my mother and her darling son, then aged ten or thereabouts, standing tall in that tiny backyard, his innocent child's head encircled by what appears to be . . . a saintly nimbus. Coincidences continue to abound. Some forty years later, Jerzy Kosinski, author of that haunting autobiographical novel of the Holocaust, *The Painted Bird,* and sometime student of sociology at

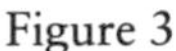

Figure 3

Figure 4 (Photo: Jerzy Kosinski)

Columbia who also happened to be a prize-winning photographer, takes a snapshot of his sometime teacher, with this result [figure 4]. As you see, my older, rather less innocent head is again nearly encircled by what surely can no longer be a *saintly* nimbus.)

Those early appearances notwithstanding, I was not greatly deprived during the rest of my fourteen years in that urban village. Thanks to its great array of institutional riches close at hand, I soon began to discover the larger world. From the start, I had a private library of some ten thousand volumes, located just a few blocks from our house (figure 5), a library thoughtfully bestowed upon me by that ultimately beneficent robber baron, Andrew Carnegie. The neighborhood was secure enough for me to make my way alone to that library of mine from the tender age of five or six. From then on, I spent countless hours there, having been adopted by the dedicated librarians—all women, of course—who indulged and guided my interest in literature, science, and history, especially in biographies and autobiographies.

It was not at school but there in the Carnegie library that I was introduced to *Tristram Shandy* which, read and reread over the years, often to cope with bouts of melancholy, eventually found expression in my Shandean Postscript, *On the Shoulders of Giants*. It was there also that I came upon James Gibbons Huneker, the Philadelphia-born-and-reared music, drama, and literary critic who introduced my teen-

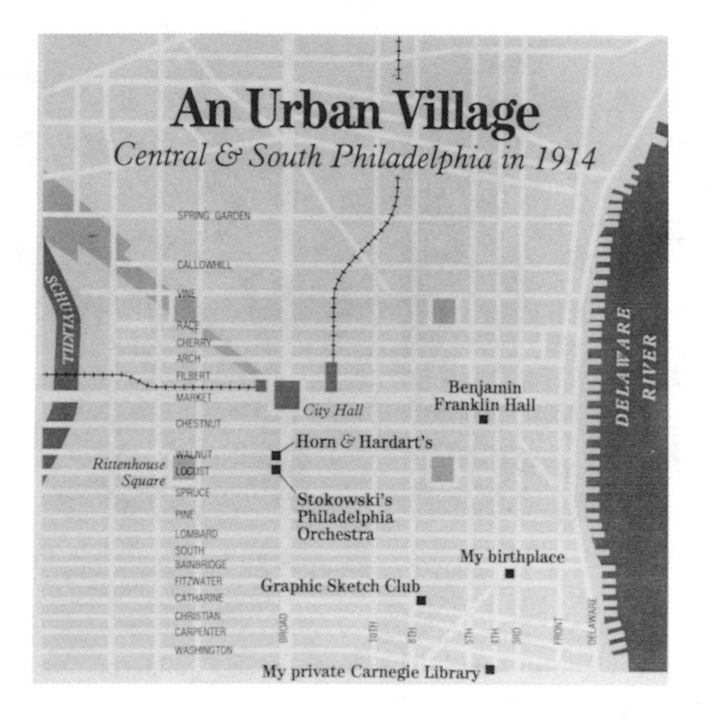

Figure 5 (Photo: Bill Marsh)

age self to new aspects of European culture: to the French symbolists, Baudelaire, Verlaine, Mallarmé, and Rimbaud, for example, and to Ibsen and George Bernard Shaw who, more than any other critic of his time, Huneker brought into the American consciousness. To say nothing of that "Beethoven of French prose," Gustave Flaubert. I still treasure the half-dozen Huneker volumes I later acquired at Leary's grand four-story bookstore, located, as I seem to remember, next to Gimbel's at Ninth and Market.

Evidently, the child was engaged in becoming father of the man as my presumably slum-bound self managed to travel widely in time and space. It may also have been in the Carnegie library that I first read David Brewster's engrossing and Victorian *Life of Newton* although I have no documents to support that conjectured memory. In any case, those early years turned out to be prelude to the years I lived in seventeenth-century England where, thanks to Harvard's Widener Library and archives, I hobnobbed with the likes of Newton, Boyle, and Christopher Wren. Just as that early addiction to biographies may have been prelude to a quantitative analysis in my doctoral dissertation of some six thousand entries in the *Dictionary of National Biography*, a mode of analysis which, I learned only much later from a paper by the Princeton historian Lawrence Stone, contributed to the research art of "historical prosopography": "the investigation of the common background characteristics of a group of actors in history by means of a collective study of their lives."

Those sojourns in libraries exemplify the Bernard-Bailyn-and-Lawrence-Cremin thesis that much consequential education takes place outside the walls of classrooms. In defense of the South Philadelphia High School of that time, however, I must report that it did provide some of us with four years of Latin, two of French, and several years of physics, chemistry, and mathematics. Not quite Groton, or Exeter, or the Bronx High School of Science, or, for that matter, Philadelphia's *Gymnasium*-like Central High School, but I might easily have done worse.

Other institutional assets were there just for the asking. A few blocks from the library was the local settlement house with its Graphic Sketch Club (figure 5) ever engaged in search of artistic talent among the culturally deprived but emphatically finding no trace of such talent in me. Still, it was there that Sundays brought us chamber music, at times by members of the celebrated Philadelphia Orchestra.

The Orchestra itself was also ours since we were within easy walking distance of its Academy of Music (figure 5). First as children and then as adolescents, we had only to wait in line for hours on end to be admitted to the Saturday night concerts. The princely sum of first 25, then 50 cents would entitle us to a seat in the last six rows of the amphitheatre; that allowed us to hear and almost get to see the charismatic Leopold Stokowski taking his orchestra of world fame through his masterly and controversial renditions of Bach—this, of course, without the customary baton. Those far-up seats also permitted us to hear him scolding the Philistine audience for noisily objecting to the new complex music of a Schoenberg, Varèse, or Alban Berg. And, after the concert, we could repair to the lavish Horn & Hardart Automat where we would sit near those of Stoki's men we had come to know and eavesdrop on their talk about the concert or, on occasion, about the baseball triumphs of Connie Mack's A's. But that too was not enough to turn me into howsoever mediocre a musician, though I do detect traces of that early musicological experience in the footnotes of *OTSOG*. Our horizons were further extended in the mid-1920s by the new, rather overwhelming Central Library and monumental Museum of Art.

At this point, my fellow sociologists will have noticed how that seemingly deprived South Philadelphia slum was providing a youngster with every sort of capital—social capital, cultural capital, human capital, and, above all, what we may call public capital—that is, with every sort of capital except the personally financial. To this day, I am impressed by the wealth of public resources made available to us ostensible poor. *Ostensible* poor, of course, since we held important property-rights in the form of ready access to valued resources otherwise possessed only by the very rich. The opportunity structure of our urban village was manifestly and rapidly expanding. But it is also the case that, in the absence of capability, all manner of opportunities being presented to me—for example, in music and the graphic arts—were without visible result. As I would argue long afterwards, in elucidating the sociological concept of opportunity structure, opportunity is probabilistic, not deterministic; it opens possibilities but does not assure their being realized. Just another biographical reminder of the continuing interplay between social structure and individual agency.

My own youthful life was also expanding through an encounter of the first magnitude with Charles Hopkins—or "Hop," as he was known to his friends—the man who became my sister's husband and, in effect, my surrogate father. And a truly chance encounter it was. Soon after my father lost his job at the Navy Yard and we moved once again, we were startled by white mice racing through our newfound row house and intrigued by rabbits in our back yard. Our next-door neighbor, Hop, came by to ask if we had happened to see his pet mice or rabbits. They turned out to be part of his stock-in-trade as an avocational magician. (Only later did I discover that his accomplished craft and artful inventions had won him a secure reputation among prime professional magicians of the time.) That encounter began Hop's courtship of my sister Emma and my idolization of Hop as he began to induct me into the art of prestidigitation. The apprenticeship continued so that I became fairly adept by the time I was fourteen. Enough so, for this arcane practice to help support me through my studies when I entered Temple College three years later. I still have copies of the card which Hop, as a Ben Franklinesque printer, designed for me (figure 6).

As you see, against a background of top hat and wand etched in soft blue, it declares in flowing script that Robert K. Merton was ready to produce "Enchanting Mysteries," presumably for a modest fee; as it turned out, chiefly at children's parties, at Sunday schools and, for part of one summer, in a small and quite unsuccessful traveling circus.[1]

1. I need hardly remind this company that Vladimir Nabokov and Edmund Wilson, those closest of friends and most devoted of antagonists, also took pleasure in the esoteric art of magic, as do Persi Diaconis and Frederick Mosteller, that pair of mathematical statisticians I have quoted on the complex subject of coincidence.

Figure 6

When I began that short-lived practice as a magician, Houdini became a "role model" (if I may resort to that once well-defined sociological term now become blurred if not vacuous by frequent and indiscriminate use; a term, incidentally, which *A Supplement to the Oxford English Dictionary* maintains was first used in 1957 by my Columbia research group then at work on *The Student Physician*). But I swiftly end this tiny digression into sociological semantics to return to another consequential moment in my youth, when I seized upon Houdini as my subject for a biographical sketch required in a high school course. During research for the paper, I soon learned that names in the performing arts were routinely Americanized; that is to say, they were transmuted into largely Anglo-American forms. For this, of course, was the era of hegemonic Americanization, generations before the emergence of anything resembling today's multiculturalism. The process of symbolic renaming was then in full force as we know, for example, from Leonard Rosenberg becoming Tony Randall, Issur Danielovitch Demsky becoming Kirk Douglas, and Irving Grossberg becoming first the musician and then the artist, Larry Rivers. And so, just as Ehrich Weiss, the son of Rabbi Mayer Samuel Weiss, had become Harry Houdini, naming himself after the celebrated French magician, Robert Houdin, the fourteen-year-old Meyer R. Schkolnick fleetingly became Robert K. Merlin, after the far more celebrated magician of Arthurian legend. Merlin, in turn, soon became Merton when my mentor Hop gently observed that Merlin was a bit hackneyed. By the time I arrived at Temple College, my close friends were more often than not calling me Bob Merton and I did not discourage them. I rather liked the sound of it, no doubt because it seemed "more American" back then in the 1920s. With the warm consent of my devoted Americanizing mother—she attended night school far more religiously than the synagogue—and the bland agreement of my rather uninterested father, this was followed by the legal transformation of my name some sixty-five years ago.[2]

II

It was at Temple, a secular college established in 1884 by the Baptist minister Russell H. Conwell for "the poor boys and girls of Philadelphia," that another chance encounter changed the direction of my life. Brought there by a scholarship, I had ventured into a class in sociology given by a young instructor, George E. Simpson, and there I found my subject. Then still at work on a doctoral dissertation on *The Negro in the Philadelphia Press,* Simpson recruited me as his research assistant and soon had me doing some of the routine work: classifying, counting, measuring, and statistically summarizing all the references to Negroes over a span of decades in Philadelphia newspapers. The purpose was, of course, to gauge changes in the public imagery of Negroes (not, I recall, of "Blacks," a term which, in those days, was regarded by us white liberals as a demeaning epithet). Only years later would George Simpson and I learn that we had engaged in the research procedure which Harold Lasswell came to designate as "content analysis"—no more aware that *that* was what we were doing than Molière's Monsieur Jourdain had been aware, before the moment of epiphany, that he had actually been speaking prose all his life. It was that research experience which sealed my decision to enter upon the still fairly new and, for many, exotic and dubious field of sociology.

It was also through George Simpson that I entered into new social and cognitive networks, especially with Negroes. Through him, I came

2. Of course, Hop and I had no idea back then that the name *Merton* had been adopted by the Moses family of British and German industrialists. That I learned only in the 1970s in noticing that a biographical sketch of me in the *Encyclopedia Judaica* followed an entry for another, rather wealthier and vastly more philanthropic, Merton family. (They had founded *Metalgesellschaft,* one of the largest metallurgical firms in Germany.) Once again, coincidence reigns. For it was the philanthropic Wilhelm Merton who founded the Academy that eventually became the University of Frankfurt where the group advocating critical theory located its Institute for Social Research, later known as "the Frankfurt School" of social philosophy, sociology, politics, and economics. When Hitler came into power, members of the Frankfurt School found their way to New York and a peripheral affiliation with Columbia University and it was there that Leo Löwenthal and occasional others of the School eventually became members of the Bureau of Applied Social Research founded by my longtime collaborator, Paul Lazarsfeld. It was not until those entries in the *Encyclopedia Judaica,* however, that Löwenthal and I took note of the wholly secularized ethnic if not national coincidence of the German and the American Mertons.

to know Ralph Bunche and Franklin Frazier from the time they were instructors at Howard University, along with the Arthur Fausets and others in the reclusive Negro Philadelphia elite of physicians, lawyers, writers, artists and musicians. While at Temple, I also came to know the Philadelphia-born, Harvard-trained philosopher, Alain Locke, who had been the first black Rhodes scholar. I had invited him to address our nascent Sociology Club at Temple and several years later he invited me to join him for a summer in Paris but, to my great regret, time-and-circumstance kept me from what would have been my first direct experience of Europe. That wide array of Negro friends provided early contexts for my later assisting Kenneth Clark to put together the much-debated Social Science Brief on desegregation in the public schools for *Brown v. Board of Education* just as they provided contexts for my later studies of racism, Negro-white intermarriage, and the social perspectives of Insiders and Outsiders.

Taking his assistant in hand, George Simpson also saw to it that I would see and hear key figures at an annual meeting of the American Sociological Society. There I met Pitirim Alexandrovich Sorokin, the founding chairman of the Department of Sociology then being tardily established at Harvard. That too proved to be a consequential encounter. For I would surely not have dared apply for graduate study at Harvard had Sorokin not encouraged me to do so. After all, my college advisers had warned me that Temple was still not fully accredited. To which I replied, rather ineptly, that it was the scholar Sorokin, not the institution Harvard, that mattered most to me. For, as a rather arrogant undergraduate, I had brought myself to believe—not entirely without foundation—that I knew just about everything American sociology had to offer in the late 1920s, although I had to confess to having only peripheral knowledge of the older and, to me, more evocative European traditions of sociological thought. Sorokin had recently published his *Contemporary Sociological Theories,* a wide-ranging, contentious overview of, in the main, European sociology, and plainly he was the teacher I was looking for. Moreover, it was evident that Sorokin was not your ordinary academic sociologist. Imprisoned three times by czarists and then three times by the Bolsheviks, he had been secretary to Alexandr Kerensky, the Socialist Revolutionary Prime Minister of Russia, and had had a death sentence commuted into exile by the normally unsparing Lenin. That too was bound to matter to me since, like many another Temple College student during the Great Depression, I was a dedicated socialist. In the event, I did nervously apply to Harvard, did receive a scholarship there, and soon found myself embarked on a new phase in a life of learning.

III

Harvard proved to be a serendipitous environment, full of evocative surprises. The first definitely consequential surprise was Sorokin's inviting me to be his research assistant, this in my first year of graduate study, and then his teaching assistant as well. That meant, of course, that I became his man-of-all-work—and, as I was soon to learn, his occasional stand-in as well. Summoning me to his office one day, he announced that he had stupidly agreed to do a paper on recent French sociology for a learned society and asked if I would be good enough to take it on in his stead. Clearly, this was less a question than an unforgiving expectation. Abandoning all pretense at attending classes, I devoted days and nights to the vast *oeuvre* issuing forth from Émile Durkheim himself and from such eminences in the Durkheim school as Lévy-Bruhl, Mauss, Halbwachs, and Bouglé. This turned out to be the first of several such unpredictable and fruitful occasions provided by the expanding opportunity structure at Harvard. This one was doubly consequential, for it catapulted me at once, in my second year of graduate study, into the role of a published scholar and led to my being invited to do the first essay-review of Durkheim's newly translated *Division of Labor in Society.* The intensive work on those two papers resulted in my becoming a transatlantic Durkheimian and laid the groundwork for what would become my own mode of structural and functional analysis.

As I've said, Sorokin, not the University, was the lodestone that drew me to Harvard. But, in the event, it was not the renowned Sorokin who most influenced my sociological thinking there; instead, it was a young instructor with no public identity whatever as a sociologist. Talcott Parsons had then published only two articles, both based on his dissertation; moreover, these had appeared in the *Journal of Political Economy,* a journal, it is fair to suppose, not much read by undergraduates in sociology bent on deciding where to do their graduate work. However, those few of us who did come into Talcott Parsons's very first course in theory (despite its long, seemingly humdrum title, "Sociological Theories of Hobhouse, Durkheim, Simmel, Toennies, and Max Weber") soon experienced him as a new sociological voice. The corpus of social thought which Sorokin summarized, Parsons anatomized and synthesized. As we students could not know and as I later learned Parsons himself did not anticipate, those lectures would provide the core of his masterwork, *The Structure of Social Action.* That monumental book did not appear in print until five years later, only after having been worked and reworked in lectures and seminars.

I truly cannot say whether that experience of observing Talcott Parsons virtually write his book in the course of his teaching led me to adopt, quite self-consciously, a similar and lifelong practice of engaging in what can be described as "oral publication"—the working out of ideas in lectures, seminars, and workshops—before finally converting their developed substance into public print. For some of us, teaching is itself a mode of scholarship. Continually revised lectures amount to new if unprinted editions. At least, that has been my experience. On exceptionally good days, the effort to rethink a subject or problem in advance of a lecture or seminar session is capped by new tentative ideas emerging in the lecture or seminar itself. On bad days, I feel that such continuities in lectures over the years risk my becoming a repetitive bore. At any rate, I notice that a dozen years raced by between the time I first lectured on "manifest and latent functions" at Harvard and the time those ideas took printed form in a "paradigm for functional analysis." Just as a dozen years intervened between my 1936 paper focussed on the unintended consequences of intentional action and the paper introducing the kindred concept of "the self-fulfilling prophecy."

Although much impressed by Parsons as a master-builder of sociological theory, I found myself departing from his mode of theorizing (as well as his mode of exposition). I still recall the grace with which he responded in a public forum to my mild-mannered but determined criticism of his kind of general theory. I had argued that his formulations were remote from providing a problematics and a direction for theory-oriented empirical inquiry into the observable worlds of culture and society and I went on to state the case for "theories of the middle range" as mediating between gross empiricism and grand speculative doctrines. In typically civil fashion, Parsons paid his respects to my filial impiety and agreed that we both had cause to disagree.

However, it was not the sociologists Sorokin or Parsons but the Harvard economic historian E. F. Gay who, with no such intent, triggered my enduring sociological interest in science and technology. Gay had studied at Berlin with the economic historian Gustav Schmoller, notorious, among other things, for his sociological bent and famous for his insistence on archival research. I decided to take Gay's course rather than an alternative in sociology and that led to still another truly consequential encounter. An assignment in the course had me doing an analytical essay on A. P. Usher's recent *History of Mechanical Invention.* Gay liked the essay and suggested that I audit Harvard's sole course in the history of science given jointly by the biochemist and self-taught Paretan sociologist L. J. Henderson, and by George Sarton, the world doyen among historians of science. I did so but it was only after

I began work on a dissertation that I dared seek guidance from Sarton. For he was reputed to be a remote and awesome presence, so dedicated to his scholarship as to be wholly inaccessible. Thus do plausible but ill-founded beliefs develop into social realities through the mechanism of the self-fulfilling prophecy. Since this forbidding scholar was unapproachable, there was no point in trying to approach him. And his subsequently having very few students only went to show how inaccessible he actually was. But when in the fall of 1933 I knocked on the door of Sarton's office in Widener Library, he did not merely invite me in; he positively ushered me in. That first audition had me sketching plans for a dissertation centered on sociological aspects of the growth of science in seventeenth-century England—a problem not exactly central to sociology back then. I cannot say that Sarton greeted those plans with enthusiasm; in his knowing judgment, so large a canvas as seventeenth-century English science might be a bit much for a novice. But he did not veto the idea. Then began my intensive, sometimes unruly, apprenticeship, followed by an epistolary friendship that continued until his death twenty-five years later.

From the start, George Sarton did much to set me on a new path of learning. He proceeded methodically—he was methodical in most things—to transform me from a graduate student (figure 7), struggling with early work on a dissertation, into a tyro scholar addressing an international community of learned scholars in print. This he did first by opening the pages of his journal *Isis* to me. During the next few years, he accepted several articles of mine along with some two dozen reviews and scores of entries for the annotated critical bibliographies appearing in *Isis*. Sarton then went on to bestow a "threshold gift": the special kind of gift which, in the words of the anthropological poet-ethicist Lewis Hyde, acts as an "agent of individual transformation." Sarton offered to publish my dissertation in *OSIRIS,* the series of monographs typically written by distinguished scholars in the history and philosophy of science; but not, surely, a series designed to include monographs by newly minted Ph.D.'s at work in what was becoming the sociology of science. Half-a-century later, his daughter, the poet and novelist May Sarton, took occasion to say that were her father still with us, he would have felt renewed pleasure in that decision to publish *Science, Technology, and Science in Seventeenth-Century England* as he observed its fiftieth year being commemorated in fine Sartonian style by a symposium in *Isis,* replete with a picture of his onetime student on the cover (figure 8).

Completion of the dissertation had other consequences. Sorokin and

*Figure 7**

Figure 8

Parsons lifted my spirits by seeing to it that I was appointed an instructor and tutor in the department. Given the dismal state of the job market, that was something of an event. But only temporarily so. This was, after all, the midst of the Great Depression—and even Harvard was hurting. Its still fairly new president, James B. Conant, signalled his

*Only now does this ancient snapshot call back to mind how it was that Filene's bargain basement of world fame allowed an impecunious graduate student to indulge himself by sporting a heavy, white-linen and originally expensive suit long before it became Tom Wolfe's signature. That Harvard student's standard of living can be gauged from a segmented summary of his weekly expenses in the academic year 1931-1932 and from a sampled daily record maintained by his roommate, budgeter, and chef, Richard Deininger, during the next academic year (Figure 7A).

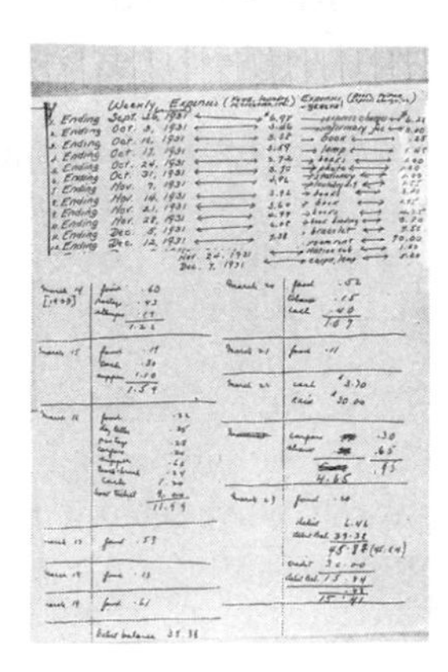

Figure 7a

intention to do away with the rank of assistant professor altogether and to limit promotions to the replacement of retiring or otherwise departing professors. That meant, of course, that a permanent post at Harvard would largely depend on the age distribution of faculty in each department. To be sure, Conant, self-described as "an amateur historian of seventeenth-century science in England," had gone out of his way to let me know "how much I enjoyed your work"—the language is his. However, the presiding elder in our fledgling department, Sorokin, was still in his forties; reason enough for me to leave the to-me indulgent yet alien Harvard before my instructorship had run its course. And so when Tulane University beckoned with a professorship in that bleak economic time, the decision was over-determined and the die was cast. Besides, for a provincial whose life had been confined to Philadelphia and Cambridge, the fanciful culture of New Orleans provided a distinct attraction. After a relaxing—and intellectually rewarding—two years at Tulane, I moved to Columbia and entered upon another, wholly unpredictable phase of learning: what turned out to be thirty-five years of an improbable collaboration with the mathematician-psychologist turned sociologist, Paul F. Lazarsfeld.

IV

I say "improbable collaboration" because Paul Lazarsfeld and I may have been the original odd couple in the domain of social science. He, the mathematically-minded methodologist, inventor of powerful techniques of social inquiry such as the panel method and latent structure analysis; I, the confirmed social theorist albeit with something of an empirical bent, insisting on the importance of sociological paradigms (in a pre-Kuhnian sense of "paradigm"); Paul, a founder of systematic empirical research on mass communications, voting behavior, opinion leadership, and individual action; I, engrossed in developing the paradigms of functional analysis and deviant behavior while trying to bring a nascent sociology of science into fuller being by exploring science as a social institution with a distinctive, historically evolving ethos, normative structure, and reward system; Paul, from his early days in Vienna, the inveterate creator of research institutes unable to imagine himself working outside of a research organization; I, the inveterate loner working chiefly in libraries and in my study at home; he, the matter-of-fact but methodologically demanding positivist; I, something of a doubting Thomas who, in my very first published paper, had dared satirize the "enlightened Boojum of Positivism." But, when I joined

Paul in his prime institutional creation, the Columbia University Bureau of Applied Social Research, presumably for just one research project, we soon discovered elective affinities and common ground. That temporary affiliation with the Bureau lasted some thirty years. Throughout that time, our shared lives of learning would center on a continuing program of theory-guided and methodologically disciplined empirical social research on a wide variety of substantive problems.

I have failed miserably in every attempt at even a meager digest of the influence Paul Lazarsfeld and I may have had on each other. Documentary evidence does testify, however, that I finally did persuade this resolute mathematician-psychologist that there really was a discipline of sociology. For eventually Paul published a little book with the engaging title *Qu'est ce que la sociologie?* which, in his private idiom, translated into the question: "What on earth *is* sociology all about?" Or, as his self-mocking inscription in my copy of the book put it: "All the questions you always wanted to have answered but never dared to ask."

Correlatively, Paul's abiding concern with research methods rubbed off on me and once resulted in a codification of what I called the focussed interview. Designed to elicit responses of groups to texts of various kinds—say, a journal article, radio program or educational film—the focussed interview took hold in academic sociology and then, after dubious sea changes, boomed its way into what we all know as the focus group. In their enthusiasm for the new ubiquitous focus group, marketeers and political advisers of every stripe, not excluding habitués of the White House and of Congress, often mislead themselves and others by failing to recognize or to acknowledge that such group interviews can at best only yield guesses about the current state of the public mind. Not being representative samples, focus groups cannot, of course, provide reliable knowledge about the extent and social distribution of public preferences, practices, and sentiments.

In retrospect, I am persuaded that the most consequential result of Paul's and my working together went far beyond our collaborations in print. It was of a quite different sort, one nicely summed up about a century ago by the French mining engineer and self-taught sociologist, Frédéric Le Play: The most important thing to come out of the mine, he wrote, is the miner. In much the same spirit, it can be said that the most important thing to come out of Columbia sociology back then was the student. Owing in no small part to the war's end and to the GI Bill, successive cohorts of brilliant students brightened our Department and Research Bureau in the 1940s and '50s and did much to bring about the intellectual excitement that then brought us a continuing flow of new talent. Paul Lazarsfeld and I had no doubt that a good many of these students would go on to leave an indelible imprint on sociological scholarship. As has proved to be the case. Indeed, I now find myself periodically diverted from work-in-slow-progress by writing papers designed specifically for those honorific volumes known as *festschriften,* not, as might be supposed, festschriften in honor of teachers or aged peers but in honor of onetime students. Hardly the usual pattern. Most recently, I have found myself gladly paying tribute to James S. Coleman, as I had gladly paid tribute before to Lewis Coser, Franco Ferrarotti, Peter Blau, Rose Coser, and Seymour Martin Lipset along with Alvin Gouldner and Louis Schneider though abjectly missing out on the two-volume festschrift for Juan Linz. Contemplating the extraordinary run of gifted students over that period of decades, I see more festschriften in the offing. In anticipatory celebration, I have begun work on a paper entitled "The Emergence and Evolution of the Festschrift: A Sociological Study in the Reward-System of Science and Learning." Prefaced by individualized tributes, it may serve as a template for contributions to future festschriften honoring onetime students whose scholarship has happily advanced beyond that of their onetime teachers.

V

In this retrospect on a life of learning, I have dwelt upon the private life rather more than upon the public learning. After all, the fruits of that learning are accessible in the public domain to those who care to sample them; the private life is not. But now that my time and your patience are rapidly drawing to a close, a few scattered remarks that bear variously upon the theme of the child as father of the man and upon oddities in my style of work over the years.

I first give way to the intrusive thought that age has its strange reckonings. I find it hard to believe that I was born a mere forty-five years after the Civil War and exceedingly hard to believe that I have lived through more than a third of our nation's history. All the more difficult to believe since, as a young romantic, I was convinced that the good die young and that, like Byron, Keats, and Shelley, I'd not live much beyond the age of thirty. A latter-day reminder that if age is renewed opportunity, it is also continuing obligation.

With regard to my work, I only touch upon three quite discrete matters: an almost lifelong addiction to editing, a preferred expositional style, and lastly, certain thematic orientations in social theory.

If Schopenhauer had it right in declaring that to put away one's own original ideas in order to take up the work of another is a sin against the Holy Ghost of scholarship, then indeed *peccavi, peccavi.* I have truly and chronically sinned. For almost as soon as sociology became my vocation, editing became my avocation. This began as early as my student days. Following upon a moderately effective editing of Sorokin's Russified English prose, I agreed to try my hand at editing Parsons's classic *Structure of Social Action.* Although kindly appreciated in the Preface, that editorial effort plainly had an indifferent effect. But this failure was evidently not enough to stay my editor's pen. For, based on some sample lists, a back-of-the-envelope estimate has me editing some 250 books and 2,000 articles over the course of the past sixty years. Behavior hardly in accord with the Schopenhauer canon.

My preferred style of exposition also emerged from the start. As in the 1936 paper on the "Unanticipated Consequences of Purposive Social Action," the 1938 paper on "Social Structure and Anomie," and the 1948 paper on "The Self-Fulfilling Prophecy," I have generally set out my sociological ideas in the form of highly condensed paradigmatic essays, typically running to few more than a dozen-or-so pages. By adopting the relatively discursive form of the essay, I have no doubt irked some sociologist-peers by departing from the tidy format long since prescribed for the scientific paper. Designed to instruct fellow scientists about a potential new contribution to a field of knowledge, the stylized scientific paper presents an immaculate appearance that tells little or nothing of the intuitive leaps, false starts, loose ends, opportunistic adaptations, and happy accidents that actually cluttered up the inquiry. After all, the scientific paper is not designed as a clinical or biographical account of the reported research. In contrast, the essay provides scope for asides and correlatives of a kind that interest historians and sociologists of science and is, in any case, better suited to my ungovernable preference for linking humanistic and scientific aspects of social knowledge.

However, those sociological essays of mine are not wholly discursive. They are disciplined by being "paradigmatic" in, as I've said, a pre-Kuhnian sense of the term "paradigm." That is to say, the analytical paradigm identifies the basic assumptions, problems, concepts, and hypotheses incorporated in the sociological idea in order to generate researchable questions and to provide for continuities of theoretical and empirical inquiry. Thus, the "paradigm of anomie-and-opportunity structure" laid out in a set of essays has been put to use by successive generations of scholars over the past half-century, first in the sociological and criminological study of deviant behavior and then in continuing researches in a variety of other disciplines, just as the "paradigm of the self-fulfilling prophecy," which was first applied to the sociological problem of ethnic and racial discrimination, has since led to traditions of theoretical and empirical inquiry in social psychology, political science, anthropology, economics, and public administration.

Reflecting briefly on thematic orientations emerging in my theoretical work, I take note of a prime aversion, a prime preference, and a prime indulgence.

My prime theoretical aversion is to any extreme sociological, economic, or psychological reductionism that claims to account uniquely and exhaustively for patterns of social behavior and social structure. By way of rationale for this aversion, I confine myself to the William James parable about the reductionist fallacy: "A Beethoven string-quartet is truly . . . a scraping of horses' tails on cats' bowels, and may be exhaustively described in such terms, but the application of this description in no way precludes the simultaneous applicability of an entirely different description."

As I have intimated, my prime theoretical preference is for sociological theories of the middle range which, I hasten to say in accord with Arthur Stinchcombe, can be shown to derive in principle from a more general theory if they are worth their salt in providing an improved understanding of social behavior, social structure, and social change.

And my prime theoretical indulgence finds its fullest expression in my one avowedly humanist and self-winding book, *On the Shoulders of Giants,* which adopts a nonlinear, divagating Shandean mode for examining the enduring tension between tradition and originality in the transmission and growth of knowledge along with a variety of related themes.

And now, as befits a short essay on an improbable life of learning, a final brief thought about autobiography, that mode of self-reflection in historical contexts which has held my interest since those distant days in the Carnegie library. But not, of course, with reference to myself—until recent decades. For it happens that ever since the publication in 1961 of the *New Yorker* profile, with its condensed South Philadelphia story, kindly disposed friends, colleagues, and publishers have been urging me to write an autobiography or, at least, a longish memoir. Would that I could. But, as all of us know, God is in the details. Without thick, textured detail, an autobiography is bound to be weary, flat, stale, and unprofitable. But the sinful fact is that I simply haven't access to the needed detail. Cursed my life long by a scant and episodic memory, I dare not rely on vagrant memories without visible means of documentary support. But, alas, I've not kept a diary or a journal, with documentation thus confined to notebooks and voluminous but still inadequate files of letters. And so, when asked to venture upon an autobiography, I have only to recall the caustic review of a memoir by the prolific novelist and playwright, Heinrich Böll. The reviewer notes Böll's many tiresome passages lamenting his inability to remember and concludes that the author "seems almost to boast of his mnemonic failures." For me, that review amounts to a preview. It provides timely warning that any memoir of mine would surely display an even more humiliating amnesia. But perhaps, just perhaps, this slight remembrance of things past will serve in its stead.[3]

3. Since my long-term memory is distinctly limited, this essay draws freely upon reminiscent passages in previous publications.

Publications by and about Robert K. Merton: A Select Bibliography

A bibliography of Merton's works can be found in "Bibliography: Robert K. Merton, 1934–1975," compiled by Mary Wilson Miles in *The Idea of Social Structure: Papers in Honor of Robert K. Merton,* edited by Lewis A. Coser (New York: Harcourt Brace Jovanovich, 1995), 497–522; supplemented by "Writings of Robert K. Merton: Publications 1975–1989," compiled by Mary Wilson Miles and Rosa Haritos in *Robert K. Merton: Consensus and Controversy,* edited by Jon Clark et al. (New York: Falmer Press, 1990), 451–75. An asterisk (*) indicates whole or partial inclusion in this volume.

Books

1938. [1970]. *Science, Technology, and Society in Seventeenth Century England.* In *OSIRIS,* ed. George Sarton, 362–632. Bruges: St. Catherine Press. 2d ed. New York: Howard Fertig, 1970; reprinted 1993.

1946. *Mass Persuasion* (with M. Fiske and A. Curtis). New York: Harper and Brothers. Reprint, Stamford, Conn.: Greenwood Press, 1971.

1956 [1990]. *The Focused Interview* (with M. Fiske and P. L. Kendall). Glencoe, Ill.: Free Press. 2d ed., 1990.

1957. *The Student-Physician: Introductory Studies in the Sociology of Medical Education* (with G. G. Reader and P. L. Kendall). Cambridge: Harvard University Press.

1965 [1985, 1993]. *On the Shoulders of Giants.* New York: Harcourt Brace Jovanovich. "Vicennial Edition," 1985. "Post-Italianate Edition," Chicago: University of Chicago Press, 1993.

1967. *On Theoretical Sociology.* New York: Free Press.

1968. [1949, 1957]. *Social Theory and Social Structure.* 3d ed., rev. and enl. New York: Free Press. Originally published Glencoe, Ill.: Free Press, 1949; rev. ed., 1957.

1973. *The Sociology of Science: Theoretical and Empirical Investigations.* Ed. N. W. Storer. Chicago: University of Chicago Press.

1976. *Sociological Ambivalence and Other Essays.* New York: Free Press.

1979. *The Sociology of Science: An Episodic Memoir.* Carbondale, Ill.: Southern Illinois University Press.

1982. *Social Research and the Practicing Professions.* Ed. A. Rosenblatt and T. F. Gieryn. Cambridge, Mass.: Abt Books.

Edited Volumes

1950. *Continuities in Social Research* (with Paul F. Lazarsfeld). Glencoe, Ill.: Free Press.

1952. *Reader in Bureaucracy* (with Ailsa Gray, Barbara Hockey, and Hanan Selvin). 2d ed. Glencoe, Ill.: Free Press.

1959. *Sociology Today: Problems and Prospects* (with Leonard Broom and Leonard Cottrell, Jr.). New York: Basic Books.

1961 [1966, 1971, 1976]. *Contemporary Social Problems* (with Robert A. Nisbet). New York: Harcourt Brace Jovanovich, 1961; 2d ed., 1966; 3d ed., 1971; 4th ed., 1976.

1975. *The Sociology of Science in Europe* (with Jerry Gaston). Carbondale, Ill.: Southern Illinois Press.

1978. *Toward a Metric of Science: Thoughts Occasioned by the Advent of Science Indicators* (with Harriet Zuckerman, Arnold Thackray, Joshua Lederberg, and Yehuda Elkana). New York: Wiley.

1979. *Genesis and Development of a Scientific Fact,* by Ludwik Fleck (ed. and trans. with Thaddeus J. Trenn). Chicago: University of Chicago Press.

1979. *Qualitative and Quantitative Social Research: Papers in Honor of Paul F. Lazarsfeld* (with James S. Coleman and Peter H. Rossi). New York: Free Press.

1980. *Sociological Traditions from Generation to Generation: Glimpses of the American Experience* (with Matilda White Riley). Norwood, N.J.: Ablex Corporation.

1981. *Continuities in Structural Inquiry* (with Peter M. Blau). London: Sage Publications.

1990. *Social Science Quotations: Who Said What, When, and Where* (with David L. Sills). Vol. 19 of the *International Encyclopedia of the Social Sciences.* New York: Macmillan Publishing Company.

Compilations

1974. *Perspectives in Social Inquiry: Classics, Staples, and Precursors in Sociology.* 40 vols. New York: Arno Press.

1975. *History, Philosophy, and Sociology of Science: Classics, Staples, and Precursors* (with Harriet Zuckerman, Arnold Thackray, and Yehuda Elkana). 60 vols. New York: Arno Press.

1980. *Dissertations in Sociology* (with Harriet Zuckerman). 61 vols. New York: Arno Press.

Selected Articles

1934a. "Recent French Sociology." *Social Forces* 12:537–45.

1934b. "Durkheim's Division of Labor in Society." *American Journal of Sociology* 40:319–28.

* 1936a. "The Unanticipated Consequences of Purposive Social Action." *American Sociological Review* 1:894–904.
1936b. "Civilization and Culture." *Sociology and Social Research* 21: 103–13.
1937. "Social Time: A Methodological and Functional Analysis" (with Pitirim A. Sorokin). *American Journal of Sociology* 42:619–29.
* 1938. "Social Structure and Anomie." *American Sociological Review* 3: 672–82.
1939. "Bureaucratic Structure and Personality." *Social Forces* 18:560–68.
1945. "Sociological Theory." *American Journal of Sociology* 50:462–73.
1946. "The Focused Interview" (with Patricia L. Kendall). *American Journal of Sociology* 51:541–57.
1948a. "The Position of Sociological Theory." *American Sociological Review* 13:164–68.
* 1948b. "The Self-Fulfilling Prophecy." *Antioch Review* summer:193–210.
1950. "Contributions to the Theory of Reference Group Behavior" (with Alice Kitt Rossi). In *Continuities in Social Research*. Ed. Robert K. Merton and Paul F. Lazarsfeld, 40–105. Glencoe, Ill.: Free Press.
1954. "Friendship as a Social Process: A Substantive and Methodological Analysis" (with Paul F. Lazarsfeld). In *Freedom and Control in Modern Society*. Ed. M. Berger, T. Abel, and C. Page, 18–66. New York: Van Nostrand.
* 1957a. "The Role-Set: Problems in Sociological Theory." *British Journal of Sociology* 2:106–20.
* 1957b. "Priorities in Scientific Discovery: A Chapter in the Sociology of Science." *American Sociological Review* 22:635–59.
1959. "Social Conformity, Deviation, and Opportunity-Structure." *American Sociological Review* 24:177–89.
* 1963. "Sociological Ambivalence" (with Elinor Barber). In *Sociological Theory, Values, and Sociocultural Change*. Ed. E. A. Tiryakian, 91–120. Glencoe, Ill.: Free Press.
1964. "Anomie, Anomia, and Social Interaction: Contexts of Deviant Behavior." In *Anomie and Deviant Behavior: Discussion and Critique*. Ed. M. Clinard, 213–42. Glencoe, Ill.: Free Press.
* 1968. "The Matthew Effect in Science." *Science* 5 (January): 55–63.
1971. "Patterns of Evaluation in Science: Institutionalization, Structure, and Functions of the Referee System" (with Harriet Zuckerman). *Minerva* 9: 66–100.
* 1972. "Insiders and Outsiders: A Chapter in the Sociology of Knowledge." *American Journal of Sociology* 78:9–47.
* 1975. "Structural Analysis in Sociology." In *Approaches to the Study of Social Structure*. Ed. Peter Blau, 21–52. New York: Free Press.
1980a. "On the Oral Transmission of Knowledge." In *Sociological Traditions from Generation to Generation*. Ed. Robert K. Merton and Matilda W. Riley, 1–35. Norwood, N.J.: Ablex.

1980b. "Remembering the Young Talcott Parsons." *The American Sociologist* 15:68–71.

1980c. "George Sarton" (with Arnold Thackray). In *Dictionary of American Biography*. Ed. John A. Garraty, 564–66. New York: Charles Scribner's Sons, 1980.

1983a. "Florian Znaniecki: A Short Reminiscence." *Journal of the History of the Behavioral Sciences* 19:123–6.

1983b. "Client Ambivalence in Professional Relationships: The Problem of Seeking Help from Strangers" (with V. Merton and E. Barber). In *New Directions in Helping*. Ed. B. M. DePaulo et al., 2:13–44. New York: Academic Press.

* 1984a. "Socially Expected Durations: A Case Study of Concept Formation in Sociology." In *Conflict and Consensus: In Honor of Lewis A. Coser*. Ed. W. W. Powell and R. Robbins, 262–83. New York: Free Press.

1984b. "The Fallacy of the Latest Word: The Case of Pietism and Science." *American Journal of Sociology* 89:1091–1121.

1985. "George Sarton: Episodic Recollections by an Unruly Apprentice." *Isis* 76:477–86.

* 1987. "Three Fragments from a Sociologist's Notebooks: Establishing the Phenomenon, Specified Ignorance, and Strategic Research Sites." *Annual Review of Sociology* 13:1–28.

* 1988a. "The Matthew Effect in Science, II: Cumulative Advantage and the Symbolism of Intellectual Property." *Isis* 79:606–23.

1988b. "Reference Groups, Invisible Colleges, and Deviant Behavior in Science." In *Surveying Social Life: Papers in Honor of Herbert H. Hyman*. Ed. H. J. O'Gorman, 174–89. Middletown, Conn.: Wesleyan University Press.

1988c. "Sociological Resonances: The Early Franco Ferrarotti and a Transatlantic Colleague." In *Omaggio a Franco Ferrarotti*. Ed. R. Cipriani and M. I. Macioti, 83–91. Rome: Siares.

1988d. "Some Thoughts on the Concept of Sociological Autobiography." In *Sociological Lives*. Ed. M. W. Riley, 17–21. Newbury Park, Calif.: Sage Publications.

1989a. "Le molteplici origini e il carattere epiceno de termine inglese *Scientist*." . . . *Scientia* 123:279–93.

1989b. "Unanticipated Consequences and Kindred Sociological Ideas: A Personal Gloss." In *L'Opera di R. K. Merton e la sociologia contemporanea*. Ed. Carlo Mongardini and Simonetta Tabboni, 307–29. Genoa: ECIG.

1990a. "Epistolary Notes on the Making of a Sociological Dissertation Classic: The Dynamics of Bureaucracy." In *Structures of Power and Constraint: Papers in Honor of Peter M. Blau*. Ed. C. Calhoun, M. W. Meyer and W. Richard Scott, 37–66. Cambridge: Cambridge University Press.

1990b. "On Becoming an Honorand of Jagiellonian University: Social Time and Socio-Cognitive Networks." *International Sociology* 5:5–10.

1990c. "*STS*: Foreshadowings of an Evolving Research Program in the Sociology of Science." In *Puritanism and the Rise of Modern Science: The Merton Thesis.* Ed. I. B. Cohen, 334–71. New Brunswick, N.J.

1992. "Social Science Quotations" (with David L. Sills). *Current Contents* 24: 4–8.

1993. "Genesis of the Field of Science, Technology, and Society (STS)." *Journal of Science Policy and Research Management* 8:200–203.

1994a. "Durkheim's *Division of Labor in Society:* A Sexagenarian Postscript." *Sociological Forum* 9:27–36.

* 1994b. "A Life of Learning." American Council of Learned Societies Occasional 25. New York: ACLS.

* 1995a. "Opportunity Structure: The Emergence, Diffusion, and Differentiation of a Sociological Concept, 1930s–1950s." In *The Legacy of Anomie Theory: Advances in Criminological Theory.* Ed. F. Adler and W. S. Laufer, 3-78. New Brunswick, N.J.: Transaction Publishers.

1995b. "The Thomas Theorem and the Matthew Effect." *Social Forces* 74: 379–424.

1996a. "Teaching James S. Coleman." In *James S. Coleman.* Ed. Jon Clark, 351–56. New York: Falmer Press.

1996b. "The Cultural and Social Incorporation of Sociological Knowledge" (with Alan Wolfe). *American Sociologist,* 26: 15-38.

Selected Commentaries on Robert K. Merton's Work

Arimoto, Akira. 1987. *Merton Kagaku-Shakaigaku no Kenkyu* (A study of Merton's sociology of science: Formation and development of its paradigm). Tokyo: Fukumura.

Ashworth, William B. 1988. "In the Lap of Nature: A Mertonian Postscript." Paper presented at the annual meeting of the American Historical Association/History of Science Society, 28 December, Cincinnati, Ohio.

Bazerman, Charles. 1981. "What Written Knowledge Does: Three Examples of Academic Discourse." *Philosophy of the Social Sciences* 11:361–87.

Bennett, James. 1990. "Merton's 'Social Structure and Anomie': Suggestions for Rhetorical Analysis." In *The Rhetoric of Social Research: Understood and Believed.* Ed. Albert Hunter. New Brunswick and London: Rutgers University Press.

Boudon, Raymond. 1991. "What Middle-Range Theories Are." *Contemporary Sociology* 20 (July): 519–22.

Bourgeois III, L. J. 1979. "Toward a Method of Middle-Range Theorizing." *The Academy of Management Review* 4:443–47.

Cataño, Gonzalo. 1991. "Un examen de la obra de Robert K. Merton." *Revista Paraguaya de Sociologia* 28 (January–April): 155–60.

Clark, Jon, C. Modgil, and S. Modgil, eds. 1990. *Robert K. Merton: Consensus and Controversy.* London and Philadelphia: Falmer Press.

Cohen, I. Bernard, ed. 1990. *Puritanism and the Rise of Modern Science: The Merton Thesis.* New Brunswick: Rutgers University Press.

Collins, Randall. [1977] 1981. "Merton's Functionalism." In *Sociology Since Midcentury,* 197–203. New York: Academic Press.

Coser, Lewis A., ed. 1975. *The Idea of Social Structure: Papers in Honor of Robert K. Merton.* New York: Harcourt Brace Jovanovich.

Crothers, Charles. 1987. *Robert K. Merton: A Key Sociologist.* London and New York: Tavistock.

Cuzzort, R. P. and Edith King. 1995. "The Unanticipated Consequences of Human Action: The Functional Analysis of Robert K. Merton." In *Twentieth-Century Social Thought* [. . .]. 5th ed., 249–73. Fort Worth, Texas: Harcourt Brace.

DeMott, Benjamin. 1991. *The Imperial Middle: Why Americans Can't Think Straight About Class,* chapter 11. New York: William Morrow and Co.

Eden, Dov. 1990. *Pygmalion in Management: Productivity as a Self-Fulfilling Prophecy,* passim. Lexington, Mass.: D.C. Heath and Company.

Feldhay, Rivka and Yehuda Elkana. 1989. "'After Merton': Protestant and Catholic Science in Seventeenth-Century Europe." *Science in Context.* Vol. 3, 3–302.

Garcia, Jesus L. 1979. *Merton: La estructura precaria: Orden y conflicto en la sociedad moderna.* Mexico: Editorial Edicol.

Gieryn, Thomas F., ed. 1980. *Science and Social Structure: A Festschrift for Robert K. Merton.* New York: New York Academy of Sciences.

Gross, Alan G. 1990. "The Emergence of a Social Norm." In *The Rhetoric of Science,* chapter 11. Cambridge: Harvard University Press.

Gvoic', Ljubivoj. 1990. *Robert Merton's Functionalism: The Theoretical Conception and Fundamental Fields of Application.* Beograd: N.p.

Henshel, Richard L. 1990. "Unanticipated Consequences of Intervention: A Conservative and Radical Notion." In *Thinking About Social Problems,* chapter 8, passim. San Diego and New York: Harcourt Brace Jovanovich.

Hilbert, Richard A. 1989. "Durkheim and Merton on Anomie: An Unexplored Contrast and Its Derivatives." *Social Problems* 36, no. 3 (June): 242–50.

Hollinger, David A. 1983. "The Defense of Democracy and Robert K. Merton's Formulation of the Scientific Ethos." In *Knowledge and Society: Studies in the Sociology of Culture.* Vol. 4. Ed. R. A. Sones and H. Kuklick, 1–15. Greenwich, Conn.: JAI Press.

Jaworski, Gary Dean. 1990. "Robert K. Merton's Extension of Simmel's *Uebersehbar.*" *Sociological Theory* 8 (spring): 99–105.

———. 1990. "Robert K. Merton as Postwar Prophet." *The American Sociologist* 21 (fall): 209–16.

Knorr-Cetina, Karin. 1991. "Merton's Sociology of Science: The First and the Last Sociology of Science?" *Contemporary Sociology* 20 (July): 522–26.

Kuvaicic, Ivan. 1989. "Development, Expansion, and Contestation of Functionalism." *Sociologija* 31, no. 4 (November–December): 653–72.

Martin, Randy, Robert J. Mutchnick, and W. Timothy Austin. 1990. "Robert K. Merton." In *Criminological Thought: Pioneers Past and Present,* 206–35. New York: Macmillan.

Mendelsohn, Everett. 1989. "Robert K. Merton: The Celebration and Defence of Science." *Science in Context* 3 (July): 269–90.

Mongardini, Carlo and Simonetta Tabboni, eds. 1989. *L'Opera di R. K. Merton e la sociologia contemporanea*. Genoa: ECIG.

Passas, Nikos. 1990. *Merton's Theory of Anomie and Deviance: An Elaboration*. Edinburgh University: Dissertation Abstracts International. A. The Humanities and Social Sciences 50, no. 7.

Patel, Praven J. 1976. *Sociology of Robert K. Merton*. Ahmedebad: Gujarat University Press.

Poggi, Gianfranco. 1991. "A View from the Amalfi Coast." *Contemporary Sociology* 20 (July): 514–16.

Pokrovsky, Nikita E. 1992. "Eleven Commandments of the Robert Merton Functionalism." *Sotsiol Iss: Sociological Research* 2:114–17.

———. 1992. "Early Evening on Morningside Heights: Personal Notes on Meeting Robert K. Merton." *SOCIS: Sociological Research* 5 (May): 11–19.

Ramsøy, Natalie Rogoff. 1971. "Merton og funksjonalisme." In *Sosiologiens klassikere*. Oslo: Cappelen forlag.

Ritzer, George. 1992. *Sociological Theory*, 252–59, passim. New York: McGraw-Hill.

Rosenfield, Richard. 1989. "Robert Merton's Contributions to the Sociology of Deviance." *Sociological Inquiry* 59:453–66.

Sivin, Nathan. 1991. "Science, Religion, and Boundary Maintenance." *Contemporary Sociology* 20 (July): 526–30.

Smelser, Neil J. 1991. "An Olio on Merton." *Contemporary Sociology* 20 (July): 510–14.

Sørenson, Aage B. 1991. "Merton and Methodology." *Contemporary Sociology* 20 (July): 516–19.

Sperber, Irwin. 1990. "Sorokin, Merton, and the Fashion Process." *Fashions in Science: Opinion Leaders and Collective Behavior in the Social Sciences*, 154–90. Minneapolis: University of Minnesota Press.

Steffenhagen, R. A. 1984. "Self-Esteem and Anomie: An Integration of Adler and Merton as a Theory of Deviance." *Deviant Behavior* 5:23–30.

Struik, Dirk. 1989. "Further Thoughts on Merton in Context." *Science in Context* 3:227–38.

Stuhlhofer, Franz. 1987. *Lohn und Strafe in der Wissenschaft: Naturforscher im Urteil der Geschichte*, passim. Vienna: Böhlau Verlag.

Sztompka, Piotr. 1986. *Robert K. Merton: An Intellectual Profile*. New York: St. Martin's Press.

Tatsis, Nicholas C. 1987. "Robert K. Merton's Theory of Anomie: A Contribution to the Sociology of Deviance." *Aremopoulos* 8:187–201.

Taylor, James B. 1983. "Bureaucratic Structure and Personality: The Merton Model Revisited." In *Bureaucracy as a Social Problem*. Ed. W. Boyd Littrell, Gideon Sjoberg, and Louis A. Zurcher. Greenwich, Conn.: JAI Press.

Thompson, Michael, Richard Ellis, and Aaron Wildavsky. 1990. "Merton,

Stinchcombe, and Elster." In *Cultural Theory,* chapter 11. Boulder, San Francisco, and Oxford: Westview Press.

Tiryakian, Edward A., ed. 1991. "Symposium: Robert K. Merton in Review." *Contemporary Sociology* 20 (July): 506–30. See especially Tiryakian, "Evaluating the Standard: An Introduction to Sociological Metrology," 506–10.

Turner, Bryan S. 1990. "The Anatomy Lesson: A Note on the Merton Thesis." *The Sociological Review* 38 (February).

Twenhöfel, Ralf. 1991. "Zur Wissenschaftssoziologie Robert K. Mertons." In *Wissenschaftliches Handeln: Aspekte und Bestimmungsgründe der Forschung,* 77–130. Berlin and New York: Walter de Gruyter.

Vaughan, Diane. 1983. *Controlling Unlawful Organizational Behavior: Social Structure and Corporate Misconduct,* 55–62. Chicago: University of Chicago Press.

———. 1992. "Theory Elaboration: The Heuristics of Case Analysis." In *What Is a Case? Issues in the Logic of Social Inquiry.* Ed. Charles Ragan and Howard S. Becker, passim. Chicago: University of Chicago Press.

Vranken, Jan, ed. 1990. *Wetenschap, technologie en maatschappij in het 17 deeeuwse Engeland.* Lewven/Amersfoot: Acco.

Wallace, Walter. 1983. *Principles of Scientific Sociology,* passim. New York: Aldine Publishing Co.

Wells, Alan. 1979. "Conflict Theory and Functionalism." *Teaching Sociology* 6 (July): 429–37.

Yan, Ming. 1992. "Genuine Knowledge Transcends National Boundaries: An Interview with Robert K. Merton," *Sociological Studies* 3 (May): 1–6.

Zuckerman, Harriet. 1989. "Accumulation of Advantage and Disadvantage: The Theory and Its Intellectual Biography." In *L'Opera di R. K. Merton e la sociologia contemporanea.* Ed. Carlo Mongardini and Simonetta Tabboni, 153–76. Genoa: ECIG.

Zuckerman, Harriet. 1989. "The Other Merton Thesis." *Science in Context* 3: 239–67.

Name Index

The indexes to this volume were prepared by Martin White.

Subject Index